VOLUME ONE HUNDRED AND SIXTEEN

ADVANCES IN CANCER RESEARCH

VOLUME ONE HUNDRED AND SIXTEEN

ADVANCES IN CANCER RESEARCH

Histone Deacetylase Inhibitors as Cancer Therapeutics

Series Editors

KENNETH D. TEW

Cell and Molecular Pharmacology,
John C. West Chair of Cancer Research,
Department of Cell and Molecular
Pharmacology & Experimental Therapeutics
at MUSCCharleston, South Carolina, USA

PAUL B. FISHER

Department of Human & Molecular Genetics,
VCU Institute of Molecular Medicine
Richmond, Virginia, USA

Edited by

STEVEN GRANT

Division of Hematology/Oncology,
Virginia Commonwealth University
Health Sciences Center,
P.O. Box 980035,
Richmond, Virginia, USA

AMSTERDAM • BOSTON • HEIDELBERG • LONDON
NEW YORK • OXFORD • PARIS • SAN DIEGO
SAN FRANCISCO • SINGAPORE • SYDNEY • TOKYO
Academic Press is an imprint of Elsevier

Academic Press is an imprint of Elsevier
525 B Street, Suite 1900, San Diego, CA 92101-4495, USA
225 Wyman Street, Waltham, MA 02451, USA
32 Jamestown Road, London, NW1 7BY, UK
The Boulevard, Langford Lane, Kidlington, Oxford, OX51GB, UK
Radarweg 29, PO Box 211, 1000 AE Amsterdam, The Netherlands

First edition 2012

ISBN: 978-0-12-394387-3
ISSN: 0065-230X

For information on all Academic Press publications
visit our website at store.elsevier.com

Printed and bound by CPI Group (UK) Ltd, Croydon, CR0 4YY
Transferred to digital print 2012

CONTENTS

CONTRIBUTORS

Kapil N. Bhalla
The University of Kansas Cancer Center, Kansas City, Kansas, USA

Mellissa Brown
The University of Queensland Diamantina Institute, Princess Alexandra Hospital, Brisbane, Queensland, Australia

Megan L. Choy
Department of Cell Biology, Sloan-Kettering Institute, Memorial Sloan-Kettering Cancer Center, New York, USA

Yun Dai
Division of Hematology/Oncology, Virginia Commonwealth University Health Sciences Center, P.O. Box 980035, Richmond, Virginia, USA

Warren Fiskus
The University of Kansas Cancer Center, Kansas City, Kansas, USA

Brian Gabrielli
The University of Queensland Diamantina Institute, Princess Alexandra Hospital, Brisbane, Queensland, Australia

Siddhartha Ganguly
The University of Kansas Cancer Center, Kansas City, Kansas, USA

Steven Grant
Division of Hematology/Oncology, Virginia Commonwealth University Health Sciences Center, P.O. Box 980035, Richmond, Virginia, USA

Ricky W. Johnstone
Cancer Therapeutics Program, Gene Regulation Laboratory, The Peter MacCallum Cancer Centre, St. Andrews Place, East Melbourne, and The Sir Peter MacCallum Department of Oncology, University of Melbourne, Parkville, Victoria, Australia

Suman Kambhampati
The University of Kansas Cancer Center, Kansas City, Kansas, USA

Ju-Hee Lee
Department of Cell Biology, Sloan-Kettering Institute, Memorial Sloan-Kettering Cancer Center, New York, USA

Paul A. Marks
Department of Cell Biology, Sloan-Kettering Institute, Memorial Sloan-Kettering Cancer Center, New York, USA

Geoffrey M. Matthews
Cancer Therapeutics Program, Gene Regulation Laboratory, The Peter MacCallum Cancer Centre, St. Andrews Place, East Melbourne, and The Sir Peter MacCallum Department of Oncology, University of Melbourne, Parkville, Victoria, Australia

David J. McConkey
Department of Urology, and Department of Cancer Biology, U.T. M.D. Anderson Cancer Center, Houston, Texas, USA

Andrea Newbold
Cancer Therapeutics Program, Gene Regulation Laboratory, The Peter MacCallum Cancer Centre, St. Andrews Place, East Melbourne, and The Sir Peter MacCallum Department of Oncology, University of Melbourne, Parkville, Victoria, Australia

Rekha Rao
The University of Kansas Cancer Center, Kansas City, Kansas, USA

Feyruz V. Rassool
Department of Radiation Oncology and Greenebaum Cancer Center, University of Maryland School of Medicine, Baltimore, Maryland, USA

Carine Robert
Department of Radiation Oncology and Greenebaum Cancer Center, University of Maryland School of Medicine, Baltimore, Maryland, USA

Matthew White
Department of Urology, and Department of Cancer Biology, U.T. M.D. Anderson Cancer Center, Houston, Texas, USA

Wudan Yan
Department of Urology, and Department of Cancer Biology, U.T. M.D. Anderson Cancer Center, Houston, Texas, USA

PREFACE

In many respects, histone deacetylase inhibitors (HDACIs) represent prototypical "epigenetic" agents which act by modifying gene expression to restore the normal differentiation or death programs of transformed cells. In fact, the ability of HDACIs such as sodium butyrate to alter histone acetylation and induce maturation of leukemia cells was discovered over 30 years ago. Over the ensuing decades, a large amount of data emerged concerning the mechanism(s) of action of these agents, culminating in the approval of two HDACIs, vorinostat (Zolinza) and romidpesin (Istodax), for the treatment of patients with cutaneous T-cell lymphoma. However, despite this large body of information, and intriguing evidence, both preclinical and clinical, suggesting a role for HDACIs in other malignancies (e.g., acute myeloid leukemia and multiple myeloma), there is a general sense that HDACIs have not fully realized their potential as antineoplastic agents. One of the major barriers to this goal is residual uncertainty about the mechanism of action by which these agents in fact trigger transformed cell death. Complicating efforts to resolve this issue has been the emerging realization that HDACIs are truly pleiotropic agents which act through a wide variety of disparate and mutually interactive mechanisms. In this context, evidence that HDACIs modify gene expression, that is, by altering chromatin structure, acetylating promoter regions, or disabling corepressors, is undisputed. However, HDACIs have many other nonhistone protein targets which may, either directly or indirectly, influence cell fate, and it is highly unlikely that such actions would not intersect with those mediated by the canonical effects of HDACIs on gene expression. This volume provides a summary of recently evolving insights into the mechanism of action of HDACIs, with the hope that such information will lay the foundation for expanding use of these agents in cancer therapy. For example, it stands to reason that HDACI actions would be closely intertwined with their effects on members of the Bcl-2 family of pro- and antiapoptotic proteins, and recent studies shed significant light on this important topic. In addition, HDACIs modify, through multiple mechanisms, various cell cycle checkpoints in transformed cells, and this capacity may be related to the putative selectivity of this class of agents. In this regard, the basis for preferential toxicity of HDACIs toward transformed cells has long been a subject of interest, and recent evidence suggests that this feature is intimately related to the question of the development of resistance to these agents. Furthermore, the selective toxicity of HDACIs has been attributed to their ability to induce oxidative injury and DNA damage through multiple mechanisms, including acetylation and disabling of DNA damage repair processes. HDACIs can also induce various forms of proteotoxic stress, in part through their capacity to disrupt the function of chaperone proteins. In addition, HDACIs may act, at least in part, by interfering with multiple systems responsible for responding to such stresses, including proteasomal or autophagic degradation, or the unfolded protein response. Finally, the broad range of actions of HDACIs render them, in many respects, ideal candidates for rational combination strategies, particularly when paired with other "targeted" agents. It is hoped that a better understanding of the mechanism(s) by which HDACIs will lead to tangible improvement in their established activity in certain malignancies, and extension to other indications, including both hematologic and nonhematologic malignancies.

STEVEN GRANT

CHAPTER ONE

Histone Deacetylase Inhibitors Disrupt the Mitotic Spindle Assembly Checkpoint By Targeting Histone and Nonhistone Proteins

Brian Gabrielli[1], Mellissa Brown[2]

The University of Queensland Diamantina Institute, Princess Alexandra Hospital, Brisbane, Queensland, Australia

[1]Corresponding author: e-mail address: brianG@uq.edu.au

[2]Current address: Division of Blood Cancers, Department of Clinical Haematology, Australian Centre for Blood Diseases, Monash University, Prahran, Victoria, Australia

Contents

Advances in Cancer Research, Volume 116
ISSN 0065-230X
http://dx.doi.org/10.1016/B978-0-12-394387-3.00001-X

Abstract

Histone deacetylase inhibitors exhibit pleiotropic effects on cell functions, both *in vivo* and *in vitro*. One of the more dramatic effects of these drugs is their ability to disrupt normal mitotic division, which is a significant contributor to the anticancer properties of these drugs. The most important feature of the disrupted mitosis is that drug treatment overcomes the mitotic spindle assembly checkpoint and drives mitotic slippage, but in a manner that triggers apoptosis. The mechanism by which histone deacetylase inhibitors affect mitosis is now becoming clearer through the identification of a number of chromatin and nonchromatin protein targets that are critical to the regulation of normal mitotic progression and cell division. These proteins are directly regulated by acetylation and deacetylation, or in some cases indirectly through the acetylation of essential partner proteins. There appears to be little contribution from deacetylase inhibitor-induced transcriptional changes to the mitotic effects of these drugs. The overall mitotic phenotype of drug treatment appears to be the sum of these disrupted mechanisms.

ABBREVIATIONS

CPC chromosomal passenger complex
HDACi histone deacetylase inhibitor
HP1 heterochromatin protein 1
SAC spindle assembly checkpoint

1. INTRODUCTION

Histone deacetylase inhibitors (HDACi) affect a wide range of biological functions and have been proposed or trialed for the treatment of a diverse range of diseases, including muscular dystrophies and neurodegenerative diseases, autoimmunity, inflammatory and immune responses, and more recently, as potential anti-HIV agents (Choi & Reddy, 2011; Grabiec, Tak, & Reedquist, 2011; Li, Jiang, Chang, Xie, & Hu, 2011; Matalon, Rasmussen, & Dinarello, 2011; Shakespear, Halili, Irvine, Fairlie, & Sweet, 2011). In the majority of these cases, the therapeutic effects of HDACi treatment are thought to occur through the epigenetic remodeling of the target organ genome. These therapeutic applications are relatively recent developments, and the main impetus for the clinical development of these drugs as therapeutic agents has been driven by their demonstrated anticancer effects.

The majority of HDACi are pan-isoforms inhibitors of two classes of histone deacetylases (HDACs): class I HDACs (HDAC1-3 and HDAC8) and class II HDACs (HDAC4-7, HDAC9, and HDAC10). HDACs have been shown to regulate the acetylation state of nuclear histones and an increasing number of nonhistone proteins. Inhibition of HDACs leads to changes in expression of genes involved in the regulation of apoptosis, proliferation, and the cell cycle. Likewise, class I HDACs are found to be overexpressed in a range of tumor types: HDAC1 in gastric cancer, HDAC2 in colorectal cancer, and HDAC3 in colon cancer (Glozak & Seto, 2007; Ozdag et al., 2006; Wilson et al., 2006; Zhu et al., 2004). Depletion and knockout of individual HDACs have uncovered the unique biological roles of the specific HDACs. HDAC1 and HDAC3 appear to be involved in regulating proliferation (Bhaskara et al., 2008; Wilson et al., 2006), while HDAC2 appears to negatively regulate apoptosis (Senese et al., 2007; Weichert et al., 2008). On the other hand, class II HDACs do not appear to regulate cell proliferation and are instead primarily involved in cellular development and differentiation (Verdin et al., 2003).

HDACi are an emerging class of anticancer drugs which possess tumor selective cytotoxicity. These drugs produce a range of effects on tumor cells including promoting the expression of differentiation markers, cell cycle effects, and induction of apoptosis in tumor cells but have little cytotoxic effect on normal cells (Bolden et al., 2006; Lindemann et al., 2004). Progression through the HDACi-induced aberrant mitosis induces rapid cell death (Blagosklonny et al., 2002; Dowling et al., 2005; Warrener et al., 2003), and this appears to be a significant contributor to the cytotoxicity of these drugs. They have also been shown to have antiangiogenic properties (Ellis, Hammers, & Pili, 2009). These drugs are currently in clinical trials as either monotherapies or in combination with other anticancer agents. The recent approval of Vorinostat (SAHA) for the treatment of cutaneous T-cell lymphoma highlights the potential of these drugs as anticancer therapeutics. HDACi treatment induces a range of transcriptional changes, although these are dependent on the HDACi and the cell line used (Glaser et al., 2003; Peart et al., 2005). The antitumor effects observed in response to HDACi are commonly thought to be the direct consequence of transcriptional changes. However, the proposed targets and mechanisms of the anticancer action of these drugs are widely varied. These include altering the expression of apoptotic regulators (Frew, Johnstone, & Bolden, 2009; Schrump, 2009) and telomerase (Rahman & Grundy, 2011), immunological recognition of cancers (Leggatt & Gabrielli, 2012), and disruption of the cell cycle through various mechanisms. However, HDACs are associated with a range of other transcription-independent functions, including the regulation of protein chaperone HSP90 activity which ultimately affects a range of cellular processes (Aoyagi & Archer, 2005; Schrump, 2009; Xu, Parmigiani, & Marks, 2007) and DNA damage recognition in association with Ku70 (Subramanian et al., 2005).

The ability of different HDACi to inhibit proliferation is largely attributed to their capacity to inhibit cell cycle progression at different phases of the cell cycle. HDACi can act at multiple points in the cell cycle to block progression; most notably in G1, G2, and mitosis, although each cell-phase block is associated with a distinct outcome. This review will focus on the G2/M phase effects of these drugs and the molecular mechanism by which the drugs trigger these outcomes, although for completeness a brief summary of the G1 phase mechanism will also be provided.

2. HDACi DYSREGULATE CELL CYCLE PROGRESSION

2.1. HDACi-induced G1 phase arrest

Progression through the cell cycle is controlled by the ordered activation of a family of protein kinases known as the Cyclin-dependent kinases (Cdks). Individual Cdks associate with a limited repertoire of Cyclins at specific stages of the cell cycle, and the activity of discrete Cyclin/Cdks complexes controls progression through each cell cycle stage. Cell cycle progression from G1 into S phase requires the sequential activation of Cyclin D/Cdk4 or Cdk6, and Cyclin E/Cdk2, although Cyclin E/Cdk2 can perform both functions in cells that lack Cyclin D/Cdk4 activity (Gray-Bablin et al., 1996). These two Cdk complexes phosphorylate Rb to allow full activation of the Rb-bound E2F, a transcription factor required for the expression of an array of genes required for progression into and through S phase (reviewed in Sherr, 2000). Cyclin E/Cdk2 and Cyclin A/Cdk2 also have critical roles in the initiation and maintenance of DNA replication (Ohtsubo, Theodoras, Schumacher, Roberts, & Pagano, 1995; Pagano, Pepperkok, Verde, Ansorge, & Draetta, 1992; Strausfeld et al., 1996).

The HDACi-induced G1 phase arrest is a consequence of the inhibition or loss of G1 phase Cdk/Cyclin complexes responsible for Rb phosphorylation, which ultimately blocks progression into S phase and DNA replication (Qiu et al., 2000; Sambucetti et al., 1999; Sandor et al., 2000). The mechanism involves HDACi-mediated upregulation of genes that negatively control G1/S phase progression, the most prominent being the Cdk inhibitor *CDKN1A* ($p21^{WAF1/CIP1}$), which occurs through increased acetylation of the chromatin in the *CDKN1A* promoter region, leading to increased transcription of the *CDKN1A* locus (Huang, Sowa, Sakai, & Pardee, 2000; Richon, Sandhoff, Rifkind, & Marks, 2000; Sambucetti et al., 1999; Xiao, Hasegawa, & Isobe, 2000). Additional modifications to the transcriptional machinery may also be required for upregulated *CDKN1A* expression (Gui, Ngo, Xu, Richon, & Marks, 2004; Kim et al., 2003). The increased $p21^{WAF1/CIP1}$ protein binds to and inhibits the critical S phase regulators, Cyclin E and Cyclin A/Cdk2, and deletion of the *CDKN1A* gene or in the few cell lines where HDACi treatment fails to upregulate $p21^{WAF1/CIP1}$ expression alleviates this G1 phase arrest (Archer, Meng, Shei, & Hodin, 1998; Burgess et al., 2001; Sandor et al., 2000).

HDACi treatment has also been reported to upregulate the expression of other Cdk inhibitors, such as the Cdk4- and Cdk6-specific $p15^{INK4B}$ and $p19^{INK4D}$, and the $p21^{WAF1/CIP1}$-related $p57^{KIP2}$ (Hitomi, Matsuzaki, Yokota, Takaoka, & Sakai, 2003; Mitsiades et al., 2004; Sato et al., 2003; Yokota et al., 2004). The INK4 family members inhibit S phase entry through inhibiting the Cdk4/Cyclin D complexes, while the $p21^{WAF1/CIP1}$-related $p57^{KIP2}$ protein inhibits the Cdk2 complexes (Satyanarayana & Kaldis, 2009). The INK4 proteins can also promote the translocation of $p21^{WAF1/CIP1}$ family proteins from Cdk4 complexes where they have little inhibitory activity *in vivo* and onto Cdk2 complexes of which they are potent inhibitors, thereby reinforcing the G1 arrest (McConnell, Gregory, Stott, Hara, & Peters, 1999). The HDACi-induced upregulation of $p21^{WAF1/CIP1}$ is widely observed in both normal and tumor cell lines (Burgess et al., 2001; Qiu et al., 2000) and is one of the most commonly upregulated genes in response to various HDACi across multiple cell lines (Glaser et al., 2003). The G1 phase arrest can deliver a cytostatic effect, but this can be readily reversed with removal of the drug. HDACi are an unusual class of anticancer drugs in that they have almost equivalent anticancer activity against proliferating and nonproliferating cancer cells, and normal tissue is reportedly resistant to the cytotoxic effects of HDACi, particularly when quiescent (Burgess et al., 2004). In the proliferating population, one of the drivers of the cytotoxicity of these drugs is their effects on G2/M phase progression. This chapter will focus on the effects that HDACi treatment has on this cell cycle transition. It will examine the literature on the effects on G2 phase progression, how lack of G2 phase arrest appears to underlie the selective cytotoxicity reported for these drugs, and the effects on mitosis—which in the absence of a G2 phase arrest deliver the cytotoxic insult to the HDACi-sensitive tumor cells.

2.2. HDACi-induced G2/M phase responses

The HDACi-induced G2/M phase effects are commonly reported, although in many cases the evidence for G2/M phase arrest is based on DNA profiles obtained through flow cytometry, which show an increased proportion of cells with 4N DNA content following HDACi treatment. These results may not be a true indication of a G2 phase arrest as is often claimed, but rather a measure of failed mitosis and cytokinesis, which have been well documented as a common response to these drugs (Magnaghi-Jaulin, Eot-Houllier, Fulcrand, & Jaulin, 2007; Stevens, Beamish, Warrener, & Gabrielli, 2008). Where the presence of a G2 phase arrest has been more comprehensively demonstrated, it has only been reported to occur in response to concentrations of HDACi that produce maximal levels of increased histone acetylation (Burgess et al., 2001; Nome et al., 2005; Prystowsky et al., 2009; Richon et al., 2000; Sandor et al., 2000), and in comparatively few cell lines (Qiu et al., 2000), whereas the G1 phase arrest is induced with lower HDACi concentrations. Interestingly, cell lines that arrest in G2 phase with HDACi treatment are resistant to the cytotoxic effects of these drugs (Krauer et al., 2004; Qiu et al., 2000; Richon et al., 2000), and loss of the G2 phase arrest in response to HDACi treatment sensitizes cells to apoptosis, whereas imposing a G2 phase arrest has a protective effect (Krauer et al., 2004; Qiu et al., 2000). Cells that are incompetent for the G2 phase arrest and proceed into the mitotic phase undergo an aberrant mitosis (Blagosklonny et al., 2002; Cimini, Mattiuzzo, Torosantucci, & Degrassi, 2003; Qiu et al., 2000; Robbins et al., 2005; Shin et al., 2003; Taddei, Maison, Roche, & Almouzni, 2001; Warrener et al., 2003) that triggers failed cytokinesis and apoptosis (Qiu et al., 2000; Stevens et al., 2008). The protection that is provided by the G2 phase arrest is likely to result from inhibiting the cells from entering mitosis until the effects of HDACi have been degraded or reversed after removal of the drugs (Fig. 1.1). In the remainder of this review, we will present an outline of our understanding of the mechanisms that underlie the protective G2 phase arrest and the cytotoxic aberrant mitosis triggered by HDACi.

2.3. Mechanisms governing normal G2/M phase progression

Progression from G2 phase into mitosis is controlled by a complex network of pathways. However at a basic level, entry into mitosis is controlled through the activation of Cyclin B/Cdk1, which is either regulated by

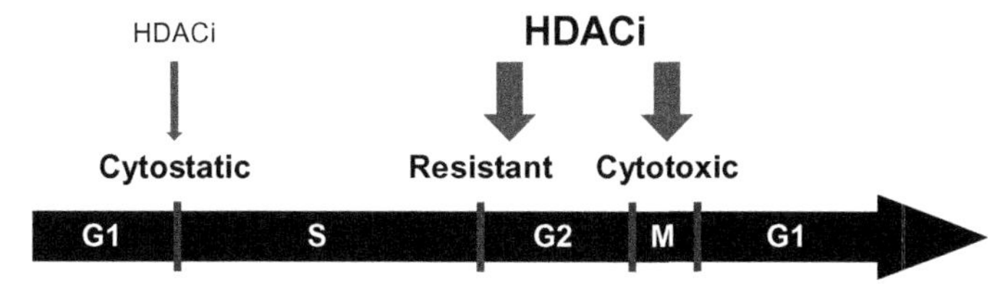

Figure 1.1 Relatively low concentrations of HDACi that do not produce maximal histone acetylation will produce a G1 phase arrest, whereas higher concentrations that produce maximal histone hyperacetylation trigger a G2 phase arrest in a subset of cell lines, particularly normal cells and aberrant mitosis in the majority of cell lines that fail to G2 phase arrest. (For color version of this figure, the reader is referred to the online version of this chapter.)

factors that directly influence this activation through positive regulation, or indirectly through the negative regulation of Cyclin B/Cdk1-activators. In mammalian cells, there are two mechanisms that independently influence Cyclin B/Cdk1 activation. The first occurs directly through positive regulation via the Cdc25 family of dual specificity phosphatases (Boutros, Dozier, & Ducommun, 2006), and the second of which occurs indirectly through Plk1-dependent regulation of negative mitotic regulators such as Wee1 (Lindqvist, Rodriguez-Bravo, & Medema, 2009). Upstream of Plk1 is Aurora A kinase which cooperates with hBora to regulate Plk1 activation (Macurek, Lindqvist, & Medema, 2009). The G2 phase Cyclin A/Cdk2 complex also has a role in regulating the timing of Cyclin B/Cdk1 activation and thus entry into mitosis (De Boer et al., 2008; Fung, Ma, & Poon, 2007). Although the mechanism by which this occurs is not fully understood, Cyclin A/Cdk2 has been shown to regulate Cyclin B levels (Lindqvist et al., 2009) and to negatively regulate Wee1 function (Fung et al., 2007; Li, Andrake, Dunbrack, & Enders, 2010; Fig. 1.2).

2.4. HDACi-sensitive G2 phase arrest

In cell lines that are competent for the HDACi-induced G2 phase arrest, the arrest is triggered by the continuous presence of HDACi throughout S phase (Lallemand et al., 1999; Qiu et al., 2000; Richon et al., 2000). The arrest operates independently of the well-studied DNA damage-induced G2 checkpoint arrest mechanisms, as DNA damaging agents can initiate a G2 phase checkpoint arrest in cells that failed to arrest in response to HDACi treatment (Qiu et al., 2000). The mechanism by which the HDACi impose the G2 phase arrest is unclear. A range of different mechanisms have been demonstrated to trigger a G2 phase arrest including: downregulation of

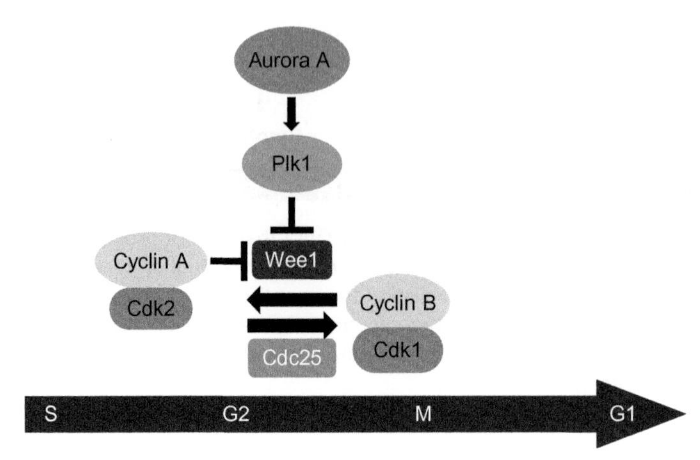

Figure 1.2 Model for the normal mechanisms controlling G2/M progression. Activation of Cyclin B/Cdk1 drives entry into mitosis. This is activated by Cdc25 and inhibited by Wee1. Cyclin A/Cdk2 controls the timing of Cyclin B/Cdk1 activation in part through regulating Wee1. Aurora A-Plk1 also regulates Wee1 to control mitotic entry. (For color version of this figure, the reader is referred to the online version of this chapter.)

G2/M phase Cyclins (Maity et al., 1996), increased expression of Cdk inhibitor proteins such as p21 (Bunz et al., 1998), and activation of checkpoint signaling through either ATM-Chk2, ATR-Chk1 (Niida & Nakanishi, 2006) or p38MAPK–MK2 (Manke et al., 2005). Expression of Epstein–Barr virus latent proteins disables ATM/ATR-dependent checkpoint signaling and the HDACi-induced G2 phase arrest (Krauer et al., 2004), which suggests the involvement of the canonical ATM/ATR checkpoint signaling pathway. However, there is no evidence that either ATM/ATR or downstream Chk1 or Chk2 signaling is activated by HDACi treatment (Lee, Choy, Ngo, Venta-Perez, & Marks, 2011; Mikhailov, Shinohara, & Rieder, 2004).

HDACi treatment has been shown to block the activation of the G2/M complexes, Cyclin A/Cdk2 and Cyclin B/Cdk1, and decrease Cyclin B1 protein levels (Lallemand et al., 1999; Qiu et al., 2000). Acetylation of Cyclin A reportedly destabilizes the protein (Mateo et al., 2009), which is likely to be responsible for reducing the levels of this critical G2 phase regulator. The protein level of Aurora A has also been reported to be reduced with HDACi treatment, possibly as a consequence of HSP90 inhibition by decreased association with HDAC6 (Cha et al., 2009). Similarly, the expression of Plk1 itself has also been shown to be down regulated in response to HDACi treatment (Lallemand et al., 1999;

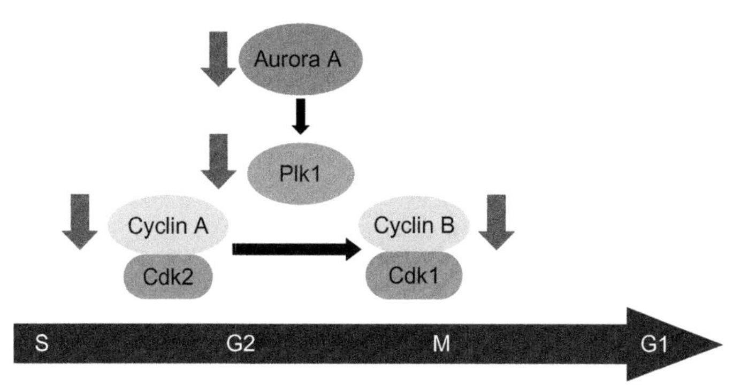

Figure 1.3 With HDACi treatment, the levels of the indicated mitotic regulators are reduced, either through reduced mRNA levels or protein stability, producing a G2 phase arrest in the subset of cell lines capable of triggering a G2 phase arrest in response to HDACi. The molecular basis determining the ability to impose this arrest is unclear, but is present in normal cell types. (For color version of this figure, the reader is referred to the online version of this chapter.)

Prystowsky et al., 2009; Fig. 1.3). Finally, the expression of Gadd45, a growth arrest and DNA damage-inducible gene that can cause a G2/M arrest by inhibiting cdc2/Cyclin B activity (Jin et al., 2000), has also been reported to be upregulated following HDACi treatment, and may also participate in the G2 arrest (Hirose et al., 2003).

A second G2 checkpoint mechanism is induced upon HDACi treatment when cells are in antephase, which occurs immediately prior to nuclear membrane breakdown. This arrest utilizes the p38MAPK–MAPKAPK2 pathway to inhibit entry into mitosis (Manke et al., 2005; Mikhailov et al., 2004). This mechanism, the antephase checkpoint, appears to be distinct from the G2 phase checkpoint initiated in response to S phase HDACi addition.

3. HDACi DISRUPT NORMAL MITOSIS

3.1. Mitotic effects of HDACi treatment

In the majority of immortalized, virally transformed or tumor cell lines, the HDACi-sensitive G2 phase arrest is defective (Burgess et al., 2001; Qiu et al., 2000). Nonimmortalized primary cells with a functional HDACi-induced G2 phase arrest are resistant to killing by even high doses of HDACi, whereas the cell lines which lack this G2 phase arrest response are sensitive to killing by these drugs. The consequence of failing to arrest in G2 phase with HDACi treatment is that cells proceed through an aberrant mitosis. Defective mitosis is not itself sufficient to drive cells into death, as aberrant mitosis is a relatively common occurrence in even normally cycling cells. Cells have a mitotic checkpoint mechanism, known as the spindle assembly checkpoint (SAC), that responds to mitotic defects and arrests cells in mitosis until such defects are repaired. If repair cannot be achieved, then the apoptosis is triggered after an extended mitotic arrest, which in many cases is >16 h (Huang, Shi, Orth, & Mitchison, 2009). This checkpoint is in place to allow for minor defects in partitioning the replicated genome to be rectified, or where the defects are beyond repair to prevent the survival of daughter cells that have either gained or lost parts or entire chromosomes. Thus, whereas the presence of a functional HDACi-induced G2 phase arrest is protective, its absence leads to cells entering an aberrant mitosis which can trigger cell death, and therefore represents the selective cytotoxicity reported for this class of drugs (Burgess et al., 2001; Qiu et al., 2000; Qiu et al., 1999). The mechanism through which the aberrant mitosis is induced by HDACi treatment is not immediately clear. It does not appear to be dependent on transcriptional changes induced by HDACi treatment. HDACi can induce aberrant mitosis in *Drosophila* early embryos which run off maternal mRNA and protein stores, thus providing evidence that HDACi disrupt mitosis in a transcription-independent manner (Warrener, Chia, Warren, Brooks, & Gabrielli, 2010). Further, no consistent changes in the expression of mitotic regulators were found in microarray analyses of various HDACi-treated cell lines (Glaser et al., 2003; Peart et al., 2005), whereas the mitotic defects are a common feature in the HDACi treatment of a wide range of cell lines (Dowling et al., 2005; Qiu et al., 2000; Robbins et al., 2005; Stevens et al., 2008; Taddei et al., 2001; Warrener et al., 2003).

To gain a better understanding of the molecular effects of these drugs on mitosis, in particular, SAC function and chromosome partitioning, a brief introduction to the SAC, the centromeric chromatin upon which it is set, the cohesin complex which maintains sister chromatid cohesion, and the connection of SAC activation with apoptotic signaling is warranted.

3.2. The SAC

The SAC ensures that the replicated genome is partitioned with fidelity to produce two genetically identical daughter cells. It does this by sensing the attachment of kinetochores to both poles of the microtubule spindle, with equal tension across the kinetochores ensuring that one copy of each of the replicated chromosomes is physically retracted to poles of the cells during anaphase. Failure to detect correct kinetochore attachments or tension signals a mitotic arrest by blocking the activity of the anaphase-promoting complex/cyclosome (APC/C), a proteasomal complex which catalyzes the proteolytic degradation key mitotic proteins. Prominent targets of the APC/C include Cyclin B, which is the regulatory component of the key mitotic kinase Cyclin B/Cdk1, and securin, an inhibitor of the separase protease that cleaves the Rad21 subunit of the cohesin complex resulting in sister chromatid separation (Nezi & Musacchio, 2009; Pines, 2011). Mitotic exit only requires the loss of Cyclin B/Cdk1 activity (Potapova et al., 2006), but the APC/C-dependent destruction of Cyclin B, securin, and other mitotic regulators, is essential for the correct partitioning of the genome (Potapova, Daum, Byrd, & Gorbsky, 2009; Skoufias, Indorato, Lacroix, Panopoulos, & Margolis, 2007; Fig. 1.4).

The kinetochore-associated SAC machinery is the key sensor of spindle integrity and function. It consists of the budding uninhibited by benzimidazole (BUB) and mitotic-arrest deficient (MAD) genes, which were originally identified in yeast genetic screens of spindle function. These proteins localize to unattached or incorrectly tensioned kinetochores in mitosis and their localization is indicative of SAC activation blocking APC/C activity and anaphase onset (Musacchio & Salmon, 2007; Przewloka & Glover, 2009). The function of the kinetochores during mitosis is to provide an anchor point for the binding of the kinetochore microtubules (k-microtubules) and to detect the equal tension and attachment of bipolar microtubules. Kinetochores transmit this information via a series of kinases to block APC/C function, essentially by regulating the conformational change in MAD2 from the inactive open (O-Mad2) to the active closed (C-Mad2) conformation, which forms a complex with Bub3, BubR1, and Cdc20, collectively called the mitotic checkpoint complex (MCC) that inhibits APC/C activity. C-Mad2 forms a complex with Mad1, which localizes to unattached kinetochores during mitosis where the C-Mad2 acts as a template to convert O-Mad2 to C-Mad2, which then binds and sequesters Cdc20 into the MCC, and ultimately inhibits the APC/C (Musacchio & Salmon, 2007; Pines, 2011). When all of the kinetochores have achieved bipolar k-microtubule attachment, the Mad1-C-Mad2 complex and other kinetochore SAC components are stripped from the kinetochore (Howell et al., 2001; Musacchio & Salmon, 2007), and together with other factors including the Mad2 binding protein, p31Comet, remove Mad2-dependent

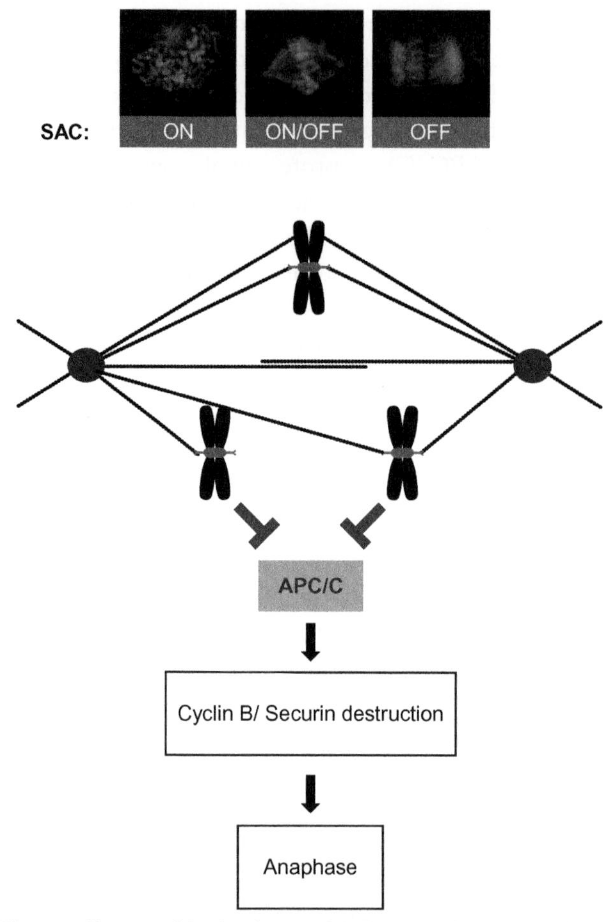

Figure 1.4 The spindle assembly checkpoint (SAC) responds to unattached kinetochore and uneven tension across the centromeres to inhibit APC/C activity and thereby mitotic exit. The SAC remains active until all kinetochores attain bipolar microtubule attachment and tension. (See Page 1 in Color Section at the back of the book.)

MCC inhibition of APC/C activity, allowing anaphase to proceed and promote mitotic exit (Fava, Kaulich, Nigg, & Santamaria, 2011; Varetti, Guida, Santaguida, Chiroli, & Musacchio, 2011; Westhorpe, Tighe, Lara-Gonzalez, & Taylor, 2011).

In addition to the core SAC components mentioned above, several other kinases are involved in mitotic progression, most notably Plk1 and Aurora B. Aurora B associates with Survivin, INCENP, and Borealin as part of the chromosomal passenger complex (CPC), which is named for its distinctive change in localization during mitosis, where it translocates from the centromeres in prophase to the forming mid-spindle in anaphase, and then to the midbody in cytokinesis (Vagnarelli & Earnshaw, 2004). Inhibition or depletion of any of these components is sufficient to mislocalize the complex, which disrupts kinetochore localization of the core SAC components (Ditchfield et al., 2003) and involves the Aurora B substrate, Zwint1 (Kasuboski et al., 2011). Aurora B has a range of substrates in mitosis that includes histone H3, the centromere-specific H3 homolog CenpA (Zeitlin, Shelby, & Sullivan, 2001), and MCAK that regulates microtubule binding (Zhang, Lan, Ems-McClung, Stukenberg, & Walczak, 2007).

The other components of the CPC are involved in specific localization and activation of the Aurora B. Survivin binds to the Thr3-phosphorylated histone H3, which is specifically localized to centromeres and responsible for the accumulation of the CPC at the centromeres in mitosis (Kelly et al., 2010; Yamagishi, Honda, Tanno, & Watanabe, 2010); this localization that is sufficient for the SAC function of Aurora B (Becker, Stolz, Ertych, & Bastians, 2010). Aurora B is responsible for regulating the biorientation of k-microtubules and for responding to low tension across the sister centromeres. In the former role, it is responsible for releasing syntelic microtubule attachment through the phosphorylation of multiple kinetochore proteins (Chan, Jeyaprakash, Nigg, & Santamaria, 2012; Maresca, 2011; Maresca & Salmon, 2010). Recently, it has been reported to be responsible for the centromere-specific activation of Plk1, which is responsible for stabilizing microtubule-kinetochore attachments (Carmena et al., 2012). Aurora B forms a diffusible gradient across the centromere, having its highest concentration at the inner centromere and reducing out to the kinetochore where it is involved in regulating microtubule attachment (Maresca & Salmon, 2010; Wang, Ballister, & Lampson, 2011). Plk1 has a number of critical roles in mitosis including the unloading of the cohesion complex from the chromosome arms (Hauf et al., 2005), and the upregulation of APC/C activity and SAC function (Petronczki, Lenart, & Peters, 2008; van de Weerdt et al., 2005).

3.3. The centromere: the constitutive heterochromatin underlying the kinetochore

DNA is packaged around nucleosomes, and it is the posttranslational modification of the nucleosome histone proteins that dictate the degree of compaction of the chromatin. In general, transcriptionally active euchromatin is organized into a more "open" state of compaction which allows ready access to the chromatin associated DNA, whereas transcriptionally silenced DNA is held in a highly condensed heterochromatic or "closed" state. The compacted heterochromatin is marked by specific methylations of histone proteins, particularly histone H3 Lys9, which is trimethylated by the SUV39H methyltranferases and forms the binding site for the chromodomain-containing protein, heterochromatin protein 1 (HP1; Bannister et al., 2001; Jacobs et al., 2001; Peters et al., 2001). Additional histone modifications found in heterochromatin include trimethylation of H4 Lys20 and H3 Lys 27, which are recognized by a different set of proteins (Martin & Zhang, 2005). Dimethylated H3 Lys9 is widely spread across the chromosomes, however, the trimethylated H3Lys9 and HP1 binding are specifically associated with heterochromatin regions of the chromosome and accumulate at the constitutive heterochromatin surrounding the centromere (Schmiedeberg, Weisshart, Diekmann, Meyer Zu Hoerste, & Hemmerich, 2004; Warrener et al., 2010).

The centromere is the region of the chromosome that is readily observed in mitotic chromosomes as the primary constriction where the sister chromatids remain attached. The physical appearance of this constriction suggests that the chromatin in this region is highly condensed. The kinetochore sits directly on the outer face of the centromere, with the inner face contacting the sister chromatid. Centromere structure is critical to the function of the kinetochore, and disruption of centromere structure is known to dramatically affect the kinetochore function. The centromere is generally based on long stretches of repetitive α-satellite DNA, although the presence of this is neither sufficient nor necessary to specify centromere location. Specification of the centromeric region starts with the incorporation of a specialized variant of histone H3 known as CenpA or CenH3 into the nucleosomes. During replication of the centromeric DNA, preexisting nucleosomes are partitioned between the template and daughter strand, and new CenpA-containing nucleosomes are not loaded until early G1 phase (Allshire & Karpen, 2008; Jansen, Black, Foltz, & Cleveland, 2007; Torras-Llort, Moreno-Moreno, & Azorin, 2009). The loading of CenpA-containing nucleosomes is dependent on multiprotein complexes comprising the chaperone RbAp48 and/or Mis18 protein (Allshire & Karpen, 2008; Fujita et al., 2007; Przewloka & Glover, 2009). CenpA nucleosomes are interspersed with canonical H3-containing nucleosomes in the centromere, such that folding of DNA results in the CenpA nucleosomes aligning in the outer centromere where the kinetochore binds on the

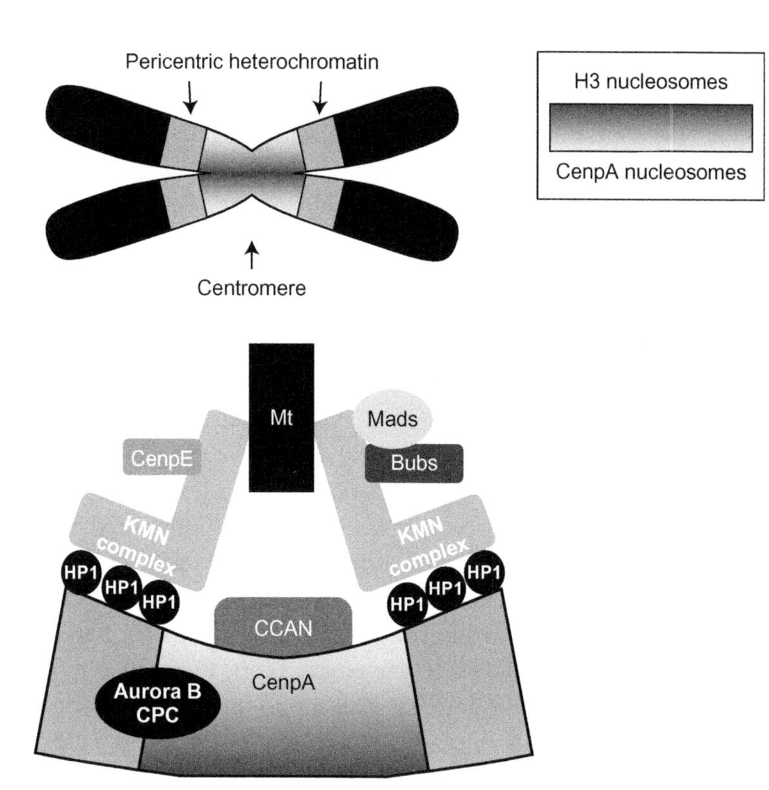

Figure 1.5 (A) The centromere is organized such that CenpA-containing nucleosomes are placed on the outer face of the centromere immediately adjacent to the kinetochore, while H3 containing nucleosomes form the inner face of the sister chromatids. (B) The CenpA-containing centromere is flanked by pericentric heterochromatin which accumulates HP1 binding to the trimethylated H3 Lys9. The CCAN complex associates with the centromere constitutively. The HP1 binding assists in localizing the Mis12 complex to the kinetochores which in turn localizes the other components of the KMN complex, Blinkin, Knl1, Ndc80, and facilitates microtubule (Mt) binding. KMN complex localization also facilitates binding of SAC components (Mads and Bubs) and CenpE. The Aurora B containing chromosomal passenger complex (CPC) localizes to the pericentric heterochromatin and operates in a gradient diffusing to the outer kinetochore. (See Page 2 in Color Section at the back of the book.)

outer face of sister chromatids, while the H3 nucleosomes form the inner centromere where the sister chromatids attach (Blower, Sullivan, & Karpen, 2002; Fig. 1.5A). The centromere is flanked by heterochromatin regions (Sullivan & Karpen, 2004), and the pericentric heterochromatin appears to be required to specify CenpA-loading regions and hence centromeres (Gonzalez-Barrios, Soto-Reyes, & Herrera, 2012; Stimpson

& Sullivan, 2010). In addition to CenpA, the constitutive centromere associated network (CCAN) complex, which consists of Cenp proteins binding to the outer centromere throughout the cell cycle, acts as a bridging interface between the centromere and kinetochore and is required for correct kinetochore function (Hori et al., 2008; Torras-Llort et al., 2009). The k-microtubules directly associate with kinetochores through binding the KMN (KLN1, Mis12, and NDC80 complexes) network (Gascoigne & Cheeseman, 2011; Przewloka & Glover, 2009). Localization of the Mis12 complex is regulated by HP1 binding (Gonzalez-Barrios et al., 2012; Obuse et al., 2004), the role of HP1 appears to be to localize the Mis12 complex to centromere during interphase; however, this binding is dispensable during mitosis (Kiyomitsu, Iwasaki, Obuse, & Yanagida, 2010). The Mis12 complex (Mis12, Mis13, Mis14, and hNnf1/PMF1) binds to Blinkin (hSpc105/hKNL1) which is, in turn, responsible for binding and localizing the SAC signaling proteins Bub1 and BubR1 and the microtubule-binding NDC80 complex to the kinetochores (Kiyomitsu et al., 2010; Kiyomitsu, Obuse, & Yanagida, 2007; Kline, Cheeseman, Hori, Fukagawa, & Desai, 2006). Another complex binding to Blinkin is the Zwint1-containing RZZ complex involved in microtubule attachment to the kinetochore (Kiyomitsu et al., 2010; Kline et al., 2006; Przewloka & Glover, 2009). The kinetochore motor protein CenpE is another protein that is dependent on the Mis12 complex for loading onto the kinetochores (Gascoigne & Cheeseman, 2011; Kline et al., 2006; Fig. 1.5B). Therefore, disruption of expression or HP1-dependent binding of the Mis12 complex to the inner centromere CCAN complex can severely disrupt SAC function, resulting in failure of chromosome congression to the metaphase plate and segregation at cytokinesis.

3.4. Cohesins: maintaining sister chromatid cohesion until anaphase

Cohesin is a complex of four proteins: SMC1, SMC3, and kleisin/Rad21, which form a ring-like structure that encompasses the sister chromatids and SA1/2 subunits that associate with kleisin/Rad21 (Losada, 2008; Nasmyth, 2011). The cohesion complex is loaded onto the replicated sister chromatids during S phase and is responsible for their tight association until early prophase, when the cohesion subunits along the chromosome arms are removed by Plk1-mediated phosphorylation of the SA1/2 subunit (Hauf et al., 2005). The SMC3 subunit of cohesion is acetylated upon loading onto the chromatin in S phase. This acetylation is maintained until the cohesion complex is unloaded from the chromosome arms in prophase, at that time SMC3 is deacetylated by the class1 deacetylase HDAC8 in humans, or Hos1 in yeast. The acetylation of SMC3 appears to have a role in locking the cohesion ring to maintain sister chromatid cohesion (Beckouet et al., 2010; Nasmyth, 2011). The centromeric cohesion is protected from Plk1-dependent unloading by the centromeric localization of Shugoshin (Sgo1), which relocalizes protein phosphatase 2A (PP2A) to block phosphorylation of the cohesin and other centromeric proteins (Xu et al., 2009). Sgo1 is localized to the chromatin by Bub1-mediated phosphorylation of histone H2A Ser120 and restricted to the centromeres by the Aurora B CPC (Boyarchuk, Salic, Dasso, & Arnaoutov, 2007; Kawashima, Yamagishi, Honda, Ishiguro, & Watanabe, 2010; Yamagishi et al., 2010). The CPC is localized to the centromere by Haspin/GSG2-mediated phosphorylation of histone H3 Thr3, which acts as binding site for the Survivin subunit of the CPC to localize Aurora B (Dai, Sullivan, & Higgins, 2006; Kelly et al., 2010; Wang et al., 2010; Yamagishi et al., 2010). Therefore, the consequence of loss of the Mis12 complex from the centromeres (which also results in Bub1 failing to localize to the kinetochore) is decreased Sgo1 localization, which would contribute to a lack of sister chromatid cohesion (Kiyomitsu et al., 2010).

3.5. SAC activation is required for signaling apoptosis after mitotic failure

A common outcome of extended SAC-induced mitotic arrest is apoptosis. The connection between the SAC and apoptotic mechanisms in mitosis or early G1 phase is unsurprising, as there must be a mechanism to destroy cells that have failed to undergo high fidelity cell division. SAC function is required for this apoptosis (Nitta et al., 2004; Sudo, Nitta, Saya, & Ueno, 2004; Vogel, Hager, & Bastians, 2007). Inhibiting the spindle checkpoint with mitotic kinase inhibitors, or inactivation of kinetochore components causes rapid exit from spindle checkpoint arrest with little apoptosis (Chan, Jablonski, Sudakin, Hittle, & Yen, 1999; Ditchfield et al., 2003; Dobles, Liberal, Scott, Benezra, & Sorger, 2000; Hauf et al., 2003; Taylor & McKeon, 1997; Wassmann & Benezra, 2001). Extended SAC-dependent mitotic arrest using antimitotic drugs such as paclitaxel can lead to mitotic slippage, where cells prematurely exit mitosis, leading to either aneuploidy or apoptosis (Castedo, Perfettini, Roumier,

Andreau, et al., 2004; Gascoigne & Taylor, 2008; Rieder & Maiato, 2004; Tao et al., 2005). In addition, the bypassing of the G2 phase checkpoint arrest in response to a range of different stresses, including HDACi, leads to aberrant mitosis detected by the SAC and eventual mitotic slippage (Castedo, Perfettini, Roumier, Valent, et al., 2004; Gabrielli et al., 2007; Stevens et al., 2008; Vogel, Kienitz, Muller, & Bastians, 2005). The signal driving apoptosis appears to be dependent upon the degree of APC/C activation, with the major target possibly being Cyclin B and inactivation of Cyclin B/Cdk1. Depletion of Cdc20, the APC/C cofactor that targets Cyclin B destruction during mitotic exit, enhances cell killing in combination with SAC-activating agents, and this degree of cell killing was also observed in response to overexpression of a indestructible Cyclin B mutant (Huang et al., 2009). Cyclin B/Cdk1 has roles in regulating components of the apoptotic machinery in mitotic-arrested cells. The antiapoptotic proteins Bcl-2 and Bcl-Xl are strongly phosphorylated and inhibited by Cyclin B/Cdk1 in SAC-arrested cells, which potentially negates their ability to protect the mitochondria from proapoptotic Bcl-2-related family members (Terrano, Upreti, & Chambers, 2010). Other pathways also appear to contribute to determining the balance of anti- and proapoptotic signaling in mitotically arrested cells. The p21-activated kinase (PAK1) appears to have an antiapoptotic role in SAC-arrested cells and inhibiting PAK1 in mitotically arrested cells has been shown to initiate rapid apoptosis (Gabrielli et al., 2007). This is possibly due to the loss of the PAK1 phosphorylation of BimL that blocks its proapoptotic function (Vadlamudi et al., 2004). In addition, the Bcl-2-related Mcl-1 is destabilized in mitotically arrested cells, as is the inhibitor of apoptosis XIAP, which ultimately removes their antiapoptotic influence (Shi, Orth, & Mitchison, 2008; Tunquist, Woessner, & Walker, 2010; Wertz et al., 2011). The combined effect of these changes is inactivation or loss of antiapoptotic components of the apoptotic mechanism which normally bind and inhibit the proapoptotic BH3-only proteins (Fletcher & Huang, 2008; Willis et al., 2007). However, apoptosis is not initiated until a signal provides a proapoptotic stimulus, possibly through increased levels of these proapoptotic proteins.

Thus the spindle checkpoint not only blocks cell cycle progression in response to mitotic defects, but it also regulates apoptotic signaling during mitotic arrest to ensure that cells that cannot fulfill the requirements for normal partitioning of the replicated genome are destroyed.

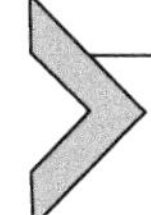

4. MECHANISM OF HDACi-INDUCED ABERRANT MITOSIS

4.1. HDACi treatment overcomes the SAC by disrupting CPC localization to the centromere in mitosis

The molecular basis of the mitotic defect that is observed following HDACi treatment is becoming clearer, with an increasing repertoire of proteins whose activity and/or function is regulated by acetylation, and thus likely to be either hyperacetylated or their acetylation stabilized by HDACi treatment. The effects of HDACi on the function and/or stability of some of these target proteins have been reported, but for many others there is currently no direct evidence of acetylation or acetylation of essential partner proteins that would mechanistically account for the drug effects observed.

The aberrant mitosis that arises from HDACi treatment results in initiation of the SAC, which blocks anaphase chromosome segregation until all of the chromosomes have achieved bipolar attachment to the mitotic spindle (see Section 3.2). There is some controversy in the literature as to whether the SAC is activated in HDACi-treated cells. Where cells have been followed by time-lapse microscopy or in synchronized populations, a mitotic delay is clearly evident, but it is only 2.5-fold the length of normal mitosis (<3 h) compared to the extended mitotic arrest observed with many of the usual antimitotic drugs that can exceed 10 h. Moreover, cells exit mitosis without having entered anaphase, and as a result, they fail to partition the replicated genome (Ma et al., 2008; Stevens et al., 2008). This is clear evidence of SAC failure and mitotic slippage (Weaver & Cleveland, 2005). Other reports have also shown that HDACi treatment causes an increase in the proportion of mitotic cells, as a direct consequence of extended mitosis (Blagosklonny et al., 2002; Cha et al., 2009; Chia, Beamish, Jafferi, & Gabrielli, 2010; Dowling et al., 2005). The kinetochore SAC signaling components: Mad2, Bub1, and BubR1, have been reported to load normally on unaligned chromosomes in HDACi-treated cells (Stevens et al., 2008), but others have reported a loss of Bub1 and BubR1 staining to kinetochores (Dowling et al., 2005; Ma et al., 2008; Robbins et al., 2005). This may be explained by the loading of these proteins onto kinetochores during the delay in prophase after HDACi treatment (Stevens et al., 2008), which would also explain the mitotic delay that is reported. However, the arrest cannot be sustained, and exit occurs with the SAC components detaching from the kinetochore, which allows mitotic exit and the normal loss of localization

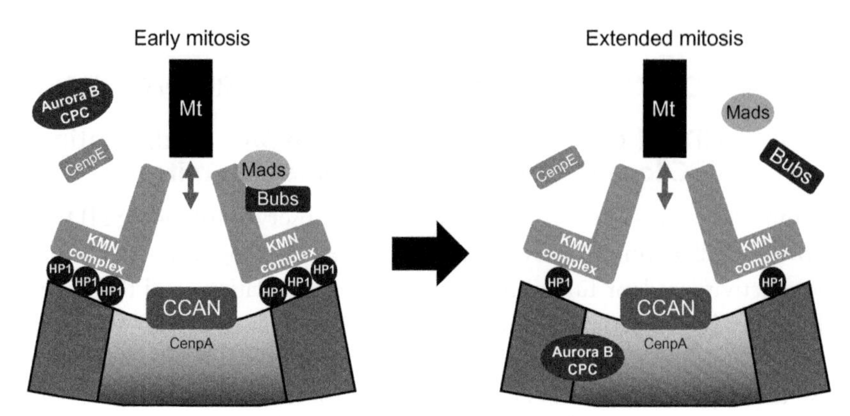

Figure 1.6 In HDACi-treated cells, the reduced levels of centromere-localized Aurora B/CPC initially have little effect on the loading of SAC components and CenpE or the KMN complex, although there is reduced Mt attachment. This triggers the SAC-dependent mitotic arrest. Cells remain arrested for 2- to 3-fold the normal period of mitosis, but as a consequence of reduced levels of Aurora B/CPC at the centromere and in the gradient out to the kinetochore, SAC protein and CenpE association with the kinetochore is lost, SAC signaling is reduced or terminated and cells prematurely exit mitosis. HP1 binding to the pericentric heterochromatin may be reduced but this does not affect KMN complex association with the centromere. (See Page 3 in Color Section at the back of the book.)

of these SAC components from the kinetochore (Dowling et al., 2005; Ma et al., 2008; Robbins et al., 2005; Stevens et al., 2008; Fig. 1.6). Further evidence that HDACi treatment effectively disrupts SAC function is demonstrated by the lack of stable mitotic arrest with either taxol or nocodazole treatment (Dowling et al., 2005; Magnaghi-Jaulin et al., 2007; Warrener et al., 2003). However, even in the presence of these antimicrotubule drugs, cells cotreated with HDACi exhibit a mitotic delay of 2–3 h, which is considerably less than the >10 h arrest that is observed with the antimicrotubule drugs alone. Further, the mitotic delay is similar in duration to HDACi treatment alone, indicating that the SAC is activated but cannot sustain the checkpoint arrest in the presence of HDACi (Stevens et al., 2008).

The mechanisms by which HDACi disrupt normal SAC are complex. HDACi treatment resulted in loss of CenpE and CenpF localization to the mitotic kinetochore (Ma et al., 2008; Robbins et al., 2005), whereas the inner centromere proteins—CenpA, CenpC (a component of CCAN), and the components of the KMN complex—Mis12 and NDC80/Hec1, are all retained at the centromere at their normal levels (Ma et al., 2008).

There is some dispute about the association of HP1 with the centromere after HDACi treatment. A number of reports have indicated a general loss of HP1 binding in mitotic cells occurring in response to HDACi treatment, with little effect of HP1 binding in G2 phase cells (Ma et al., 2008; Robbins et al., 2005; Warrener et al., 2010). There was also little reduction in the level of trimethylated histone H3 Lys9, consistent with retention of HP1 binding in the G2 phase cells (Warrener et al., 2010). Further, the lack of effect on Mis12 centromere localization, which is dependent on interphase HP1 binding to the pericentric heterochromatin (Kiyomitsu et al., 2010; Obuse et al., 2004), supports a relative lack of effect on HP1 binding (Fig. 1.6).

Despite HDACi treatment having no obvious effect on the levels or centromere localization of the microtubule-binding KMN complex, the attachment of the microtubules to the kinetochores is severely reduced (Ma et al., 2008; Robbins et al., 2005; Stevens et al., 2008). The lack of stable interactions between the microtubules and kinetochores explains the defects in spindle structure observed with HDACi treatment and is likely to be a consequence of a lack of Aurora B activity at the centromere. In normal cells, Aurora B/CPC accumulates at the centromere in mitosis, but this accumulation is absent in HDACi-treated cells (Ma et al., 2008; Stevens et al., 2008). The consequence of this reduction in centromeric Aurora B/CPC is loss of Aurora B and Plk1-dependent microtubule attachment to the kinetochore (Carmena et al., 2012; Chan et al., 2012; Maresca, 2011; Maresca & Salmon, 2010), and the loss of CenpE and SAC proteins from the kinetochore (Becker et al., 2010; Kim, Holland, Lan, & Cleveland, 2010; Maresca, 2011). The observation that they initially localize to produce an attenuated mitotic arrest may be explained by the fact that the concentration of Aurora B/CPC at the centromere is reduced, and thus the concentration of the Aurora B diffusing to the kinetochore lower still, but the complex remains active. Aurora B phosphorylation of CenpA is normal in HDACi-treated cells, demonstrating that the reduced levels of Aurora B are still capable of catalyzing centromeric phosphorylation events (Stevens et al., 2008). The regulation of Aurora B activation and CPC localization to the chromosome arms utilizes a different mechanism which is unaffected by HDACi treatment, and this is demonstrated by the lack of effect on histone H3 Ser10 phosphorylation on the chromosome arms (Cimini et al., 2003; Eot-Houllier, Fulcrand, Watanabe, Magnaghi-Jaulin, & Jaulin, 2008; Robbins et al., 2005; Stevens et al., 2008; Fig. 1.7).

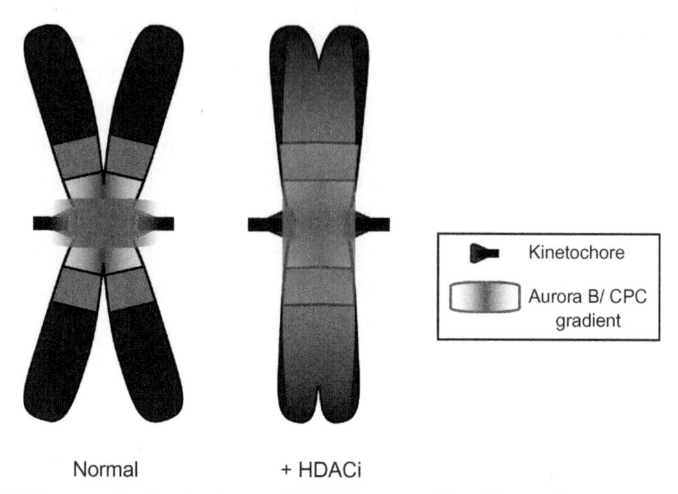

Figure 1.7 In normal mitosis, Aurora B/CPC forms a diffusible gradient centered on the centromere out to the kinetochore, regulating both centromeric and kinetochore functions. With HDACi treatment, the centromere accumulation is reduced, severely reducing the level of Aurora B/CPC that can regulate kinetochore function. The complex appears to localize along the chromosome arms. (See Page 3 in Color Section at the back of the book.)

4.2. HDACi disrupts Aurora B localization to the centromere

The mechanism controlling Aurora B/CPC centromeric localization involves the binding of the Survivin subunit of the CPC to phosphorylated histone H3 Thr3 via its conserved BIR domain (Kelly et al., 2010; Wang et al., 2010; Yamagishi et al., 2010). Blocking H3 Thr3 phosphorylation by depleting the responsible kinase, Haspin, or expressing a BIR mutant of survivin, produced an equivalent loss of CPC localization to the centromere. Loss of centromeric accumulation of CPC by inhibiting Haspin-dependent histone H3 Thr3 phosphorylation results in the CPC relocalizing along the chromosome arms (Dai et al., 2006), and similar chromosomal localization of CPC was also observed with HDACi treatment (Stevens et al., 2008). The mechanism by which HDACi affects CPC localization is not likely to occur through changes in Haspin levels, as HDACi treatment has not been reported to affect Haspin expression. Instead, HDACi treatment appears to mimic the loss of H3 Thr3 phosphorylation through increased trimethylation of the adjacent histone H3 Lys4 residue. HDACi treatment has been shown to rapidly increase levels of H3 Lys4 trimethylation, which is detectable within a few hours following drug addition in S phase, and prior to mitotic entry (Nightingale et al., 2007; Warrener et al., 2010). Trimethylation of H3 Lys4 can inhibit Survivin binding to phosphorylated H3 Thr3 and prevents Haspin phosphorylation of H3 Thr3 *in vitro* (Eswaran et al., 2009; Wang et al., 2010). As yet, there have been no reports investigating the effects of HDACi treatment on H3 Thr3 phosphorylation *in vivo*, although the reduction in CPC localization to the centromere suggests that the increased H3 Lys4 trimethylation blocks CPC binding rather than affecting Thr3 phosphorylation. The increased H3 Lys4 trimethylation is also likely to be related to increased localization of the H3 Lys4 methyltransferase MLL4, which binds acetylated histone H3 Lys9 (Nightingale et al., 2007). Elevated H3 Lys9 acetylation is rapidly induced by HDACi treatment. In parallel, HDACi treatment decreases the expression of H3 Lys4 demethylases such as RBP2 (Huang et al., 2011), and together these could readily account for the increased H3Lys4 trimethylation that is observed.

The loss of histone H3 Thr3 phosphorylation and CPC centromeric localization can account for many, but not all of the mitotic effects induced by HDACi. One of the most obvious differences is the ability of HDACi to overcome SAC activity in both nocodazole- and taxol-arrested cells (Dowling et al., 2005; Magnaghi-Jaulin et al., 2007; Stevens et al., 2008; Warrener et al., 2003). In contrast, Haspin depletion only affects the taxol-induced mitotic checkpoint arrest (Wang et al., 2010), and in this way, resembles the effect of Aurora B inhibition (Ditchfield et al., 2003; Hauf et al., 2003). A plausible explanation for this could be that while depletion of Haspin blocks Aurora B/CPC association with the centromere, and Aurora B inhibitors abrogate Aurora B activity, HDACi reduce but do not abolish Aurora B accumulation at the centromere. This clearly affects the kinetochore function of Aurora B as a direct result of reduced diffusible complex at the centromere but has little detectible effect on its centromeric functions such as CenpA phosphorylation (Stevens et al., 2008). This suggests that the maintenance of Aurora B activity at the centromere may be necessary to overcome SAC signaling in response to lack of microtubule binding and lack of tension across the centromere (Maresca & Salmon, 2010).

The other major difference between Haspin depletion and HDACi treatment is that HDACi treatment causes the sister chromatid arms to remain tightly associated in the "closed" conformation which results in failed separation of the sister chromatids in mitosis (Cimini et al., 2003; Stevens

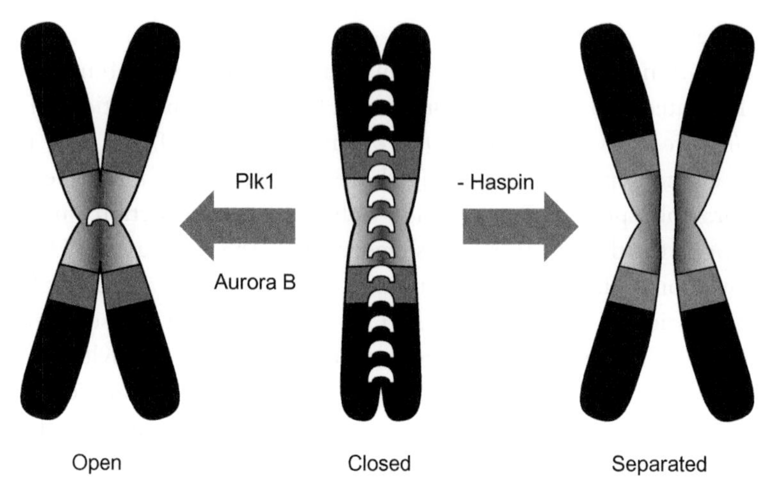

Figure 1.8 At entry into mitosis, the cohesin complexes along the chromosome arms are in the closed conformation with cohesin holding the sister chromatids closely associated. Early in normal mitosis, cohesin along the chromosome arms is unloaded by a mechanism involving Aurora B and Plk1, cohesin at the centromere is protected by Sgo1. Depletion/inhibition of Haspin causes loss of cohesion. HDACi treatment blocks the chromosomes in the closed arm conformation, blocking unloading of cohesin from the chromosome arms. (For color version of this figure, the reader is referred to the online version of this chapter.)

et al., 2008; Warrener et al., 2010). This contrasts with Haspin inhibition/depletion which results in loss of sister chromatid cohesion ("separated" conformation; Fig. 1.8) (Dai et al., 2006). Thus, the difference is not due to HDACi decreasing Haspin activity, histone H3 Thr3 phosphorylation, and/or Survivin/CPC binding, as cohesion is maintained along the chromosome arms, indicating the retention of the cohesin complex along the chromosome arms in mitosis. The effect of HDACi treatment phenocopies overexpression of Haspin in maintaining the closed arm conformation (Dai et al., 2006).

A probable target of HDACi inhibition that accounts for at least a subset of these mitotic effects is HDAC3. Short term depletion (1–2 days) of HDAC3 also mimics the effect of Haspin overexpression, with an increase in histone H3 Thr3 phosphorylation observed across the chromosome arms (Eot-Houllier et al., 2008). Longer term (5 days) depletion of HDAC3 resulted in a complete loss of sister chromatid attachment, which is identical to the depletion of Sgo1, a negative regulator of H3 Thr3 phosphorylation. Several reports indicate that HDAC3 appears to have a specific mitotic role. HDAC3 selectively binds the chromosome-associated AKAP95 in mitosis and has been suggested to affect Aurora B function (Li et al., 2006), and depletion of HDAC3 (but not other HDAC isoforms) has been demonstrated to produce similar effects on mitosis as HDACi, including premature mitotic exit from defective mitosis (Ishii, Kurasawa, Wong, & Yu-Lee, 2008; Magnaghi-Jaulin et al., 2007; Warrener et al., 2010). In addition, Class I selective HDACi produce the full spectrum of mitotic defects observed with pan-isoform inhibitors (Chia et al., 2010; Park et al., 2004), indicating that inhibition of HDAC3 is likely to be a significant contributor to the mitotic defects that is observed with HDACi treatment.

Another Class I HDAC, HDAC8, may also contribute to the closed arm phenotype described above. HDAC8 deacetylates the SMC3 subunit of cohesin in prophase when the cohesion complex is unloaded from the chromosome arms (Beckouet et al., 2010). It has been proposed that the acetylation may lock the cohesion ring in place (Nasmyth, 2011). Thus HDACi which inhibit Class I HDACs, which includes the majority of HDACi, would also block SMC3 deacetylation, ultimately inhibiting cohesion unloading. In addition to the direct effects on HDACs, HDACi treatment has also been reported to reduce Plk1 levels, or at least disrupt its normal chromosome and kinetochore localization in mitosis (Lallemand et al., 1999; Ma et al., 2008). Plk1 has a critical role in the unloading of the cohesion complex from the chromosome arms in prophase (Hauf et al., 2005; Losada, 2008; Nasmyth, 2011) and in k-microtubule binding in mitosis (Petronczki et al., 2008); therefore providing another potential mechanism through which HDACi may block sister chromatid separation in mitosis.

4.3. Nonhistone targets HDACi can contribute to aberrant mitosis

In addition to disrupting the chromatin modifications that are required for normal mitotic progression, HDACi affect the normal activity and/or function of other nonhistone proteins, which can also contribute to the defective mitosis observed. A prominent example is the molecular chaperone, HSP90. HDACi treatment stabilizes acetylation of HSP90 by inhibiting its associated HDAC6 activity, thus inhibiting HSP90 chaperone function (Drysdale, Brough, Massey, Jensen, & Schoepfer, 2006). Inhibition of HSP90 activity using selective HSP90 inhibitors such as 17-AAG results in mitotic defects similar to HDACi treatment, but the outcome is generally an extended mitotic arrest. The mitotic defects appear to be a consequence of the destabilization of various mitosis-specific HSP90 clients including: Aurora A,

Chk1, and Plk1, although the effects that have been reported were highly dependent on the cell line treated (de Carcer, 2004; Lyman et al., 2011; Zajac, Moneo, Carnero, Benitez, & Martinez-Delgado, 2008). The effect of HDACi treatment on the stability of the HSP90 client protein Aurora A, appears to be selective with destabilization of the cytoplasmic fraction and not the centrosome-localized fraction (Park et al., 2008; unpublished observations). This selectivity may account for the absence of monopolar spindles (which is a feature of Aurora A inhibition) in HDACi-treated cells (Andrews, Knatko, Moore, & Swedlow, 2003; Robbins et al., 2005; Stevens et al., 2008). Additionally, the HSP90 effects will be selective for HDACi capable of inhibiting HDAC6. A number of common HDACi, including sodium butyrate and depsipeptide (romidepsin, FK228), are unable to inhibit HDAC6-dependent α-tubulin deacetylation but produce similar mitotic defects as other pan-HDAC inhibitors (Warrener et al., 2003). This suggests that reduced levels of HSP90 client proteins are likely to be minor contributors to HDACi-induced mitotic defects.

4.4. HDACi kill tumor cells by driving premature exit from aberrant mitosis

The consequence of HDACi-induced aberrant mitosis is the rapid onset of apoptosis. Premature exit from the aberrant mitosis is a requirement for HDACi-induced apoptosis, as blocking mitotic exit has been reported to inhibit apoptosis (Warrener et al., 2003). The mechanism of HDACi-induced premature mitotic exit has been outlined in Section 3.1. The outcome of the premature exit is that cells are unlikely to completely disengage the proapoptotic, or reactivate the antiapoptotic mechanisms that establish the proapoptotic gearing in SAC mitotic-arrested cells. It is possible that this is related to the incomplete destruction of Cyclin B (Warrener et al., 2003), or other substrates of the APC/C that are normally destroyed during mitotic exit. The effect of this proapoptotic state may also be exacerbated by increased expression of proapoptotic effectors and reduced expression of antiapoptotic signaling components that are commonly associated with HDACi treatment, although the genes that are affected are often cell line or lineage-dependent (Frew et al., 2009; Henderson & Brancolini, 2003; Lindemann et al., 2007; Peart et al., 2003). Thus, the cell death that is observed upon HDACi treatment is the consequence of the combination of a lack of the HDACi-sensitive G2 phase arrest that is an intrinsic feature of the cancer cells, and HDACi-induced bypass of the compensating SAC. Normal cells are protected from cell death by their intact G2 phase arrest, and do not proceed into mitosis until the drug has been removed and the hyperacetylation reversed.

5. CONCLUSIONS

In summary, HDACi affect a number of components of the mitotic machinery, which combine to disrupt mitosis in tumor cells. These appear to include: inhibiting Survivin binding to phosphorylated H3 Thr3 which results in the mislocalization of Aurora B/CPC, maintenance of acetylated cohesin loading on the chromosome arms, and destabilization of Aurora A and Plk1; the latter also likely to contribute to the continued cohesion loading. These combine to allow for the establishment of a SAC-dependent mitotic arrest that cannot be maintained and results in premature mitotic exit. The premature exit triggers the rapid apoptosis that is observed. Normal tissue is protected from the deleterious mitotic effects of these drugs by triggering an HDACi-sensitive G2 phase arrest that persists until the drug is removed.

ACKNOWLEDGMENTS

This work was supported by funding from Cancer Council Queensland. B. G. is a National Health and Medical Research Council Senior Research Fellow.

REFERENCES

Allshire, R. C., & Karpen, G. H. (2008). Epigenetic regulation of centromeric chromatin: Old dogs, new tricks? *Nature Reviews. Genetics*, *9*, 923–937.

Andrews, P. D., Knatko, E., Moore, W. J., & Swedlow, J. R. (2003). Mitotic mechanics: The auroras come into view. *Current Opinion in Cell Biology*, *15*, 672–683.

Aoyagi, S., & Archer, T. K. (2005). Modulating molecular chaperone Hsp90 functions through reversible acetylation. *Trends Cell Biology*, *15*, 565–567.

Archer, S. Y., Meng, S., Shei, A., & Hodin, R. A. (1998). p21(WAF1) is required for butyrate-mediated growth inhibition of human colon cancer cells. *Proceedings of the National Academy of Sciences of the United States of America*, *95*, 6791–6796.

Bannister, A. J., Zegerman, P., Partridge, J. F., Miska, E. A., Thomas, J. O., & Allshire, R. C. (2001). Selective recognition of methylated lysine 9 on histone H3 by the HP1 chromo domain. *Nature*, *410*, 120–124.

Becker, M., Stolz, A., Ertych, N., & Bastians, H. (2010). Centromere localization of INCENP-Aurora B is sufficient to support spindle checkpoint function. *Cell Cycle*, *9*, 1360–1372.

Beckouet, F., Hu, B., Roig, M. B., Sutani, T., Komata, M., Uluocak, P., et al. (2010). An Smc3 acetylation cycle is essential for establishment of sister chromatid cohesion. *Molecular Cell*, *39*, 689–699.

Bhaskara, S., Chyla, B. J., Amann, J. M., Knutson, S. K., Cortez, D., Sun, Z. W., et al. (2008). Deletion of histone deacetylase 3 reveals critical roles in S phase progression and DNA damage control. *Molecular Cell*, *30*, 61–72.

Blagosklonny, M. V., Robey, R., Sackett, D. L., Du, L., Traganos, F., Darzynkiewicz, Z., et al. (2002). Histone deacetylase inhibitors all induce p21 but differentially cause tubulin acetylation, mitotic arrest, and cytotoxicity. *Molecular Cancer Therapeutics*, *1*, 937–941.

Blower, M. D., Sullivan, B. A., & Karpen, G. H. (2002). Conserved organization of centromeric chromatin in flies and humans. *Developmental Cell*, *2*, 319–330.

Bolden, J. E., Peart, M. J., & Johnstone, R. W. (2006). Anticancer activities of histone deacetylase inhibitors. *Nature Reviews Drug Discovery*, *5*, 769–784.

Boutros, R., Dozier, C., & Ducommun, B. (2006). The when and wheres of CDC25 phosphatases. *Current Opinion in Cell Biology*, *18*, 185–191.

Boyarchuk, Y., Salic, A., Dasso, M., & Arnaoutov, A. (2007). Bub1 is essential for assembly of the functional inner centromere. *The Journal of Cell Biology*, *176*, 919–928.

Bunz, F., Dutriaux, A., Lengauer, C., Waldman, T., Zhou, S., Brown, J. P., et al. (1998). Requirement for p53 and p21 to sustain G2 arrest after DNA damage. *Science*, *282*, 1497–1501.

Burgess, A. J., Pavey, S., Warrener, R., Hunter, L. J., Piva, T. J., Musgrove, E. A., et al. (2001). Up-regulation of p21(WAF1/CIP1) by histone deacetylase inhibitors reduces their cytotoxicity. *Molecular Pharmacology*, *60*, 828–837.

Burgess, A., Ruefli, A., Beamish, H., Warrener, R., Saunders, N., Johnstone, R., et al. (2004). Histone deacetylase inhibitors specifically kill nonproliferating tumour cells. *Oncogene*, *23*, 6693–6701.

Carmena, M., Pinson, X., Platani, M., Salloum, Z., Xu, Z., Clark, A., et al. (2012). The chromosomal passenger complex activates Polo kinase at centromeres. *PLoS Biology*, *10*, e1001250.

Castedo, M., Perfettini, J. L., Roumier, T., Andreau, K., Medema, R., & Kroemer, G. (2004). Cell death by mitotic catastrophe: A molecular definition. *Oncogene*, *23*, 2825–2837.

Castedo, M., Perfettini, J. L., Roumier, T., Valent, A., Raslova, H., Yakushijin, K., et al. (2004). Mitotic catastrophe constitutes a special case of apoptosis whose suppression entails aneuploidy. *Oncogene*, *23*, 4362–4370.

Cha, T. L., Chuang, M. J., Wu, S. T., Sun, G. H., Chang, S. Y., Yu, D. S., et al. (2009). Dual degradation of aurora A and B kinases by the histone deacetylase inhibitor LBH589 induces G2-M arrest and apoptosis of renal cancer cells. *Clinical Cancer Research*, *15*, 840–850.

Chan, G. K., Jablonski, S. A., Sudakin, V., Hittle, J. C., & Yen, T. J. (1999). Human BUBR1 is a mitotic checkpoint kinase that monitors CENP-E functions at kinetochores and binds the cyclosome/APC. *The Journal of Cell Biology*, *146*, 941–954.

Chan, Y. W., Jeyaprakash, A. A., Nigg, E. A., & Santamaria, A. (2012). Aurora B controls kinetochore-microtubule attachments by inhibiting Ska complex-KMN network interaction. *The Journal of Cell Biology*, *196*, 563–571.

Chia, K., Beamish, H., Jafferi, K., & Gabrielli, B. (2010). The histone deacetylase inhibitor MGCD0103 has both deacetylase and microtubule inhibitory activity. *Molecular Pharmacology*, *78*, 436–443.

Choi, S., & Reddy, P. (2011). HDAC inhibition and graft versus host disease. *Molecular Medicine*, *17*, 404–416.

Cimini, D., Mattiuzzo, M., Torosantucci, L., & Degrassi, F. (2003). Histone hyperacetylation in mitosis prevents sister chromatid separation and produces chromosome segregation defects. *Molecular Biology of the Cell*, *14*, 3821–3833.

Dai, J., Sullivan, B. A., & Higgins, J. M. (2006). Regulation of mitotic chromosome cohesion by Haspin and Aurora B. *Developmental Cell*, *11*, 741–750.

De Boer, L., Oakes, V., Beamish, H., Giles, N., Stevens, F., Somodevilla-Torres, M., et al. (2008). Cyclin A/cdk2 coordinates centrosomal and nuclear mitotic events. *Oncogene*, *27*, 4261–4268.

de Carcer, G. (2004). Heat shock protein 90 regulates the metaphase-anaphase transition in a polo-like kinase-dependent manner. *Cancer Research*, *64*, 5106–5112.

Ditchfield, C., Johnson, V. L., Tighe, A., Ellston, R., Haworth, C., Johnson, T., et al. (2003). Aurora B couples chromosome alignment with anaphase by targeting BubR1, Mad2, and Cenp-E to kinetochores. *The Journal of Cell Biology*, *161*, 267–280.

Dobles, M., Liberal, V., Scott, M. L., Benezra, R., & Sorger, P. K. (2000). Chromosome missegregation and apoptosis in mice lacking the mitotic checkpoint protein Mad2. *Cell*, *101*, 635–645.

Dowling, M., Voong, K. R., Kim, M., Keutmann, M. K., Harris, E., & Kao, G. D. (2005). Mitotic spindle checkpoint inactivation by trichostatin a defines a mechanism for increasing cancer cell killing by microtubule-disrupting agents. *Cancer Biology & Therapy*, *4*, 197–206.

Drysdale, M. J., Brough, P. A., Massey, A., Jensen, M. R., & Schoepfer, J. (2006). Targeting Hsp90 for the treatment of cancer. *Current Opinion in Drug Discovery & Development*, *9*, 483–495.

Ellis, L., Hammers, H., & Pili, R. (2009). Targeting tumor angiogenesis with histone deacetylase inhibitors. *Cancer Letters*, *280*, 145–153.

Eot-Houllier, G., Fulcrand, G., Watanabe, Y., Magnaghi-Jaulin, L., & Jaulin, C. (2008). Histone deacetylase 3 is required for centromeric H3K4 deacetylation and sister chromatid cohesion. *Genes & Development*, *22*, 2639–2644.

Eswaran, J., Patnaik, D., Filippakopoulos, P., Wang, F., Stein, R. L., Murray, J. W., et al. (2009). Structure and functional characterization of the atypical human kinase haspin. *Proceedings of the National Academy of Sciences of the United States of America*, *106*, 20198–20203.

Fava, L. L., Kaulich, M., Nigg, E. A., & Santamaria, A. (2011). Probing the in vivo function of Mad1:C-Mad2 in the spindle assembly checkpoint. *The EMBO Journal*, *30*, 3322–3336.

Fletcher, J. I., & Huang, D. C. (2008). Controlling the cell death mediators Bax and Bak: Puzzles and conundrums. *Cell Cycle*, *7*, 39–44.

Frew, A. J., Johnstone, R. W., & Bolden, J. E. (2009). Enhancing the apoptotic and therapeutic effects of HDAC inhibitors. *Cancer Letters*, *280*, 125–133.

Fujita, Y., Hayashi, T., Kiyomitsu, T., Toyoda, Y., Kokubu, A., Obuse, C., et al. (2007). Priming of centromere for CENP-A recruitment by human hMis18alpha, hMis18beta, and M18BP1. *Developmental Cell*, *12*, 17–30.

Fung, T. K., Ma, H. T., & Poon, R. Y. (2007). Specialized roles of the two mitotic cyclins in somatic cells: Cyclin A as an activator of M phase-promoting factor. *Molecular Biology of the Cell*, *18*, 1861–1873.

Gabrielli, B., Chau, Y. Q., Giles, N., Harding, A., Stevens, F., & Beamish, H. (2007). Caffeine promotes apoptosis in mitotic spindle checkpoint-arrested cells. *The Journal of Biological Chemistry*, *282*, 6954–6964.

Gascoigne, K. E., & Cheeseman, I. M. (2011). Kinetochore assembly: If you build it, they will come. *Current Opinion in Cell Biology*, *23*, 102–108.

Gascoigne, K. E., & Taylor, S. S. (2008). Cancer cells display profound intra- and interline variation following prolonged exposure to antimitotic drugs. *Cancer Cell*, *14*, 111–122.

Glaser, K. B., Staver, M. J., Waring, J. F., Stender, J., Ulrich, R. G., & Davidsen, S. K. (2003). Gene expression profiling of multiple histone deacetylase (HDAC) inhibitors: Defining a common gene set produced by HDAC inhibition in T24 and MDA carcinoma cell lines. *Molecular Cancer Therapeutics*, *2*, 151–163.

Glozak, M. A., & Seto, E. (2007). Histone deacetylases and cancer. *Oncogene*, *26*, 5420–5432.

Gonzalez-Barrios, R., Soto-Reyes, E., & Herrera, L. A. (2012). Assembling pieces of the centromere epigenetics puzzle. *Epigenetics*, *7*, 3–13.

Grabiec, A. M., Tak, P. P., & Reedquist, K. A. (2011). Function of histone deacetylase inhibitors in inflammation. *Critical Reviews in Immunology, 31*, 233–263.

Gray-Bablin, J., Zalvide, J., Fox, M. P., Knickerbocker, C. J., DeCaprio, J. A., & Keyomarsi, K. (1996). Cyclin E, a redundant cyclin in breast cancer. *Proceedings of the National Academy of Sciences of the United States of America, 93*, 15215–15220.

Gui, C. Y., Ngo, L., Xu, W. S., Richon, V. M., & Marks, P. A. (2004). Histone deacetylase (HDAC) inhibitor activation of p21WAF1 involves changes in promoter-associated proteins, including HDAC1. *Proceedings of the National Academy of Sciences of the United States of America, 101*, 1241–1246 Epub 2004 Jan 20.

Hauf, S., Cole, R. W., LaTerra, S., Zimmer, C., Schnapp, G., Walter, R., et al. (2003). The small molecule Hesperadin reveals a role for Aurora B in correcting kinetochore-microtubule attachment and in maintaining the spindle assembly checkpoint. *The Journal of Cell Biology, 161*, 281–294.

Hauf, S., Roitinger, E., Koch, B., Dittrich, C. M., Mechtler, K., & Peters, J. M. (2005). Dissociation of cohesin from chromosome arms and loss of arm cohesion during early mitosis depends on phosphorylation of SA2. *PLoS Biology, 3*, e69.

Henderson, C., & Brancolini, C. (2003). Apoptotic pathways activated by histone deacetylase inhibitors: Implications for the drug-resistant phenotype. *Drug Resistance Updates, 6*, 247–256.

Hirose, T., Sowa, Y., Takahashi, S., Saito, S., Yasuda, C., Shindo, N., et al. (2003). p53-independent induction of Gadd45 by histone deacetylase inhibitor: Coordinate regulation by transcription factors Oct-1 and NF-Y. *Oncogene, 22*, 7762–7773.

Hitomi, T., Matsuzaki, Y., Yokota, T., Takaoka, Y., & Sakai, T. (2003). p15(INK4b) in HDAC inhibitor-induced growth arrest. *FEBS Letters, 554*, 347–350.

Hori, T., Amano, M., Suzuki, A., Backer, C. B., Welburn, J. P., Dong, Y., et al. (2008). CCAN makes multiple contacts with centromeric DNA to provide distinct pathways to the outer kinetochore. *Cell, 135*, 1039–1052.

Howell, B. J., McEwen, B. F., Canman, J. C., Hoffman, D. B., Farrar, E. M., Rieder, C. L., et al. (2001). Cytoplasmic dynein/dynactin drives kinetochore protein transport to the spindle poles and has a role in mitotic spindle checkpoint inactivation. *The Journal of Cell Biology, 155*, 1159–1172.

Huang, P. H., Chen, C. H., Chou, C. C., Sargeant, A. M., Kulp, S. K., Teng, C. M., et al. (2011). Histone deacetylase inhibitors stimulate histone H3 lysine 4 methylation in part via transcriptional repression of histone H3 lysine 4 demethylases. *Molecular Pharmacology, 79*, 197–206.

Huang, H. C., Shi, J., Orth, J. D., & Mitchison, T. J. (2009). Evidence that mitotic exit is a better cancer therapeutic target than spindle assembly. *Cancer Cell, 16*, 347–358.

Huang, L., Sowa, Y., Sakai, T., & Pardee, A. B. (2000). Activation of the p21WAF1/CIP1 promoter independent of p53 by the histone deacetylase inhibitor suberoylanilide hydroxamic acid (SAHA) through the Sp1 sites. *Oncogene, 19*, 5712–5719.

Ishii, S., Kurasawa, Y., Wong, J., & Yu-Lee, L. Y. (2008). Histone deacetylase 3 localizes to the mitotic spindle and is required for kinetochore-microtubule attachment. *Proceedings of the National Academy of Sciences of the United States of America, 105*, 4179–4184.

Jacobs, S. A., Taverna, S. D., Zhang, Y., Briggs, S. D., Li, J., Eissenberg, J. C., et al. (2001). Specificity of the HP1 chromo domain for the methylated N-terminus of histone H3. *The EMBO Journal, 20*, 5232–5241.

Jansen, L. E., Black, B. E., Foltz, D. R., & Cleveland, D. W. (2007). Propagation of centromeric chromatin requires exit from mitosis. *The Journal of Cell Biology, 176*, 795–805.

Jin, S., Antinore, M. J., Lung, F. D., Dong, X., Zhao, H., Fan, F., et al. (2000). The GADD45 inhibition of Cdc2 kinase correlates with GADD45-mediated growth suppression. *The Journal of Biological Chemistry, 275*, 16602–16608.

Kasuboski, J. M., Bader, J. R., Vaughan, P. S., Tauhata, S. B., Winding, M., Morrissey, M. A., et al. (2011). Zwint-1 is a novel Aurora B substrate required for the assembly of a dynein-binding platform on kinetochores. *Molecular Biology of the Cell, 22*, 3318–3330.

Kawashima, S. A., Yamagishi, Y., Honda, T., Ishiguro, K., & Watanabe, Y. (2010). Phosphorylation of H2A by Bub1 prevents chromosomal instability through localizing shugoshin. *Science, 327*, 172–177.

Kelly, A. E., Ghenoiu, C., Xue, J. Z., Zierhut, C., Kimura, H., & Funabiki, H. (2010). Survivin reads phosphorylated histone H3 threonine 3 to activate the mitotic kinase Aurora B. *Science, 330*, 235–239.

Kim, Y. K., Han, J. W., Woo, Y. N., Chun, J. K., Yoo, J. Y., Cho, E. J., et al. (2003). Expression of p21(WAF1/Cip1) through Sp1 sites by histone deacetylase inhibitor apicidin requires PI 3-kinase-PKC epsilon signaling pathway. *Oncogene, 22*, 6023–6031.

Kim, Y., Holland, A. J., Lan, W., & Cleveland, D. W. (2010). Aurora kinases and protein phosphatase 1 mediate chromosome congression through regulation of CENP-E. *Cell, 142*, 444–455.

Kiyomitsu, T., Iwasaki, O., Obuse, C., & Yanagida, M. (2010). Inner centromere formation requires hMis14, a trident kinetochore protein that specifically recruits HP1 to human chromosomes. *The Journal of Cell Biology, 188*, 791–807.

Kiyomitsu, T., Obuse, C., & Yanagida, M. (2007). Human Blinkin/AF15q14 is required for chromosome alignment and the mitotic checkpoint through direct interaction with Bub1 and BubR1. *Developmental Cell, 13*, 663–676.

Kline, S. L., Cheeseman, I. M., Hori, T., Fukagawa, T., & Desai, A. (2006). The human Mis12 complex is required for kinetochore assembly and proper chromosome segregation. *The Journal of Cell Biology, 173*, 9–17.

Krauer, K. G., Burgess, A., Buck, M., Flanagan, J., Sculley, T. B., & Gabrielli, B. (2004). The EBNA-3 gene family proteins disrupt the G2/M checkpoint. *Oncogene, 23*, 1342–1353.

Lallemand, F., Courilleau, D., Buquet-Fagot, C., Atfi, A., Montagne, M. N., & Mester, J. (1999). Sodium butyrate induces G2 arrest in the human breast cancer cells MDA-MB-231 and renders them competent for DNA rereplication. *Experimental Cell Research, 247*, 432–440.

Lee, J. H., Choy, M. L., Ngo, L., Venta-Perez, G., & Marks, P. A. (2011). Role of checkpoint kinase 1 (Chk1) in the mechanisms of resistance to histone deacetylase inhibitors. *Proceedings of the National Academy of Sciences of the United States of America, 108*, 19629–19634.

Leggatt, G. R., & Gabrielli, B. (2012). Histone deacetylase inhibitors in the generation of the anti-tumour immune response. *Immunology and Cell Biology, 90*, 33–38.

Li, C., Andrake, M., Dunbrack, R., & Enders, G. H. (2010). A bifunctional regulatory element in human somatic Wee1 mediates cyclin A/Cdk2 binding and Crm1-dependent nuclear export. *Molecular and Cellular Biology, 30*, 116–130.

Li, G., Jiang, H., Chang, M., Xie, H., & Hu, L. (2011). HDAC6 alpha-tubulin deacetylase: A potential therapeutic target in neurodegenerative diseases. *Journal of the Neurological Sciences, 304*, 1–8.

Li, Y., Kao, G. D., Garcia, B. A., Shabanowitz, J., Hunt, D. F., Qin, J., et al. (2006). A novel histone deacetylase pathway regulates mitosis by modulating Aurora B kinase activity. *Genes & Development, 20*, 2566–2579.

Lindemann, R. K., Gabrielli, B., & Johnstone, R. W. (2004). Histone-Deacetylase Inhibitors for the Treatment of Cancer. *Cell Cycle, 3*, 779–788.

Lindemann, R. K., Newbold, A., Whitecross, K. F., Cluse, L. A., Frew, A. J., Ellis, L., et al. (2007). Analysis of the apoptotic and therapeutic activities of histone deacetylase inhibitors by using a mouse model of B cell lymphoma. *Proceedings of the National Academy of Sciences of the United States of America, 104*, 8071–8076.

Lindqvist, A., Rodriguez-Bravo, V., & Medema, R. H. (2009). The decision to enter mitosis: Feedback and redundancy in the mitotic entry network. *The Journal of Cell Biology, 185*, 193–202.

Losada, A. (2008). The regulation of sister chromatid cohesion. *Biochimica et Biophysica Acta, 1786*, 41–48.

Lyman, S. K., Crawley, S. C., Gong, R., Adamkewicz, J. I., McGrath, G., Chew, J. Y., et al. (2011). High-content, high-throughput analysis of cell cycle perturbations induced by the HSP90 inhibitor XL888. *PLoS One, 6*, e17692.

Ma, Y., Cai, S., Lu, Q., Lu, X., Jiang, Q., Zhou, J., et al. (2008). Inhibition of protein deacetylation by trichostatin A impairs microtubule-kinetochore attachment. *Cellular and Molecular Life Sciences, 65*, 3100–3109.

Macurek, L., Lindqvist, A., & Medema, R. H. (2009). Aurora-A and hBora join the game of Polo. *Cancer Research, 69*, 4555–4558.

Magnaghi-Jaulin, L., Eot-Houllier, G., Fulcrand, G., & Jaulin, C. (2007). Histone deacetylase inhibitors induce premature sister chromatid separation and override the mitotic spindle assembly checkpoint. *Cancer Research, 67*, 6360–6367.

Maity, A., Hwang, A., Janss, A., Phillips, P., McKenna, W. G., & Muschel, R. J. (1996). Delayed cyclin B1 expression during the G2 arrest following DNA damage. *Oncogene, 13*, 1647–1657.

Manke, I. A., Nguyen, A., Lim, D., Stewart, M. Q., Elia, A. E., & Yaffe, M. B. (2005). MAPKAP kinase-2 is a cell cycle checkpoint kinase that regulates the G2/M transition and S phase progression in response to UV irradiation. *Molecular Cell, 17*, 37–48.

Maresca, T. J. (2011). Cell division: Aurora B illuminates a checkpoint pathway. *Current Biology, 21*, R557–R559.

Maresca, T. J., & Salmon, E. D. (2010). Welcome to a new kind of tension: Translating kinetochore mechanics into a wait-anaphase signal. *Journal of Cell Science, 123*, 825–835.

Martin, C., & Zhang, Y. (2005). The diverse functions of histone lysine methylation. *Nature Reviews. Molecular Cell Biology, 6*, 838–849.

Matalon, S., Rasmussen, T. A., & Dinarello, C. A. (2011). Histone deacetylase inhibitors for purging HIV-1 from the latent reservoir. *Molecular Medicine, 17*, 466–472.

Mateo, F., Vidal-Laliena, M., Canela, N., Busino, L., Martinez-Balbas, M. A., Pagano, M., et al. (2009). Degradation of cyclin A is regulated by acetylation. *Oncogene, 28*, 2654–2666.

McConnell, B. B., Gregory, F. J., Stott, F. J., Hara, E., & Peters, G. (1999). Induced expression of p16(INK4a) inhibits both CDK4- and CDK2-associated kinase activity by reassortment of cyclin-CDK-inhibitor complexes. *Molecular and Cellular Biology, 19*, 1981–1989.

Mikhailov, A., Shinohara, M., & Rieder, C. L. (2004). Topoisomerase II and histone deacetylase inhibitors delay the G2/M transition by triggering the p38 MAPK checkpoint pathway. *The Journal of Cell Biology, 166*, 517–526.

Mitsiades, C. S., Mitsiades, N. S., McMullan, C. J., Poulaki, V., Shringarpure, R., Hideshima, T., et al. (2004). Transcriptional signature of histone deacetylase inhibition in multiple myeloma: Biological and clinical implications. *Proceedings of the National Academy of Sciences of the United States of America, 101*, 540–545.

Musacchio, A., & Salmon, E. D. (2007). The spindle-assembly checkpoint in space and time. *Nature Reviews. Molecular Cell Biology, 8*, 379–393.

Nasmyth, K. (2011). Cohesin: A catenase with separate entry and exit gates? *Nature Cell Biology, 13*, 1170–1177.

Nezi, L., & Musacchio, A. (2009). Sister chromatid tension and the spindle assembly checkpoint. *Current Opinion in Cell Biology, 21*, 785–795.

Nightingale, K. P., Gendreizig, S., White, D. A., Bradbury, C., Hollfelder, F., & Turner, B. M. (2007). Cross-talk between histone modifications in response to histone

deacetylase inhibitors: MLL4 links histone H3 acetylation and histone H3K4 methylation. *The Journal of Biological Chemistry, 282*, 4408–4416.
Niida, H., & Nakanishi, M. (2006). DNA damage checkpoints in mammals. *Mutagenesis, 21*, 3–9.
Nitta, M., Kobayashi, O., Honda, S., Hirota, T., Kuninaka, S., Marumoto, T., et al. (2004). Spindle checkpoint function is required for mitotic catastrophe induced by DNA-damaging agents. *Oncogene, 23*, 6548–6558.
Nome, R. V., Bratland, A., Harman, G., Fodstad, O., Andersson, Y., & Ree, A. H. (2005). Cell cycle checkpoint signaling involved in histone deacetylase inhibition and radiation-induced cell death. *Molecular Cancer Therapeutics, 4*, 1231–1238.
Obuse, C., Iwasaki, O., Kiyomitsu, T., Goshima, G., Toyoda, Y., & Yanagida, M. (2004). A conserved Mis12 centromere complex is linked to heterochromatic HP1 and outer kinetochore protein Zwint-1. *Nature Cell Biology, 6*, 1135–1141.
Ohtsubo, M., Theodoras, A. M., Schumacher, J., Roberts, J. M., & Pagano, M. (1995). Human cyclin E, a nuclear protein essential for the G1-to-S phase transition. *Molecular and Cellular Biology, 15*, 2612–2624.
Ozdag, H., Teschendorff, A. E., Ahmed, A. A., Hyland, S. J., Blenkiron, C., Bobrow, L., et al. (2006). Differential expression of selected histone modifier genes in human solid cancers. *BioMed Central Genomics, 7*, 90.
Pagano, M., Pepperkok, R., Verde, F., Ansorge, W., & Draetta, G. (1992). Cyclin A is required at two points in the human cell cycle. *The EMBO Journal, 11*, 961–971.
Park, J. H., Jong, H. S., Kim, S. G., Jung, Y., Lee, K. W., Lee, J. H., et al. (2008). Inhibitors of histone deacetylases induce tumor-selective cytotoxicity through modulating Aurora-A kinase. *Journal of Molecular Medicine, 86*, 117–128.
Park, J. H., Jung, Y., Kim, T. Y., Kim, S. G., Jong, H. S., Lee, J. W., et al. (2004). Class I histone deacetylase-selective novel synthetic inhibitors potently inhibit human tumor proliferation. *Clinical Cancer Research, 10*, 5271–5281.
Peart, M. J., Smyth, G. K., van Laar, R. K., Bowtell, D. D., Richon, V. M., Marks, P. A., et al. (2005). Identification and functional significance of genes regulated by structurally different histone deacetylase inhibitors. *Proceedings of the National Academy of Sciences of the United States of America, 102*, 3697–3702.
Peart, M. J., Tainton, K. M., Ruefli, A. A., Dear, A. E., Sedelies, K. A., O'Reilly, L. A., et al. (2003). Novel mechanisms of apoptosis induced by histone deacetylase inhibitors. *Cancer Research, 63*, 4460–4471.
Peters, A. H., O'Carroll, D., Scherthan, H., Mechtler, K., Sauer, S., Schofer, C., et al. (2001). Loss of the Suv39h histone methyltransferases impairs mammalian heterochromatin and genome stability. *Cell, 107*, 323–337.
Petronczki, M., Lenart, P., & Peters, J. M. (2008). Polo on the rise—From mitotic entry to cytokinesis with Plk1. *Developmental Cell, 14*, 646–659.
Pines, J. (2011). Cubism and the cell cycle: The many faces of the APC/C. *Nature Reviews. Molecular Cell Biology, 12*, 427–438.
Potapova, T. A., Daum, J. R., Byrd, K. S., & Gorbsky, G. J. (2009). Fine tuning the cell cycle: Activation of the Cdk1 inhibitory phosphorylation pathway during mitotic exit. *Molecular Biology of the Cell, 20*, 1737–1748.
Potapova, T. A., Daum, J. R., Pittman, B. D., Hudson, J. R., Jones, T. N., Satinover, D. L., et al. (2006). The reversibility of mitotic exit in vertebrate cells. *Nature, 440*, 954–958.
Prystowsky, M. B., Adomako, A., Smith, R. V., Kawachi, N., McKimpson, W., Atadja, P., et al. (2009). The histone deacetylase inhibitor LBH589 inhibits expression of mitotic genes causing G2/M arrest and cell death in head and neck squamous cell carcinoma cell lines. *The Journal of Pathology, 218*, 467–477.
Przewloka, M. R., & Glover, D. M. (2009). The kinetochore and the centromere: A working long distance relationship. *Annual Review of Genetics, 43*, 439–465.

Qiu, L., Burgess, A., Fairlie, D. P., Leonard, H., Parsons, P. G., & Gabrielli, B. G. (2000). Histone deacetylase inhibitors trigger a G2 checkpoint in normal cells that is defective in tumor cells. *Molecular Biology of the Cell, 11*, 2069–2083.

Qiu, L., Kelso, M. J., Hansen, C., West, M. L., Fairlie, D. P., & Parsons, P. G. (1999). Antitumour activity in vitro and in vivo of selective differentiating agents containing hydroxamate. *British Journal of Cancer, 80*, 1252–1258.

Rahman, R., & Grundy, R. (2011). Histone deacetylase inhibition as an anticancer telomerase-targeting strategy. *International Journal of Cancer, 129*, 2765–2774.

Richon, V. M., Sandhoff, T. W., Rifkind, R. A., & Marks, P. A. (2000). Histone deacetylase inhibitor selectively induces p21WAF1 expression and gene-associated histone acetylation. *Proceedings of the National Academy of Sciences of the United States of America, 97*, 10014–10019.

Rieder, C. L., & Maiato, H. (2004). Stuck in division or passing through: What happens when cells cannot satisfy the spindle assembly checkpoint. *Developmental Cell, 7*, 637–651.

Robbins, A. R., Jablonski, S. A., Yen, T. J., Yoda, K., Robey, R., Bates, S. E., et al. (2005). Inhibitors of histone deacetylases alter kinetochore assembly by disrupting pericentromeric heterochromatin. *Cell Cycle, 4*, 717–726.

Sambucetti, L. C., Fischer, D. D., Zabludoff, S., Kwon, P. O., Chamberlin, H., Trogani, N., et al. (1999). Histone deacetylase inhibition selectively alters the activity and expression of cell cycle proteins leading to specific chromatin acetylation and antiproliferative effects. *The Journal of Biological Chemistry, 274*, 34940–34947.

Sandor, V., Senderowicz, A., Mertins, S., Sackett, D., Sausville, E., Blagosklonny, M. V., et al. (2000). P21-dependent g(1)arrest with downregulation of cyclin D1 and upregulation of cyclin E by the histone deacetylase inhibitor FR901228. *British Journal of Cancer, 83*, 817–825.

Sato, N., Fukushima, N., Maitra, A., Matsubayashi, H., Yeo, C. J., Cameron, J. L., et al. (2003). Discovery of novel targets for aberrant methylation in pancreatic carcinoma using high-throughput microarrays. Histone acetylation-mediated regulation of genes in leukaemic cells. Gene expression profiling of multiple histone deacetylase (HDAC) inhibitors: defining a common gene set produced by HDAC inhibition in T24 and MDA carcinoma cell lines. *Cancer Research, 63*, 3735–3742.

Satyanarayana, A., & Kaldis, P. (2009). Mammalian cell-cycle regulation: Several Cdks, numerous cyclins and diverse compensatory mechanisms. *Oncogene, 28*, 2925–2939.

Schmiedeberg, L., Weisshart, K., Diekmann, S., Meyer Zu Hoerste, G., & Hemmerich, P. (2004). High- and low-mobility populations of HP1 in heterochromatin of mammalian cells. *Molecular Biology of the Cell, 15*, 2819–2833.

Schrump, D. S. (2009). Cytotoxicity mediated by histone deacetylase inhibitors in cancer cells: Mechanisms and potential clinical implications. *Clinical Cancer Research, 15*, 3947–3957.

Senese, S., Zaragoza, K., Minardi, S., Muradore, I., Ronzoni, S., Passafaro, A., et al. (2007). Role for histone deacetylase 1 in human tumor cell proliferation. *Molecular Cell Biology, 27*, 4784–4795.

Shakespear, M. R., Halili, M. A., Irvine, K. M., Fairlie, D. P., & Sweet, M. J. (2011). Histone deacetylases as regulators of inflammation and immunity. *Trends in Immunology, 32*, 335–343.

Sherr, C. J. (2000). The Pezcoller lecture: Cancer cell cycles revisited. *Cancer Research, 60*, 3689–3695.

Shi, J., Orth, J. D., & Mitchison, T. (2008). Cell type variation in responses to antimitotic drugs that target microtubules and kinesin-5. *Cancer Research, 68*, 3269–3276.

Shin, H. J., Baek, K. H., Jeon, A. H., Kim, S. J., Jang, K. L., Sung, Y. C., et al. (2003). Inhibition of histone deacetylase activity increases chromosomal instability by the aberrant regulation of mitotic checkpoint activation. *Oncogene, 22*, 3853–3858.

Skoufias, D. A., Indorato, R. L., Lacroix, F., Panopoulos, A., & Margolis, R. L. (2007). Mitosis persists in the absence of Cdk1 activity when proteolysis or protein phosphatase activity is suppressed. *The Journal of Cell Biology*, *179*, 671–685.

Stevens, F. E., Beamish, H., Warrener, R., & Gabrielli, B. (2008). Histone deacetylase inhibitors induce mitotic slippage. *Oncogene*, *27*, 1345–1354.

Stimpson, K. M., & Sullivan, B. A. (2010). Epigenomics of centromere assembly and function. *Current Opinion in Cell Biology*, *22*, 772–780.

Strausfeld, U. P., Howell, M., Descombes, P., Chevalier, S., Rempel, R. E., Adamczewski, J., et al. (1996). Both cyclin A and cyclin E have S-phase promoting (SPF) activity in Xenopus egg extracts. *Journal of Cell Science*, *109*, 1555–1563.

Subramanian, C., Opipari, A. W., Jr., Bian, X., Castle, V. P., & Kwok, R. P. (2005). Ku70 acetylation mediates neuroblastoma cell death induced by histone deacetylase inhibitors. *Proceedings of the National Academy of Sciences of the United States of America*, *102*, 4842–4847.

Sudo, T., Nitta, M., Saya, H., & Ueno, N. T. (2004). Dependence of paclitaxel sensitivity on a functional spindle assembly checkpoint. *Cancer Research*, *64*, 2502–2508.

Sullivan, B. A., & Karpen, G. H. (2004). Centromeric chromatin exhibits a histone modification pattern that is distinct from both euchromatin and heterochromatin. *Nature Structural & Molecular Biology*, *11*, 1076–1083.

Taddei, A., Maison, C., Roche, D., & Almouzni, G. (2001). Reversible disruption of pericentric heterochromatin and centromere function by inhibiting deacetylases. *Nature Cell Biology*, *3*, 114–120.

Tao, W., South, V. J., Zhang, Y., Davide, J. P., Farrell, L., Kohl, N. E., et al. (2005). Induction of apoptosis by an inhibitor of the mitotic kinesin KSP requires both activation of the spindle assembly checkpoint and mitotic slippage. *Cancer Cell*, *8*, 49–59.

Taylor, S. S., & McKeon, F. (1997). Kinetochore localization of murine Bub1 is required for normal mitotic timing and checkpoint response to spindle damage. *Cell*, *89*, 727–735.

Terrano, D. T., Upreti, M., & Chambers, T. C. (2010). Cyclin-dependent kinase 1-mediated Bcl-xL/Bcl-2 phosphorylation acts as a functional link coupling mitotic arrest and apoptosis. *Molecular and Cellular Biology*, *30*, 640–656.

Torras-Llort, M., Moreno-Moreno, O., & Azorin, F. (2009). Focus on the centre: The role of chromatin on the regulation of centromere identity and function. *The EMBO Journal*, *28*, 2337–2348.

Tunquist, B. J., Woessner, R. D., & Walker, D. H. (2010). Mcl-1 stability determines mitotic cell fate of human multiple myeloma tumor cells treated with the kinesin spindle protein inhibitor ARRY-520. *Molecular Cancer Therapeutics*, *9*, 2046–2056.

Vadlamudi, R. K., Bagheri-Yarmand, R., Yang, Z., Balasenthil, S., Nguyen, D., Sahin, A. A., et al. (2004). Dynein light chain 1, a p21-activated kinase 1-interacting substrate, promotes cancerous phenotypes. *Cancer Cell*, *5*, 575–585.

Vagnarelli, P., & Earnshaw, W. C. (2004). Chromosomal passengers: The four-dimensional regulation of mitotic events. *Chromosoma*, *113*, 211–222.

van de Weerdt, B. C., van Vugt, M. A., Lindon, C., Kauw, J. J., Rozendaal, M. J., Klompmaker, R., et al. (2005). Uncoupling anaphase-promoting complex/cyclosome activity from spindle assembly checkpoint control by deregulating polo-like kinase 1. *Molecular and Cellular Biology*, *25*, 2031–2044.

Varetti, G., Guida, C., Santaguida, S., Chiroli, E., & Musacchio, A. (2011). Homeostatic control of mitotic arrest. *Molecular Cell*, *44*, 710–720.

Verdin, E., Dequiedt, F., & Kasler, H. G. (2003). Class II histone deacetylases: versatile regulators. *Trends in Genetics*, *19*, 286–293.

Vogel, C., Hager, C., & Bastians, H. (2007). Mechanisms of mitotic cell death induced by chemotherapy-mediated G2 checkpoint abrogation. *Cancer Research*, *67*, 339–345.

Vogel, C., Kienitz, A., Muller, R., & Bastians, H. (2005). The mitotic spindle checkpoint is a critical determinant for topoisomerase-based chemotherapy. *The Journal of Biological Chemistry*, *280*, 4025–4028.

Wang, E., Ballister, E. R., & Lampson, M. A. (2011). Aurora B dynamics at centromeres create a diffusion-based phosphorylation gradient. *The Journal of Cell Biology*, *194*, 539–549.

Wang, F., Dai, J., Daum, J. R., Niedzialkowska, E., Banerjee, B., Stukenberg, P. T., et al. (2010). Histone H3 Thr-3 phosphorylation by Haspin positions Aurora B at centromeres in mitosis. *Science*, *330*, 231–235.

Warrener, R., Beamish, H., Burgess, A., Waterhouse, N. J., Giles, N., Fairlie, D., et al. (2003). Tumor cell-selective cytotoxicity by targeting cell cycle checkpoints. *The FASEB Journal*, *17*, 1550–1552.

Warrener, R., Chia, K., Warren, W. D., Brooks, K., & Gabrielli, B. (2010). Inhibition of histone deacetylase 3 produces mitotic defects independent of alterations in histone h3 lysine 9 acetylation and methylation. *Molecular Pharmacology*, *78*, 384–393.

Wassmann, K., & Benezra, R. (2001). Mitotic checkpoints: From yeast to cancer. *Current Opinion in Genetics & Development*, *11*, 83–90.

Weaver, B. A., & Cleveland, D. W. (2005). Decoding the links between mitosis, cancer, and chemotherapy: The mitotic checkpoint, adaptation, and cell death. *Cancer Cell*, *8*, 7–12.

Wertz, I. E., Kusam, S., Lam, C., Okamoto, T., Sandoval, W., Anderson, D. J., et al. (2011). Sensitivity to antitubulin chemotherapeutics is regulated by MCL1 and FBW7. *Nature*, *471*, 110–114.

Weichert, W., Roske, A., Niesporek, S., Noske, A., Buckendahl, A. C., Dietel, M., et al. (2008). Class I histone deacetylase expression has independent prognostic impact in human colorectal cancer: specific role of class I histone deacetylases in vitro and in vivo. *Clinical Cancer Research*, *14*, 1669–1677.

Westhorpe, F. G., Tighe, A., Lara-Gonzalez, P., & Taylor, S. S. (2011). p31comet-mediated extraction of Mad2 from the MCC promotes efficient mitotic exit. *Journal of Cell Science*, *124*, 3905–3916.

Willis, S. N., Fletcher, J. I., Kaufmann, T., van Delft, M. F., Chen, L., Czabotar, P. E., et al. (2007). Apoptosis initiated when BH3 ligands engage multiple Bcl-2 homologs, not Bax or Bak. *Science*, *315*, 856–859.

Wilson, A. J., Byun, D. S., Popova, N., Murray, L. B., L'Italien, K., Sowa, Y., et al. (2006). Histone deacetylase 3 (HDAC3) and other class I HDACs regulate colon cell maturation and p21 expression and are deregulated in human colon cancer. *The Journal of Biological Chemistry*, *281*, 13548–13558.

Xiao, H., Hasegawa, T., & Isobe, K. (2000). p300 collaborates with Sp1 and Sp3 in p21 (waf1/cip1) promoter activation induced by histone deacetylase inhibitor. *The Journal of Biological Chemistry*, *275*, 1371–1376.

Xu, Z., Cetin, B., Anger, M., Cho, U. S., Helmhart, W., Nasmyth, K., et al. (2009). Structure and function of the PP2A-shugoshin interaction. *Molecular Cell*, *35*, 426–441.

Xu, W. S., Parmigiani, R. B., & Marks, P. A. (2007). Histone deacetylase inhibitors: Molecular mechanisms of action. *Oncogene*, *26*, 5541–5552.

Yamagishi, Y., Honda, T., Tanno, Y., & Watanabe, Y. (2010). Two histone marks establish the inner centromere and chromosome bi-orientation. *Science*, *330*, 239–243.

Yokota, T., Matsuzaki, Y., Miyazawa, K., Zindy, F., Roussel, M. F., & Sakai, T. (2004). Histone deacetylase inhibitors activate INK4d gene through Sp1 site in its promoter. *Oncogene*, *26*, 26.

Zajac, M., Moneo, M. V., Carnero, A., Benitez, J., & Martinez-Delgado, B. (2008). Mitotic catastrophe cell death induced by heat shock protein 90 inhibitor in BRCA1-deficient breast cancer cell lines. *Molecular Cancer Therapeutics*, 7, 2358–2366.

Zeitlin, S. G., Shelby, R. D., & Sullivan, K. F. (2001). CENP-A is phosphorylated by Aurora B kinase and plays an unexpected role in completion of cytokinesis. *The Journal of Cell Biology*, *155*, 1147–1157.

Zhang, X., Lan, W., Ems-McClung, S. C., Stukenberg, P. T., & Walczak, C. E. (2007). Aurora B phosphorylates multiple sites on mitotic centromere-associated kinesin to spatially and temporally regulate its function. *Molecular Biology of the Cell*, *18*, 3264–3276.

CHAPTER TWO

Mechanisms of Resistance to Histone Deacetylase Inhibitors

Ju-Hee Lee, Megan L. Choy, Paul A. Marks[1]

Department of Cell Biology, Sloan-Kettering Institute, Memorial Sloan-Kettering Cancer Center, New York, USA

[1]Corresponding author: e-mail address: marksp@mskcc.org

Contents

Abstract

Histone deacetylase (HDAC) inhibitors are a new class of anticancer agents. HDAC inhibitors induce acetylation of histones and nonhistone proteins which are involved in regulation of gene expression and in various cellular pathways including cell growth arrest, differentiation, DNA damage and repair, redox signaling, and apoptosis (Marks, 2010). The U.S. Food and Drug Administration has approved two HDAC inhibitors, vorinostat and romidepsin, for the treatment of cutaneous T-cell lymphoma (Duvic & Vu, 2007; Grant et al., 2010; Marks & Breslow, 2007). Over 20 chemically different HDAC inhibitors are in clinical trials for hematological malignancies and solid tumors. This review considers the mechanisms of resistance to HDAC inhibitors that have been identified which account for the selective effects of these agents in inducing cancer but not normal cell death. These mechanisms, such as functioning Chk1, high levels of thioredoxin, or the prosurvival BCL-2, may also contribute to resistance of cancer cells to HDAC inhibitors.

Advances in Cancer Research, Volume 116
ISSN 0065-230X
http://dx.doi.org/10.1016/B978-0-12-394387-3.00002-1

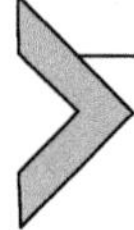

1. INTRODUCTION

1.1. Histone deacetylases

In humans, 11 members of Zn^{2+}-dependent histone deacetylases (HDACs) are classified into three groups: class I, class II, and class IV, based on their sequence homology to yeast orthologue Rpd3 and Hda1, respectively (Gregoretti, Lee, & Goodson, 2004; Witt, Deubzer, Milde, & Oehme, 2009; Table 2.1). Rpd3-like class I HDACs, HDAC1, 2, 3, and 8, are 45–55 kDa proteins ubiquitously expressed in human tissue. They have one catalytic domain and act as a deacetylase for histones. HDAC1 and HDAC2 are located in the nucleus, while HDAC3 and HDAC8 shuttle between the nucleus and the cytoplasm. Knockout of HDAC1 or HDAC3 in mice results in embryonic lethality due to impaired cell cycle

Table 2.1 Zinc-dependent histone deacetylases

Class	HDAC	Yeast homology	Localization	Chromosomal site	Size (a.a.)
Class I	HDAC1	Rpd3	Nucleus	1p34.1	483
	HDAC2[a]	Rpd3	Nucleus	6p21	488
	HDAC3	Rpd3	Nucleus	5q31	423
	HDAC8[a]	Rpd3	Nucleus	Xq13	377
Class IIa	HDAC4[a]	Hda1	Nucleus/ cytosol	2q372	1084
	HDAC5	Hda1	Nucleus/ cytosol	17q21	1122
	HDAC7[a]	Hda1	Nucleus/ cytosol	12q13	855
	HDAC9	Hda1	Nucleus/ cytosol	7p21–p15	1011
Class IIb	HDAC6	Hda1	Cytosol	Xp11.22–33	1215
	HDAC10	Hda1	Cytosol	22q13.31–33	669
Class IV	HDAC11	Rpd3/Hda1	Nucleus/ cytosol	3p25.2	347

[a]The crystal structure has been reported. See text for details and references.

(Bhaskara et al., 2008; Knutson et al., 2008; Lagger et al., 2002). Knockout of HDAC2 in mice induces cardiac abnormalities and exhibits perinatal lethality (Montgomery et al., 2007; Trivedi et al., 2007). Mice lacking HDAC8 develop perinatal lethality due to skull instability (Haberland, Mokalled, Montgomery, & Olson, 2009). Hda1-like Class II HDACs are 70–130 kDa proteins and are subdivided into class IIa, HDAC4, 5, 7, and 9, and Class IIb, HDAC6 and HDAC10. They have tissue-specific expression patterns. Class IIa HDACs shuttle between the nucleus and the cytoplasm. Class IIa HDACs have conserved binding sites for transcription factors, like the myocyte enhancing factor-2 (MEF2) and 14-3-3 protein. Binding of MEF2 or 14-3-3 to these binding sites stimulates transit of class IIa HDACs between the nucleus and the cytoplasm and regulates function of class IIa HDACs as transcription repressors. Mice lacking HDAC4 exhibit ectopic chondrocyte hypertrophy via constitutive expression of RUNX2 in chondrocyte and develop premature ossification in developing bone (Arnold et al., 2007; Vega et al., 2004). Knockout of HDAC7 results in embryonic lethality due to endothelial dysfunction (Chang et al., 2006). Mice with deleted HDAC5 or HDAC9 are viable and have minor myocardial hypertrophy (Chang et al., 2004; Zhang et al., 2002).

Class IIb HDAC6 is a cytoplasmic protein which has two active catalytic sites, a ubiquitin binding site and a C-terminal zinc-finger domain. HDAC6 is a lysine deacetylase of client proteins including alpha-tubulin and hsp90, but not histones. HDAC6-deficient mice are viable and show hyper-acetylation of alpha-tubulin in most tissues including heart, kidney, spleen, liver, and testis (Zhang et al., 2008). Only minor changes are observed in immune responses and bone mineral density. HDAC10 has two catalytic domains but one is inactive. HDAC11 belongs to class IV HDAC which has structural homology with both class I and class II HDACs. HDAC10 and HDAC11 knockout mice have not been reported.

Activation of class I, II, and IV HDACs require Zn^{2+} to catalyze the removal of acetyl groups from the lysines of histones and nonhistone proteins. Crystal structures have been reported for HDAC2, 4, 7, and 8 (Bottomley et al., 2008; Bressi et al., 2010; Schuetz et al., 2008; Somoza et al., 2004; Vannini et al., 2004) and bacterial HDAC-like proteins: HDAC-like protein, HDLP; HDAC-like amidohydrolase, HDAH; acetylpolyamine amidohydrolase, APAH (Lombardi, Cole, Dowling, & Christianson, 2011). These HDACs have a conserved catalytic pocket that is narrow and has a tube-like shape. Zn^{2+} is located 8 Å deep inside of the 12-Å pocket which is equivalent to length of four- to six-carbon straight chains (Bressi et al., 2010; Chen, Jiang, Zhou, Yu, & You, 2008).

HDAC inhibitors are anticancer agents which have tumor-selectivity. They are designed to bind to Zn^{2+} in the catalytic pocket and inhibit the enzyme activity, resulting in hyperacetylation of substrates including histones and nonhistone proteins which play a role in various cellular pathways including cell proliferation, cell survival, cell death, reactive oxygen species (ROS) signaling, DNA damage repair, and protein turnover. There are more than 50 nonhistone proteins that have been reported as HDACs substrates and 3600 acetylated lysine sites have been identified in 1750 proteins (Choudhary et al., 2009; Marks, 2010). HDAC inhibitors induce pleiotropic changes in cells which normal, but not cancer cells, can repair.

1.2. HDAC inhibitors

HDAC inhibitors can be divided into several classes based on their chemical structures including hydroxamic acids, cyclic peptides, electrophilic ketones, short-chain fatty acids, and benzamides (Table 2.2).

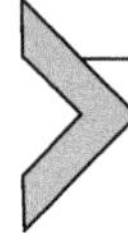

2. MECHANISMS OF RESISTANCE TO HDAC INHIBITORS

2.1. DNA damage and repair mechanisms

HDAC inhibitors have been shown to cause DNA damage as indicated by an increase in accumulation of phosphorylated H2AX, an early marker of DNA double strand breaks (DSBs), in various cancer cell lines. There is no evidence that DNA mutations are caused by HDAC inhibitors. Hyperacetylation of histones by HDAC inhibitors alters genomic stability and increases exposure of DNA to damaging agents: UV, ROS, radiation, and cytotoxic agents.

2.1.1 Alteration in expression of DNA repair proteins

HDAC inhibitors decrease the expression of genes encoding DNA repair proteins involved in DNA DSBs repair. Studies have shown that HDAC inhibitors can decrease levels of expression of genes for protein involved in the homologous recombination DNA repair including RAD51, RAD52, BRCA1/2, CtIP, Bloom syndrome gene, Nijmegen breakage syndrome1, and XRCC2 (Adimoolam et al., 2007; Kachhap et al., 2010; Lopez et al., 2009; Palmieri et al., 2009; Robert et al., 2011; Zhang et al., 2007); and components of the nonhomologous recombination end-joining DNA

Table 2.2 Histone deacetylase inhibitors

Class	Drug	Structure	HDAC's inhibited	PMID
Hydroxamic acid	Vorinostat (MK-0683, Zolinza, suberoylanilide hydroxamic acid, SAHA)		HDAC's: 1–3 6	Arts et al., 2009 and Bradner et al. (2010)
	Panobinostat (LBH589)		HDAC's: 1, 4, 5, 10, 11	Arts et al. (2009) and Bradner et al. (2010)
	Belinostat (PXD101)		HDAC's: 1–9	Bradner et al. (2010)
	Givinostat (ITF2357)		Class I	Bradner et al. (2010)
	Dacinostat (LVP-LAQ824, LAQ824)		Acetylates histones H3 and H4	Bradner et al. (2010) and Romanski et al. (2004)
	PCI-24781 (CRA-024781, CRA-02478)		HDAC's: 1–3, 6, 8, 10	Buggy et al. (2006) and Cao et al. (2006)

Continued

Table 2.2 Histone deacetylase inhibitors—cont'd

Class	Drug	Structure	HDAC's inhibited	PMID
	CHR-3996	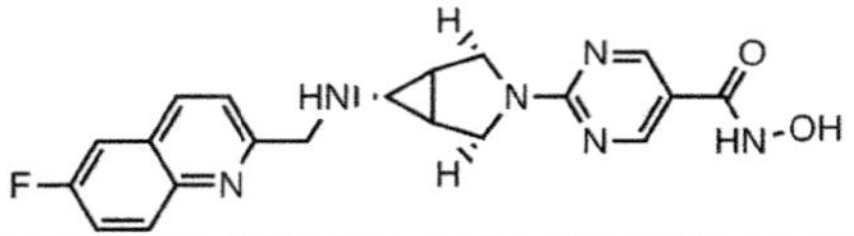	Class I: 1–3	Moffat et al. (2010)
	JNJ-26481585	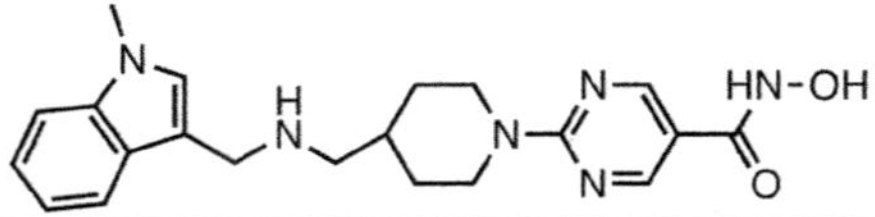	HDAC's: 1, 2, 4, 10, 11	Arts et al. (2009)
	SB939	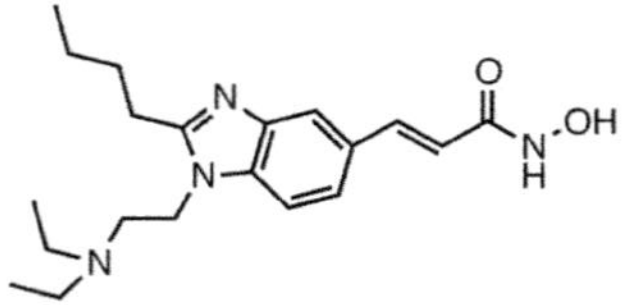	Class I, II, and IV (except HDAC6)	Novotny-Diermayr et al. (2010)
	AR-42	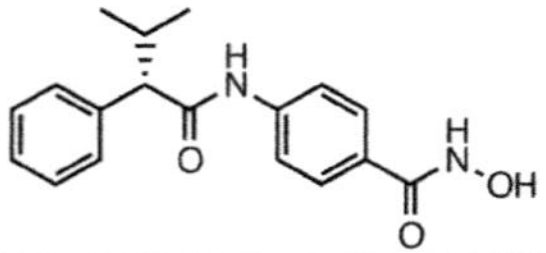	Acetylates histones H3 and H4 and alpha-tubulin	Lin et al. (2010) and Lucas et al. (2010)
	ACY-1215	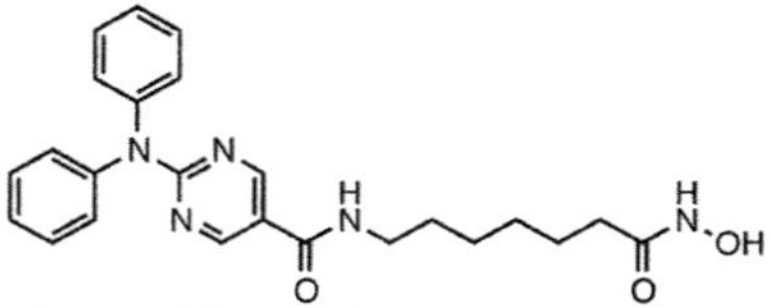	HDAC6 specific	Santo et al. (2012)

Cyclic peptide	Romidepsin (FK228, FR901228, depsipeptide)	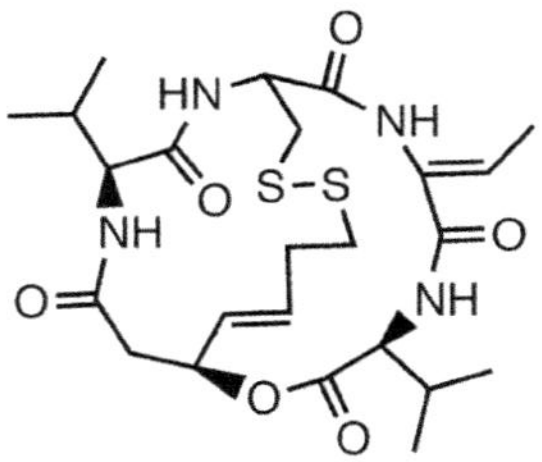	Class I: Mainly 1	Bradner et al. (2010)
Electrophilic ketone	Alpha-ketomide	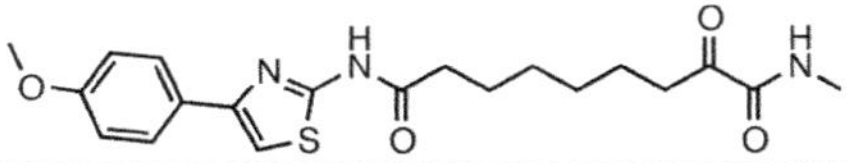	HDAC's: 1, 2	Wada et al. (2003)
	HKI 46F08	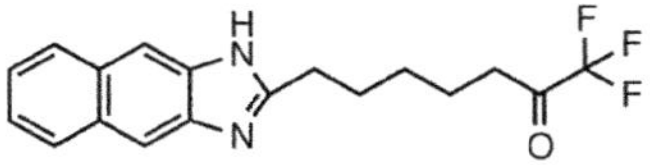	Acetylates histone H4 and alpha-tubulin	Wegener et al. (2008)
Short chain fatty acid	Valproic acid (VPA, valproate, Depakene, Valrelease)	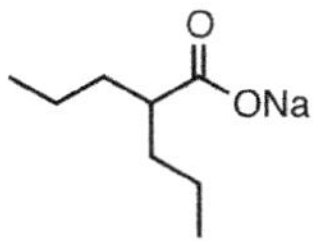	Class I and IIa	Gurvich, Tsygankova, Meinkoth, and Klein (2004)
	Butyrate (NaB, sodium, butyrate)	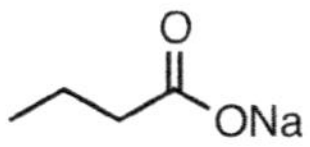	Acetylates histones	Aviram, Zimrah, Shaklai, Nudelman, and Rephaeli (1994)
	Phenylbutyrate (PB)	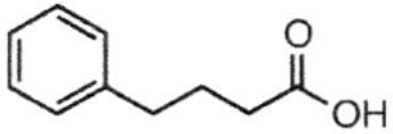	Acetylates histones	Warrell, He, Richon, Calleja, and Pandolfi (1998)
	Pivanex (AN-9)	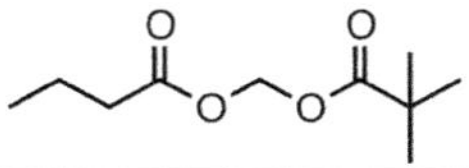	Acetylates histones	Aviram et al. (1994)

Continued

Table 2.2 Histone deacetylase inhibitors—cont'd

Class	Drug	Structure	HDAC's inhibited	PMID
Benzamide	Entinostat (MS-275, SNDX-275)	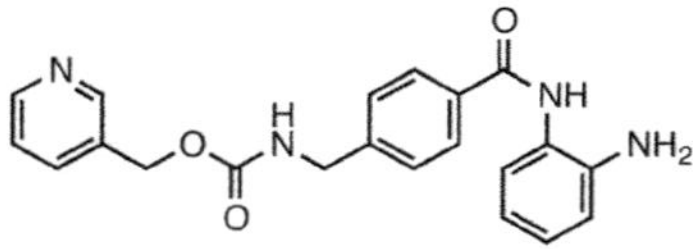	Class I: 1, 3	Hu et al., 2003, and Bradner et al. (2010)
	Mocetinostat (MGCD0103)	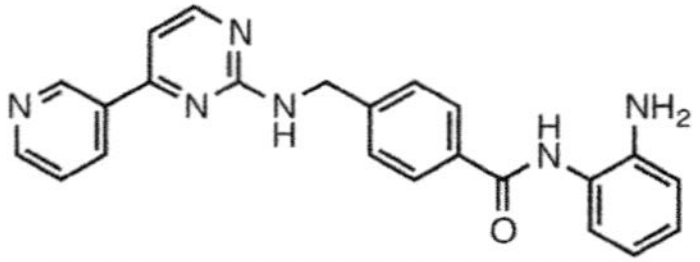	Class I: 1–3	Bradner et al. (2010)
	Tacedinaline (CI-994)	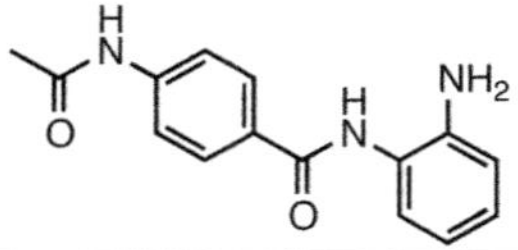	Class I: 1	Bradner et al. (2010)
Chimeric hydroxamate	CUDC-101	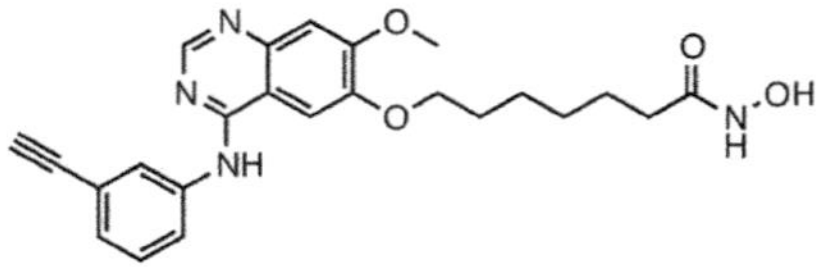	Class I and II	Lai et al. (2010)

repair including Ku70, Ku86, DNA-PKCs, XRCC4, and DNA ligase 4 (Chen et al., 2007; Munshi et al., 2005; Yaneva, Li, Marple, & Hasty, 2005).

HDAC inhibitors can induce hyperacetylation of nonhistone proteins involved in DNA repair, altering their function and causing accumulation of DNA damage. Trichostatin A, vorinostat, dacinostat (LAQ-824), and entinostat induce acetylation of Ku70 and sensitize cancer cells to DNA damage agents (Chen et al., 2007; Rosato et al., 2008).

Vorinostat induces the accumulation of DNA DSBs in both normal and transformed cells within 30 min of culture (Lee, Choy, Ngo, Foster, & Marks, 2010). Normal cells, but not cancer cells, can repair the HDAC inhibitor-induced DSBs. In transformed cells, vorinostat decreases the levels of DNA repair proteins and causes accumulation of DNA damage with consequent cell death. Transformed cells cannot repair the DNA damage and this DNA damage persists even after withdrawal of vorinostat (Lee et al., 2010).

The relative resistance of normal cells to HDAC inhibitors is abolished by inhibition of checkpoint kinase 1 (Chk1; Lee, Choy, Ngo, Venta-Perez, & Marks, 2011). Combined treatment of HDAC inhibitor (vorinostat, romidepsin, or entinostat) with Chk1 inhibitor (UCN-01, AZD-7762, or CHIR124) causes accumulation of DNA DSBs and induces cell death in normal cells. Expression of Chk1 is decreased by vorinostat in both normal and transformed cells (Brazelle et al., 2010; Lee et al., 2011). No change is observed in activity and expression of Chk2. This finding indicates that Chk1 has an important role in the resistance of normal cells to HDAC inhibitor-induced DNA DSBs.

2.1.2 Abnormalities in the regulation of cell cycle and mitotic chromosomes

HDAC inhibitors can induce cell cycle arrest in both G1 and G2/M phase with depletion of cells in S phase (Richon, Sandhoff, Rifkind, & Marks, 2000). Induction of p21 by HDAC inhibitors has been shown to cause G1 arrest in p53-dependent and p53-independent cells (Gui, Ngo, Xu, Richon, & Marks, 2004; Ju & Muller, 2003; Richon et al., 2000; Sandor et al., 2000). The HDAC inhibitors, valproic acid, PCI-24781, and vorinostat have been shown to induce a delay in DNA replication and to activate dormant origins, leading to the accumulation of replication-mediated DNA damage in transformed cells (Adimoolam et al., 2007; Conti et al., 2010; Kachhap et al., 2010; Richon et al., 2000).

Vorinostat causes mitotic chromosomal breaks and premature sister chromatin separation in transformed cells but not in normal cells (Lee et al., 2011).

The HDAC inhibitor-induced mitotic abnormalities persist in transformed cells after withdrawal of the HDAC inhibitor, indicating that transformed cells cannot recover from these mitotic defects (Lee et al., 2010; Qiu et al., 2000). In cell-based studies, it was found that inhibition of Chk1 could increase sensitivity of normal and cancer cells to HDAC inhibitors. Further, in normal mice, combined treatment with vorinostat and Chk1 inhibitor, but not treatment with either agent alone, causes severe weight loss with shrinkage of spleen and depletion of bone marrow cells. There is accumulation of chromosomal abnormalities in bone marrow cells (Lee et al., 2011). These findings indicate that Chk1 can play an important role in resistance of normal and tumor cells to HDAC inhibitors.

2.2. Reactive oxidative species and redox pathways

The redox pathway plays an important role in the resistance to HDAC inhibitors of normal and transformed cells. There is a negative correlation between the sensitivity to HDAC inhibitors (vorinostat, entinostat, and romidepsin) and the levels of thioredoxin (TRX). TRX is a thiol reductase that is a scavenger of ROS (Marks, 2006). Vorinostat increases the level of TRX expression in normal cells, but not transformed cells, which may account, in part, for the relative resistance of normal cells compared to cancer cells to HDAC inhibitors. In transformed cells, vorinostat induces upregulation of thioredoxin binding protein, a negative regulator of TRX, and accumulation of ROS, resulting in cell death.

High levels of TRX protect cells from damage caused by ROS generation, resulting in resistance to HDAC inhibitors. In peripheral blood cells from acute myeloid leukemia patients, resistance to vorinostat correlated with increased expressions of the antioxidant genes including TRX, superoxide dismutase 2, and glutathione reductase (Garcia-Manero, Yang, et al., 2008; Marks, 2006). Combination of HDAC inhibitor with ROS scavengers or glutathione precursor overcomes the resistance to HDAC inhibitor in solid tumors and human leukemia. These findings indicate that the response to HDAC inhibitor-induced oxidative stress is one determinant of resistance to HDAC inhibitors (Hu et al., 2010; Rosato et al., 2008).

2.3. Drug efflux mechanisms

Multidrug resistance is a major concern in cancer chemotherapy. P-glycoprotein (P-gp), also known as an ATP-binding cassette subfamily B member 1, is a transporter of drugs from the inside to the outside of cells.

Increased level of P-gp can be associated with resistance to HDAC inhibitors. The HDAC inhibitors, valproic acid, apicidin, romidepsin, and sodium butyrate have been found to increase levels of MDR1 mRNA, and its product P-gp expression in several cancer cells (Cerveny et al., 2007; Frommel, Coon, Tsuruo, & Roninson, 1993; Kim et al., 2008; Xiao et al., 2005; Yamada et al., 2006).

A study with trichostatin A found increased MDR1 mRNA due to transcription activation by increased interaction of NF-YA and P/CAF (p300/CREB binding protein-associated factor) on the MDR1 promoter (Jin & Scotto, 1998). Romidepsin is the only HDAC inhibitor which has been identified as a substrate of P-gp (Lee et al., 1994). Romidepsin-resistant cell lines have high levels of P-gp expression (Piekarz et al., 2004). The level of P-gp expression is reversible and is dependent on the presence of romidepsin (Yamada et al., 2006). HATs and HDACs activities are not affected (Xiao et al., 2005; Yamada et al., 2006). In cell-based studies, combined treatment of romidepsin with verapamil (P-gp inhibitor) or MK-571 (MDR1 inhibitor) reverses the resistance to romidepsin and apicidin (Okada et al., 2006).

Increased MDR1 expression has been found in normal and malignant peripheral blood mononuclear cells obtained from patients with refractory neoplasm receiving romidepsin (Robey et al., 2006). There is up to a sixfold increase in normal cells and up to an eightfold increase of MDR1 expression in circulating tumor cells after romidepsin administration. Unlike romidepsin, vorinostat and oxamflatin induce cell death in both P-gp expressing and non-P-gp expressing cancer cells, while romidepsin kills only non-P-gp expressing cells (Peart et al., 2003). Overall, this reveals that the resistance to certain HDAC inhibitors, in particular, romidepsin, may result from P-gp mediated drug efflux mechanisms.

2.4. Cell signaling-related mechanisms

Transformed cells may become resistant to HDAC inhibitors due to changes in their signaling pathways that circumvent inhibitor-induced cell death.

2.4.1 BCL-2 signaling

Alterations of the expression of antiapoptotic proteins in the intrinsic apoptosis pathway can cause resistance to HDAC inhibitors in transformed cells. BCL-2 (B-cell lymphoma-2) family proteins have a key role in HDAC inhibitor-induced cell death. BCL-2 family proteins are subdivided into three groups including antiapoptotic BCL-2 proteins, proapoptotic BAX-like proteins, and proapoptotic BH3-only proteins (Lindemann et al., 2007). Overexpression of BCL-2 or BCL-XL has been found to inhibit cell death in transformed cells treated with vorinostat, sodium butyrate, valproic acid, and entinostat (Condorelli, Gnemmi, Vallario, Genazzani, & Canonico, 2008; Duan, Heckman, & Boxer, 2005; Lucas et al., 2004; Mitsiades et al., 2003). Vorinostat-induced cell death is blocked by overexpression of BCL-2 in Eμ-Myc lymphomas bearing mice (Lindemann et al., 2007). In patients with cutaneous T-cell lymphoma, overexpression of BCL-2 has been associated with resistance to vorinostat, romidepsin, and panobinostat (Bolden, Peart, & Johnstone, 2006; Shao et al., 2010). Suppression of expression of proapoptotic BMF, BAX, or MCL-1 proteins can block apoptosis in cells treated with panobinostat, romidepsin, and m-carboxycinnamic acid bis-hydroxamide (Inoue, Walewska, Dyer, & Cohen, 2008; Shao et al., 2010; Zhang, Adachi, Kawamura, & Imai, 2006).

2.4.2 NF-κB activation

Constitutive activation of nuclear factor κB (NF-κB) has been reported in solid tumors and hematological malignancies and is related to resistance to chemotherapeutic agents (Aggarwal, 2004). Active NF-κB induces transcription of antiapoptotic genes including TNF receptor-associated factors, inhibitor of apoptosis, Mn-SOD, and BCL-XL, resulting in resistance of cells to various agents (Pahl, 1999).

RelA, a member of NF-κB family is acetylated by p300/CBP (Chen, Fischle, Verdin, & Greene, 2001). Acetylated RelA increases its DNA binding activity and impairs interaction to IκB. Vorinostat, trichostatin A, entinostat, and panobinostat activate NF-κB and attenuate HDAC inhibitor-induced cell death in various cancer cell lines (Dai, Rahmani, Dent, & Grant, 2005; Domingo-Domenech et al., 2008; Rosato et al., 2010; Rundall, Denlinger, & Jones, 2004). Combined treatment of HDAC inhibitor (trichostatin A, vorinostat, PXD101, or sodium butyrate) with either NF-κB inhibitor (BAY-11-7085 or IκB-SR) or proteosomal degradation inhibitor (MG132 or Bortezomib) can abolish resistance to the HDAC inhibitors (Domingo-Domenech et al., 2008; Dai et al., 2005; Duan et al., 2007; Rundall et al., 2004).

2.4.3 JAK and STAT signaling

Aberrant activation of the signal transducers and activators of transcription (STAT) pathways has been linked to resistance to chemotherapeutic agents in solid tumors and hematological malignancies (Yu & Jove, 2004). The effects of HDAC inhibitors have been studied in the STAT pathways.

Trichostatin A, sodium butyrate, and vorinostat inhibit transcription of STAT5 target genes without affecting its phosphorylation (active form of STAT5) (Rascle, Johnston, & Amati, 2003). Vorinostat, at a cytotoxic concentration, diminishes the level of STAT6 protein expression, but not of STAT3 protein expression, in cutaneous T-cell lymphoma cell lines and peripheral blood lymphocytes (Zhang, Richon, Ni, Talpur, & Duvic, 2005).

Despite these inhibitory effects of HDAC inhibitors on STATs-target antiapoptotic genes, overexpression and aberrant activation of STAT signaling have been related to resistance to HDAC inhibitors. Activation of STAT3 and STAT5 in PRL-3 overexpressed cell lines is associated with resistance to vorinostat (Zhou et al., 2011). Elevated expression and aberrant activation of STAT1 and STAT3 have also been associated with resistance to vorinostat in lymphoma cell lines and cutaneous T-cell lymphoma (Fantin et al., 2008). Combination of vorinostat with an inhibitor of Janus kinases (JAKs), an upstream molecule of STATs, induced apoptosis in vorinostat-resistant lymphoma cell lines (Fantin et al., 2008). Acquired mutation of Janus kinase2 (JAK2 V617P), which disrupts the autoinhibitory action of JAK2, is downregulated by treatment with panobinostat (Wang et al., 2009). Combination of panobinostat and JAK2 inhibitor TG101209 shows synergistic cytotoxic effects in myeloproliferative neoplastic cells, suggesting a possible therapeutic option to increase cell death in STAT-overexpressed tumor cells.

2.4.4 Retinoic acid signaling

Retinol (vitamin A) and retinoids have been found to have antitumor activity in cancer (Altucci & Gronemeyer, 2001). Inhibition of retinoic acid (RA) signaling causes resistance to HDAC inhibitors. A functional genetic screen identified two RA signaling-related genes, retinoic acid receptor α (RARα) and preferentially expressed antigen of melanoma (PRAME), that are involved in resistance to HDAC inhibitors (Epping et al., 2007). Ectopic expression of RARα or PRAME inhibits RA signaling and causes resistance to entinostat, vorinostat, and butyrate. Depletion of RARα or PRAME expression sensitizes cells to HDAC inhibitor-induced cell death *in vivo* and *in vitro*. UNC45A (GCUNC45) is a HSP90 co-chaperone protein and its expression is associated with malignancies in ovarian carcinomas (Bazzaro et al., 2007). Overexpression of UNC45A inhibits RA signaling, due to repression of retinoic acid receptor-dependent transcription activation, and causes resistance to HDAC inhibitors (Epping, Meijer, Bos, & Bernards, 2009).

2.5. Autophagy

Autophagy is a "self-eating" process which maintains cellular homeostasis and involves lysosome-dependent degradation of cellular components (Mizushima, Yoshimori, & Levine, 2010). The autophagosome is a unique double membrane vesicle that carries unfolded or aggregated proteins or cellular component. Fusion of autophagosomes with lysosomes degrades its delivered contents (Mizushima et al., 2010).

There are controversies regarding whether the effects of autophagy enhance antitumor activity or contribute to resistance to therapeutic agents (Carew, Giles, & Nawrocki, 2008; Janku, McConkey, Hong, & Kurzrock, 2011). Treatment with HDAC inhibitors has shown apoptotic cell death as well as autophagic cell death (Ellis et al., 2009; Fu, Shao, Chen, Ng, & Chen, 2010; Hrzenjak et al., 2008; Robert et al., 2011; Shao, Gao, Marks, & Jiang, 2004). However, autophagy has been associated with resistance to HDAC inhibitors. Accumulation of autophagosomes is increased in HDAC inhibitor-resistant malignant peripheral nerve sheath tumor cells treated with PCI-24781 (Lopez et al., 2011). Gene expression profiling in PCI-24781-resistant cells identifies three autophagy-related genes, IRGM, CXCR4, and TMEM74. Induced expression of these genes by HDAC inhibitors is observed *in vivo* and *in vitro* (Lopez et al., 2011). Knockdown of IRGM with siRNA blocks PCI-24781-induced autophagy and increases apoptosis in HDAC inhibitor-resistant cells. Blockage of autophagy with pharmacologic inhibitor (chloroquine or bafilomycin) increases cell death in HDAC inhibitor-resistant cells (Carew et al., 2007; Lopez et al., 2011).

2.6. Endoplasmic reticulum stress-related signaling

The endoplasmic reticulum (ER) is a key organelle that maintains cellular homeostasis by controlling protein synthesis, folding, delivery, and degradation (Ron & Walter, 2007). Interruptions of functions of ER by mutant proteins or excessive unfolded proteins cause ER stress which triggers unfolded protein responses (UPR).

HDAC inhibitors induce ER stress and activate UPR signaling. Glucose-regulated protein 78 (GRP78/Bip), an ATP-dependent chaperone in the ER, is involved in protein folding and assembly (Lee, 2001). Under stress conditions, GRP78 is dissociated from transmembrane ER-stress mediators, binds to unfolded proteins, and leads to proteosomal degradation. HDAC inhibitors induce GRP78 expression *in vivo* and *in vitro* (Baumeister,

Dong, Fu, & Lee, 2009; Shi, Gerritsma, Bowes, Capretta, & Werstuck, 2007; Wang, Bown, & Young, 1999). Overexpression of GRP78 is associated with resistance to HDAC inhibitor-induced cell death. The CCAAT motif at the GRP78 promoter is responsible for transcriptional activation by HDAC inhibitors (Baumeister et al., 2009).

GRP78 protein is a substrate for acetylation (Kahali et al., 2010; Rao, Nalluri, Kolhi, et al., 2010). Acetylation of GRP78 by panobinostat or vorinostat dissociates GRP78 protein from protein kinase RNA-like EF kinase (PERK), a transmembrane ER-stress mediator, and causes apoptosis (Rao, Nalluri, Kolhe, et al., 2010). Interruption of this signaling by depletion of PERK expression by siRNA leads to resistance to HDAC inhibitor-induced cell death (Kahali et al., 2010). Dissociation of PERK from GRP78 activates PERK, which upregulates proapoptotic CAAT/ enhancer binding protein homologous protein (CHOP) by ATF4 (Ron & Walter, 2007). Inhibition of CHOP decreases the sensitivity of cancer cells to HDAC inhibitor-induced cell death (Rao, Nalluri, Fiskus, et al., 2010).

2.7. HDAC expression and cancer

The levels of expression of HDACs have been found to be associated with prognosis in various cancers (Table 2.3). A high level of HDAC1 expression is correlated with poor prognosis in adenocarcinoma of lung (Minamiya et al., 2011) and endometrioid subtypes of ovarian and endometrial carcinomas (Hayashi et al., 2010; Weichert et al., 2008). A high level of HDAC2 expression is reported in advanced stages of oral cancer (Chang et al., 2009) and is associated with poor prognosis in hepatocellular carcinoma (Quint et al., 2011). Low expression of HDAC5 and HDAC10 has been associated with poor prognosis in non-small cell lung cancer (Osada et al., 2004). High expression of HDAC8 is related to poor outcomes in

Table 2.3 Expression of histone deacetylases associated with poor prognosis

HDAC	Expression levels	Tumor types
HDAC1	High	Lung, ovarian, endometrial cancer
HDAC2	High	Oral, hepatocellular cancer
HDAC5	Low	Non-small cell lung cancer
HDAC8	High	Pediatric neuroblastoma
HDAC9	Low	Glioblastoma
HDAC10	Low	Non-small cell lung cancer

pediatric neuroblastoma (Oehme, Deubzer, Lodrini, Milde, & Witt, 2009). In glioblastomas, compared to normal tissue, mRNA levels of class II and IV HDACs are downregulated and HDAC9 protein level is also downregulated (Lucio-Eterovic et al., 2008).

2.8. Acquired resistance to HDAC inhibitors

The acquisition of resistance to anticancer agents is a major problem in cancer therapy. The heterogeneity in cancer cell populations provides a survival benefit against therapeutic agents due to induction of mutations or maintaining a resistant stem cell-like population. This acquired resistance may be not only to the anticancer drug that has been used but also from other therapeutic agents to which the tumor has not been exposed (cross-resistance) (Gottesman, 2002; Kruh, 2003).

HDAC inhibitors can be effective in "reversible drug resistance" (Dannenberg & Berns, 2010). Treatment with HDAC inhibitors has been shown in some cases to reverse resistance of the drug-tolerant subpopulation. These findings suggest that an altered epigenetic state may play a role in maintaining the survival of the drug-tolerant cells (Sharma et al., 2010).

Our understanding of the mechanisms of acquired resistance to HDAC inhibitors is incomplete. Recent studies have found that acquired resistance to vorinostat causes cross-resistance to the hydroxamic acid- and the aliphatic acid-based HDAC inhibitors but not to the benzamide- and the cyclic peptide-based HDAC inhibitors (Dedes et al., 2009; Fiskus et al., 2008). Loss of acetylation of histones and loss of the G2 checkpoint activation are found in cancer cells with acquired resistance to vorinostat (Dedes et al., 2009). Dacinostat-resistant cancer cells show cross-resistance to vorinostat and panobinostat and have a lack of HDAC6 expression with an increased level of HDAC1, HDAC2, and HDAC4 expression (Fiskus et al., 2008). Dacinostat-resistant cancer cells show cross-resistance not only to other HDAC inhibitors, vorinostat and panobinostat, but also to chemotherapeutic agents including etoposide, cytarabine, and TRAIL (Fiskus et al., 2008). The acquired resistance to dacinostat is abolished by treatment with hsp90 inhibitors.

3. CLINICAL TRIALS OF HDAC INHIBITORS

HDAC inhibitors in clinical trials, either as monotherapy or combination therapy, for solid tumors and hematological malignancies are listed in Tables 2.4 and 2.5. HDAC inhibitors have been used in combination

Table 2.4 A partial list of completed clinical trials of monotherapy and combination therapy of vorinostat and romidepsin and its clinical outcomes

Author	Intervention	Tumor type	Response
Kelly et al. (2005)	Vorinostat as monotherapy	Solid tumor	1 CR, 3 PR, 2 unconfirmed PR, 15 SD
Krug et al. (2006)		Solid tumor	2 PR, 4 SD, 6 SD, 1 NE
Rubin et al. (2006)		Solid tumor	1/14 PR, 2/14 SD (≥15 months, 10.8 months)
O'Connor et al. (2006)		Hematological malignancy	5 reduction in measurable tumor 1 CR, 2 PR, 3 SD (9 months)
Olsen et al. (2007)		Hematological malignancy	ORR: 29.7 All initial responses PR, 1 CR Median TTR: 56 days (range 28–171 days) Median DOR: not reached 185 days (range 34+ to 441+ days) Pruritus improvement in responders and nonresponders
Mann et al. (2007)		Hematological malignancy	Median DOR: 168 days (end of response defined by 50% increase in SWAT score from nadir) Median TTP: 148 days Median TTP (patients stage IIB or higher): 169 days
Duvic et al. (2007)		Hematological malignancy	8 PR Median TTR: 11.9 weeks; Median DOR: 15.1 weeks Median TTP: 30.2 weeks; 14/31 Pruritus relief

Continued

Table 2.4 A partial list of completed clinical trials of monotherapy and combination therapy of vorinostat and romidepsin and its clinical outcomes—cont'd

Author	Intervention	Tumor type	Response
Luu et al. (2008)		Solid tumor	4 SD, TTP: 4, 8, 9, 14 months
Vansteenkiste et al. (2008)		Solid tumor	8 SD, Median TTP: 33.5 days (range 14–98 days)
Richardson et al. (2008)		Hematological malignancy	9 SD, 1 MR
Crump et al. (2008)		Hematological malignancy	1 CR (TTR: 85 days; DOR: $\geq$468 days), 1 SD (301 days) 16 PD (median TTP: 44 days)
Modesitt, Sill, Hoffman, and Bender (2008)		Solid tumor	1 PR, 9 SD, 14 PD within 2 months of study entry 2 PFS 6 months
Garcia-Manero, Yang, et al. (2008)		Hematological malignancy	7/41 HI (>50% decrease in blast counts with incomplete blood count recovery) 2 CR, 2 CR with incomplete blood count recovery
Blumenschein et al. (2008)		Solid tumor	3 SD, 1 PR-unconfirmed, 8 PD
Galanis et al. (2009)		Solid tumor	9/52 PFS 6 months Median TTP: 1.9 months (range 0.3 to 28+ months) Median OS: 5.7 months (range 0.7 to 28+ months)

Munster, Rubin, et al. (2009)	Vorinostat as monotherapy (continued)	Solid tumor	QTc did not prolong with single 800 mg dose
Schaefer et al. 2009		Hematological malignancy	1 CR (duration: 398 days); not enough CR in study to deem treatment effective
Bradley et al. (2009)		Solid tumor	2 SD (84 and 135 days), no decline in PSA
Fujiwara et al. (2009)		Solid tumor	9 SD, 9 PD
Traynor et al. (2009)		Solid tumor	8 SD, median TTP: 2.3 months (range 0.9–19.4 months) Median OS: 7.1 months (range 1.4–30.0 + months) 1 year OS rate: 19% (SE 10%)
Ramalingam et al. (2010)		Solid tumor	12 SD, 1 PR, 4 SD
Kirschbaum et al. (2011)		Solid tumor	5 CR, 5 PR → ORR: 29%
Doi et al. (2012)		Solid tumor	Group 1: 10 patients received vorinostat 5 SD (≥8 weeks), Median TTP: 70 days (range 21–245 days) Group 2: six patients received vorinostat 3 SD
Kirschbaum et al. (2012)		Hematological malignancy	1 PR, 7 SD (>1 year)
Ramalingam et al. (2007)	Vorinostat + carboplatin and paelitaxel	Solid tumor	11/25 PR, 7 SD
Badros et al. (2009)	Vorinostat + bortezomtb	Hematological malignancy	3/9 PR from bortezomib refractory patients, ORR: 42%

Continued

Table 2.4 A partial list of completed clinical trials of monotherapy and combination therapy of vorinostat and romidepsin and its clinical outcomes—cont'd

Author	Intervention	Tumor type	Response
Fakih et al. (2009)	Vorinostat + 5-fluorouracil, leucovorin, oxaliplatin	Solid tumor	11 SD by CT
Fouladi et al. (2010)	Vorinostat + 13-*cis*-retinoic acid	Solid tumor (pediatric)	Vorinostat alone: No objective responses Prolonged SD (median duration: 4 cycles; range 4–8 cycles) Combination therapy: 1 CR, 2 SD
Fakih et al. (2010)	Vorinostat + 5-fluorouracil, leucovorin	Solid tumor	Once-daily arm: 16 SD, 1 PR (18 months) Median PFS: 4.4 months (95%, CI: 2.1–7.7 months) Median OS: 9.2 months (95%, CI: 8.4 months-unreached) Twice-daily arm: 8 SD
Kadia et al. (2010)	Vorinostat + idarubicin	Hematological malignancy	7 (17%) responses: 2 (5%) CR 1 (2.5%) CR without platelet recovery 4 bone marrow responses
Ree et al. (2010)	Vorinostat + radiotherapy	Solid tumor	MTD: 300 mg Vorinostat q.d. with pelvic palliative radiation to 30 Gy in 3 Gy daily fractions over 2 weeks
Wilson et al. (2010)	Vorinostat + 5-fluorouracil	Solid tumor	2 SD (4 and 6 months) PD for all other evaluable patients within 2 months of study initiation

Millward et al. (2011)	Vorinostat + marizomib	Solid tumor	No objective responses (RECIST criteria) SD: 61% of evaluable patients Decrease in tumor measurement: 39% SD
Ramaswamy et al. (2011)	Vorinostat + paclitaxel, bevacizumab	Solid tumor	Intent-to-treat analysis: 2 CR, 24 PR → 26 (49%, 95%, CI: 37–60%) with objective response 16 (30%) SD lasting 24 weeks 5 (9%) PD 6 NE
Fakih, Groman, McMahon, Wilding, and Muindi (2011)	Vorinostat + 5-fluorouracil, leucovorin	Solid tumor	1 PR, 8 SD → PFS rate 60% 23/43 PFS 2 months, PFS rate; 53% Median PFS: 2.4 (95%, CI: 1.8–3.6) months Median OS: 6.5 (95%, CI: 4.8–7.8) months
Munster et al. (2011)	Vorinostat + tamoxifen	Solid tumor	9 SD (≥24 weeks)
Stathis et al. (2011)	Vorinostat + decitabine	Solid tumor and hematological malignancy	11 SD (≥4 cycles)
Dickson et al. (2011)	Vorinostat + flavopiridol	Solid tumor	8 SD
Friday et al. (2012)	Vorinostat + bortezomib	Solid tumor	1 PR Median TTP: 1.5 months (range 0.5–5.6 months) Median OS: 3.2 months Patients who had received prior bevacizumab therapy had shorter TTP and OS compared to those who had not

Continued

Table 2.4 A partial list of completed clinical trials of monotherapy and combination therapy of vorinostat and romidepsin and its clinical outcomes—cont'd

Author	Intervention	Tumor type	Response
Chinnaiyan et al. (2012)	Vorinostat + bevacizumab, CPT-11	Solid tumor	PFS: 3.6 months, OS: 7.3 months
Schneider et al. (2012)	Vorinostat + docetaxel	Solid tumor	Trial stopped due to excessive toxicity; No responses noted
Dummer et al. (2012)	Vorinostat + bexarotene	Hematological malignancy	7 Pruritus relief at low doses of each drug
Byrd et al. (2005)	Romidepsin (FK228, depsipeptide) as monotherapy	Hematological malignancy	Several patients had evidence of antitumor activity following treatment but did not meet NCI CR/PR criteria
Stadler, Margolin, Ferber, McCulloch, and Thompson (2006)		Solid tumor	1 CR, 1 sudden death → ORR 7% (95%, CI: 0.8–23%)
Klimek et al. (2008)		Hematological malignancy	1 CR (AML), 6 SD, 4 PD
Schrump et al. (2008)		Solid tumor	No objective responses observed. Nine patients with transient SD
Odenike et al. (2008)		Hematological malignancy	2 Patients had clearance of bone marrow blasts 3 Patients had > 50% decrease in bone marrow blasts
Whitehead et al. (2009)		Solid tumor	No objective responses observed by end of first stage so study permanently closed 4 Patients had SD as best response

Piekarz et al. (2009)	Romidepsin as monotherapy (continued)	Hematological malignancy	4 CR, 20 PR → ORR 34% (95%, CI: 23–46%) Median DOR: 13.7 months
Otterson, Hodgson, Pang, and Vokes (2010)		Solid tumor and hematological malignancy	3 (19%) SD Median PFS: 1.8 months Median OS: 6 months
Niesvizky et al. (2011)		Hematological malignancy	4/12 Exhibited evidence of M-protein stabilization Several other patients experienced improvement in bone pain and resolution of hypercalcemia
Whittaker et al. (2010)		Hematological malignancy	6 CR 26/68 Patients with advanced disease had response (5 CR) Median TTR: 2 months Median DOR: 15 months 28/65 Improvement in pruritus
Molife et al. (2010)		Solid tumor	2 Confirmed radiological partial response (RECIST) lasting ≥6 months with continued PSA decline ≥50%
Piekarz et al. (2011)		Hematological malignancy	8/45 CR, 9/45 PR → ORR 38% (95%, CI: 24–53%)
Iwamoto et al. (2011)	Romidepsin + enzyme inducing antiepileptic drugs	Solid tumor	Median PFS: 8 weeks, 1 patient had PFS ≥6 months 34 (97%) Patients died Median survival duration: 34 weeks

Table 2.5 A partial list of completed clinical trials of monotherapy and combination therapy of HDAC inhibitors and their clinical outcomes

Author	Intervention	Tumor type	Response
Ellis et al. (2008)	Panobinostat as monotherapy	Hematological *malignancy*	2 CR, 4 PR, *2* SD, 2 PD 23 Genes commonly regulated by panobinostat in all patients tested
Dickinson et al. (2009)		Hematological malignancy	Best response by PET: 7 PR, 5 SD Best response by CT: 5 PR, 8 SD, 6 PD
Hainsworth et al. (2011)		Solid tumor	No objective responses; all patients progressed or stopped treatment prior to 16-week reevaluation 5/20 SD at 8 weeks, 3 of these 5 had PD in following 8 weeks Accrual stopped and trial closed
Rathkopf et al. (2010)	Panobinostat + docetaxel	Solid tumor	Arm 1: all CRPC patients had PD Arm 2: 5/8 50% decline in PSA
Jones et al. (2011)	Panobinostat + gemcitabine	Solid tumor	1 Unconfirmed PR, 5 SD (duration $\geq$4 cycles)
Hamberg et al. (2011)	Panobinostat + ketoconazole	Solid tumor	37.5% SD ($\geq$3 months)
Chavez-Blanco et al. (2005)	Valproic acid as monotherapy	Solid tumor	20 mg and 40 mg/Kg VPA inhibits HDAC activity in histones of tumor tissues
Atmaca et al. (2007)		Solid tumor	2 SD (duration 3–5 months)

Wolff et al. (2008)	Valproic acid as monotherapy (continued)	Solid tumor (pediatric)	19/22 SD, 1/12 PD (patients started VPA with no measurable tumor), 2/10 PD (patients started VPA with stable measurable tumors) 22 Patients using VPA as relapse treatment: 1/14 PR, 7/14 SD, 6/14 PD (only 14 patients had MRI)
Mohammed et al. (2011)		Solid tumor	4 SD 5/7 Patients showed improvement in tumor markers
Su et al. (2011)	Valproic acid	Solid tumor (pediatric)	1 PR (duration 7 months), 1 MR (46% reduction in bidimensional measurements brainstem glioma, 5 months), 1 child (39% reduction in bidimensional measurements, removed from study due to unexpected toxicities unrelated to VPA)
Pilatrino et al. (2005)	Valproic acid + ATRA	Hematological malignancy	6/11 Evaluable patients had HI according to WHO 5 Major platelet response and consequential platelet transfusion independence
Bug et al. (2005)	Valproic acid + ATRA	Hematological malignancy	1 MR (patient with *de novo* AML), 2 PR (patients with secondary AML arising from myeloproliferative disorder) with clearance of peripheral blood blasts

Continued

Table 2.5 A partial list of completed clinical trials of monotherapy and combination therapy of HDAC inhibitors and their clinical outcomes—cont'd

Author	Intervention	Tumor type	Response
Kuendgen et al. (2005)	Valproic acid + ATRA	Hematological malignancy	18 (24%) HI according to IWG criteria for MDS
Kasperlik-Zaluska, Zgliczynski, Jeske, and Zdunowski (2005)	Valproic acid + dexamethasone and somatostatin	Solid tumor (Nelson's adenoma)	VPA decreased moderately ACTH concentration in 2 patients
Kuendgen et al. (2006)	Valproic acid + ATRA	Hematological malignancy	ORR: 5% according to IWG criteria for AML ORR: 16% according to IWG criteria for MDS
Arce et al. (2006)	Valproic acid + doxorubicin and cyclophosphamide and hydralazine	Solid tumor	5 (31%) CR, 8 (50%) PR → ORR 81% 1/15 Operated patients had pathological CR and 70% residual disease <3 cm
Garcia-Manero et al. (2006)	Valproic acid + decitabine	Hematological malignancy	10 (19%) CR, 2 (3%) CR with incomplete platelet recovery Elderly AML/MDS: 4 CR, 1 CR incomplete platelet recovery 1 Induction mortality
Candelaria et al. (2007)	Valproate + Hydralazine	Solid tumor	4 PR, 8 SD
Blum et al. (2007)	Valproic acid + decitabine	Hematological malignancy	Intent-to-treat analysis response rate: 44% (11/25) 11/21 Assessable patients: 4 CR (morphologic, cytogenetic), 4 incomplete CR, 3 PR 4/9 Assessable AML achieved CR

Soriano et al. (2007)	Valproic acid + ATRA and 5-azacitidine	Hematological malignancy	ORR: 42% Median remission duration: 26 weeks Median survival has not been reached
Munster et al. (2007)	Valproic acid + epirubicin	Solid tumor	9 (22%) PR, 16 (39%) SD
Braiteh et al. (2008)	Valproic acid + 5-azacytidine	Solid tumor	14 SD (duration 4–12 months)
Munster, Marchion, et al. (2009)	Valproic acid + epirubicin/FEC (5-FU, epirubicin, cyclophosphamide)	Solid tumor	Dose escalation phase: 9/41 patients PR
Daud et al. (2009)	Valproic acid + karenitecin	Solid tumor	7/15 Patients SD Median OS: 32.8 weeks Median TTP: 10.2 weeks
Rocca et al. (2009)	Valproic acid + dacarbazine and interferon-alpha	Solid tumor	1 CR, 2 PR, 3 SD (duration > 24 weeks)
Khanim et al. (2009)	Valproate + ATRA and theophylline	Hematological malignancy	1 CR (duration 4 months), 2 PR (duration > 5 months), 1 clearance of circulating blasts without reduction in bone marrow blasts
David, Mongan, Smith, Gudas, and Nanus (2010)	Valproate (Depakote) + ATRA-IV (ATRA analog)	Solid tumor	1 SD
Corsetti et al. (2011)	Valproic acid + low-dose Ara-C	Hematological malignancy	8 CR, 3 HI → 35% total response rate
Raffoux et al. (2010)	Valproic acid + ATRA, 5-azacitidine	Hematological malignancy	14 CR, 3 PR Median OS: 12.4 months

Continued

Table 2.5 A partial list of completed clinical trials of monotherapy and combination therapy of HDAC inhibitors and their clinical outcomes—cont'd

Author	Intervention	Tumor type	Response
Candelaria et al. (2011)	Valproate + hydralazine	Hematological malignancy	1 CR, 1 PR, 4 HI (erythroid series) → 50% ORR 2 PD (MDS → AML)
Scherpereel et al. (2011)	Valproate + doxorubicin	Solid tumor	7 PR, 2 toxic deaths (febrile neutropenia, thrombotic event)
Phuphanich et al. (2005)	Phenylbutyrate as monotherapy	Solid tumor	1 CR (duration 5 years, 5% of patients)
Camacho et al. (2007)		Solid tumor	3 SD (4–7 months)
Maslak et al. (2006)	Phenylbutyrate + 5-azacytidine	Hematological malignancy	3 PR, 2 SD
Sung and Waxman (2007)	Phenylbutyrate + 5-Fluorouracil	Solid tumor	3/4 SD (8 week treatment)
Hauschild et al. (2008)	Entinostat (MS27S) as monotherapy	Solid tumor	No objective response detected disease stabilization observed: 4 Patients arm A (3 mg biweekly) 3 Patients arm B (7 mg weekly)
Gore et al. (2008)		Solid tumor and hematological malignancy	2/27 PR 6/27 prolonged SD
Kummar et al. (2007)		Solid tumor and hematological malignancy	1 SD (over 8 months)
Gojo et al. (2007)		Hematological malignancy	Improvement in ANC lasting 1–10 weeks

Ryan et al. (2005)	Entinostat as monotherapy (continued)	Solid tumor and hematological malignancy	15 SD (duration 62–309 days)
Fandy et al. (2009)	Entinostat + 5-azacytidine	Hematological malignancy	3 CR, 4 PR, 7 HI, 16 NR
Richards et al. (2006)	CI-994 + gemcitabine	Solid tumor	C1-994 + gemcitabine ORR 12% versus placebo + gemcitabine ORR 14%: increased neutropenia and thrombocytopenia with CI-994 and gemcitabine compared to placebo and gemcitabine
Garcia-Manero, Assouline, et al. (2008)	Mocetinostat as monotherapy	Hematological malignancy	3 Complete bone marrow response lasting 1–3 cycles
Siu et al. (2008)		Solid tumor	5/32 SD (4–5 cycles)
Younes et al. (2011)		Hematological malignancy	2 CR, 12 PR, 1 SD
Blum et al. (2009)	Mocetinostat + rituximab	Hematological malignancy	20/21 SD
de Bono et al. (2008)	Dacinostat (LAQ824)	Solid tumor	3 SD (≤14 months)
Rambaldi et al. (2010)	Givinostat (ITF2357)	Hematological malignancy	PV/ET: 5 CR (reduced to 4 after 24 weeks), 1 PR MF: 3 major response (PR splenomegaly, pruritus control)
Galli et al. (2010)		Hematological malignancy	5 SD, 5 PD, 9 deaths (progressive multiple myeloma)

Continued

Table 2.5 A partial list of completed clinical trials of monotherapy and combination therapy of HDAC inhibitors and their clinical outcomes—cont'd

Author	Intervention	Tumor type	Response
Gimsing et al. (2008)	Belinostat (PXD101) as monotherapy	Hematological malignancy	5 SD (2–9 cycles)
Steele et al. (2008)	Belinostat as monotherapy (continued)	Solid tumor	18 SD
Ramalingam et al. (2009)		Solid tumor	Median survival: 5 months Median PFS: 1 month 1 Death deemed "possibly" related to therapy
Mackay et al. (2010)		Solid tumor	1 PR—unconfirmed, 10 SD, 3 NE
Giaccone et al. (2011)		Solid tumor	2 PR, 25 SD, 13 PD
Lassen et al. (2010)	Belinostat + carboplatin, paclitaxel	Solid tumor	2 PR, 1 complete CA-125 response, 6 SD ($\geq$6 months)
Reid et al. (2004)	Pivanex (AN-9)	Solid tumor	3 PR (6.4% and 95%, CI: 1.4–18.7%), 14 (30%) SD $\geq$12 weeks Median survival: 7.8 months 1-Year survival: 31%
Yong et al. (2011)	SB939 as monotherapy	Solid tumor	5 (21%) SD—4 achieved stabilization > 6 months 1 MR (maximum decrease in tumor size of 18.4%)
Razak et al. (2011)		Solid tumor	10 (32%) SD, 21 (68%) PD

DOR, duration of responses; TTR, time to response; ORR, overall response rate; CR, complete response; PR, partial response; SD, stable disease; PD, disease progression; NE, nonevaluable; PV, Polycythemia Vera; ET, Essential Thrombocythemia; MF, Myelofibrosis; PFS, progression free survival; ANC, absolute neutrophil count; HI, hematological improvement; OS, overall survival; MR, minor response; MDS, myelodysplastic syndrome; AML, acute myeloid leukemia; TTP, time to progression; ACTH, Adrenocorticotropic hormone; IWG, International Working Group; WHO, World Health Organization; PET, positron emission tomography; CT, computed tomography; CRPC, castration-resistant prostate cancer; PSA, prostate-specific antigen.

therapy with radiation, cytotoxic agents, and different targeted therapeutic agents. Vorinostat and romidepsin are approved by the U.S. Food and Drug Administration for the treatment of cutaneous T-cell lymphoma. HDAC inhibitors currently in clinical trials include vorinostat, panobinostat, valproic acid, phenylbutyrate, entinostat, CI-994, mocetinostat, dacinostat, givinostat, belinostat, pivanex, and SB939 (Table 2.5). Common adverse effects of HDAC inhibitors have been observed, including fatigue, nausea, vomiting, dehydration, diarrhea, and thrombocytopenia.

To date, the therapeutic efficacy of the HDAC inhibitors has been seen primarily in hematological malignancies. The anticancer activity of HDAC inhibitors in solid tumors has been more limited. The mechanism of this apparent resistance in certain solid tumors is not clear. Unfortunately, no marker of sensitivity or resistance to HDAC inhibitors of a cancer has been established with the possible exception of cutaneous T-cell lymphoma. HR23B, a protein that shuttles ubiquitinated cargo proteins to the proteasome for degradation, is found to be expressed at high levels in cutaneous T-cell lymphomas which are sensitive to the HDAC inhibitor, vorinostat (Stimson & La Thangue, 2009)

4. CONCLUSION

In preclinical studies, HDAC inhibitors have been shown to have a wide range of anticancer activities. The translation to clinical efficacy is being pursued in numerous clinical trials (see Tables 2.4 and 2.5). There is an increasing understanding of the mechanisms of resistance to HDAC inhibitors. Normal cells, which are relatively resistant to HDAC inhibitor-induced cell death, can repair HDAC inhibitor-induced alterations in signaling pathways, defects in DNA DSBs repair, accumulation of reactive oxygen species, and aberrant cell cycle checkpoint (Lee et al., 2010, 2011; Parmigiani et al., 2008; Ungerstedt et al., 2005). Unlike normal cells, cancer cells generally are not capable of repairing the HDAC inhibitor-induced damage. Cancers are a heterogeneous population of cells which can have varying degrees of resistance to the inhibitors. Resistance to HDAC inhibitors in cancer cells may involve both "intrinsic" and "acquired" mechanisms. Aberrant expression and modifications of signaling molecules cause intrinsic resistance to HDAC inhibitors in cancer cells. With further studies, the mechanisms of intrinsic resistance to HDAC inhibitors may be determined. Identifying markers of resistance could make for more selective and effective therapeutic strategies for HDAC inhibitors.

ACKNOWLEDGMENTS

Memorial Sloan-Kettering Cancer Center and Columbia University hold patents on vorinostat (SAHA, suberoylanilide hydroxamic acid) and related compounds that were exclusively licensed in 2001 to Aton Pharma, of which P. A. M was a founder. Aton Pharma was wholly acquired by Merck, Inc. in April 2004. P. A. M. received a royalty on the license and has no relationship with Merck.

REFERENCES

Adimoolam, S., Sirisawad, M., Chen, J., Thiemann, P., Ford, J. M., & Buggy, J. J. (2007). HDAC inhibitor PCI-24781 decreases RAD51 expression and inhibits homologous recombination. *Proceedings of the National Academy of Sciences of the United States of America, 104*, 19482–19487.

Aggarwal, B. B. (2004). Nuclear factor-kappaB: The enemy within. *Cancer Cell, 6*, 203–208.

Altucci, L., & Gronemeyer, H. (2001). The promise of retinoids to fight against cancer. *Nature Reviews. Cancer, 1*, 181–193.

Arce, C., Perez-Plasencia, C., Gonzalez-Fierro, A., de la Cruz-Hernandez, E., Revilla-Vazquez, A., Chavez-Blanco, A., et al. (2006). A proof-of-principle study of epigenetic therapy added to neoadjuvant doxorubicin cyclophosphamide for locally advanced breast cancer. *PLoS One, 1*, e98.

Arnold, M. A., Kim, Y., Czubryt, M. P., Phan, D., McAnally, J., Qi, X., et al. (2007). MEF2C transcription factor controls chondrocyte hypertrophy and bone development. *Developmental Cell, 12*, 377–389.

Arts, J., King, P., Marien, A., Floren, W., Belien, A., Janssen, L., et al. (2009). JNJ-26481585, a novel "second-generation" oral histone deacetylase inhibitor, shows broad-spectrum preclinical antitumoral activity. *Clinical Cancer Research: An Official Journal of the American Association for Cancer Research, 15*, 6841–6851.

Atmaca, A., Al-Batran, S. E., Maurer, A., Neumann, A., Heinzel, T., Hentsch, B., et al. (2007). Valproic acid (VPA) in patients with refractory advanced cancer: A dose escalating phase I clinical trial. *British Journal of Cancer, 97*, 177–182.

Aviram, A., Zimrah, Y., Shaklai, M., Nudelman, A., & Rephaeli, A. (1994). Comparison between the effect of butyric acid and its prodrug pivaloyloxymethylbutyrate on histones hyperacetylation in an HL-60 leukemic cell line. *International Journal of Cancer, 56*, 906–909.

Badros, A., Burger, A. M., Philip, S., Niesvizky, R., Kolla, S. S., Goloubeva, O., et al. (2009). Phase I study of vorinostat in combination with bortezomib for relapsed and refractory multiple myeloma. *Clinical Cancer Research: An Official Journal of the American Association for Cancer Research, 15*, 5250–5257.

Baumeister, P., Dong, D., Fu, Y., & Lee, A. S. (2009). Transcriptional induction of GRP78/BiP by histone deacetylase inhibitors and resistance to histone deacetylase inhibitor-induced apoptosis. *Molecular Cancer Therapeutics, 8*, 1086–1094.

Bazzaro, M., Santillan, A., Lin, Z., Tang, T., Lee, M. K., Bristow, R. E., et al. (2007). Myosin II co-chaperone general cell UNC-45 overexpression is associated with ovarian cancer, rapid proliferation, and motility. *American Journal of Pathology, 171*, 1640–1649.

Bhaskara, S., Chyla, B. J., Amann, J. M., Knutson, S. K., Cortez, D., Sun, Z. W., et al. (2008). Deletion of histone deacetylase 3 reveals critical roles in S phase progression and DNA damage control. *Molecular Cell, 30*, 61–72.

Blum, K. A., Advani, A., Fernandez, L., Van Der Jagt, R., Brandwein, J., Kambhampati, S., et al. (2009). Phase II study of the histone deacetylase inhibitor MGCD0103 in patients with previously treated chronic lymphocytic leukaemia. *British Journal of Haematology, 147*, 507–514.

Blum, W., Klisovic, R. B., Hackanson, B., Liu, Z., Liu, S., Devine, H., et al. (2007). Phase I study of decitabine alone or in combination with valproic acid in acute myeloid leukemia. *Journal of Clinical Oncology: Official Journal of the American Society of Clinical Oncology, 25*, 3884–3891.

Blumenschein, G. R., Jr., Kies, M. S., Papadimitrakopoulou, V. A., Lu, C., Kumar, A. J., Ricker, J. L., et al. (2008). Phase II trial of the histone deacetylase inhibitor vorinostat (Zolinza, suberoylanilide hydroxamic acid, SAHA) in patients with recurrent and/or metastatic head and neck cancer. *Investigational New Drugs, 26*, 81–87.

Bolden, J. E., Peart, M. J., & Johnstone, R. W. (2006). Anticancer activities of histone deacetylase inhibitors. *Nature Reviews. Drug Discovery, 5*, 769–784.

Bottomley, M. J., Lo Surdo, P., Di Giovine, P., Cirillo, A., Scarpelli, R., Ferrigno, F., et al. (2008). Structural and functional analysis of the human HDAC4 catalytic domain reveals a regulatory structural zinc-binding domain. *The Journal of Biological Chemistry, 283*, 26694–26704.

Bradley, D., Rathkopf, D., Dunn, R., Stadler, W. M., Liu, G., Smith, D. C., et al. (2009). Vorinostat in advanced prostate cancer patients progressing on prior chemotherapy (National Cancer Institute Trial 6862): Trial results and interleukin-6 analysis: A study by the Department of Defense Prostate Cancer Clinical Trial Consortium and University of Chicago Phase 2 Consortium. *Cancer, 115*, 5541–5549.

Bradner, J. E., West, N., Grachan, M. L., Greenberg, E. F., Haggarty, S. J., Warnow, T., et al. (2010). Chemical phylogenetics of histone deacetylases. *Nature Chemical Biology, 6*, 238–243.

Braiteh, F., Soriano, A. O., Garcia-Manero, G., Hong, D., Johnson, M. M., Silva Lde, P., et al. (2008). Phase I study of epigenetic modulation with 5-azacytidine and valproic acid in patients with advanced cancers. *Clinical Cancer Research: An Official Journal of the American Association for Cancer Research, 14*, 6296–6301.

Brazelle, W., Kreahling, J. M., Gemmer, J., Ma, Y., Cress, W. D., Haura, E., et al. (2010). Histone deacetylase inhibitors downregulate checkpoint kinase 1 expression to induce cell death in non-small cell lung cancer cells. *PLoS One, 5*, e14335.

Bressi, J. C., Jennings, A. J., Skene, R., Wu, Y., Melkus, R., De Jong, R., et al. (2010). Exploration of the HDAC2 foot pocket: Synthesis and SAR of substituted N-(2-aminophenyl)benzamides. *Bioorganic & Medicinal Chemistry Letters, 20*, 3142–3145.

Bug, G., Ritter, M., Wassmann, B., Schoch, C., Heinzel, T., Schwarz, K., et al. (2005). Clinical trial of valproic acid and all-trans retinoic acid in patients with poor-risk acute myeloid leukemia. *Cancer, 104*, 2717–2725.

Buggy, J. J., Cao, Z. A., Bass, K. E., Verner, E., Balasubramanian, S., Liu, L., et al. (2006). CRA-024781: A novel synthetic inhibitor of histone deacetylase enzymes with antitumor activity in vitro and in vivo. *Molecular Cancer Therapeutics, 5*, 1309–1317.

Byrd, J. C., Marcucci, G., Parthun, M. R., Xiao, J. J., Klisovic, R. B., Moran, M., et al. (2005). A phase 1 and pharmacodynamic study of depsipeptide (FK228) in chronic lymphocytic leukemia and acute myeloid leukemia. *Blood, 105*, 959–967.

Camacho, L. H., Olson, J., Tong, W. P., Young, C. W., Spriggs, D. R., & Malkin, M. G. (2007). Phase I dose escalation clinical trial of phenylbutyrate sodium administered twice daily to patients with advanced solid tumors. *Investigational New Drugs, 25*, 131–138.

Candelaria, M., Gallardo-Rincon, D., Arce, C., Cetina, L., Aguilar-Ponce, J. L., Arrieta, O., et al. (2007). A phase II study of epigenetic therapy with hydralazine and magnesium valproate to overcome chemotherapy resistance in refractory solid tumors. *Annals of Oncology: Official Journal of the ESMO, 18*, 1529–1538.

Candelaria, M., Herrera, A., Labardini, J., Gonzalez-Fierro, A., Trejo-Becerril, C., Taja-Chayeb, L., et al. (2011). Hydralazine and magnesium valproate as epigenetic treatment for myelodysplastic syndrome. Preliminary results of a phase-II trial. *Annals of Hematology, 90*, 379–387.

Cao, Z. A., Bass, K. E., Balasubramanian, S., Liu, L., Schultz, B., Verner, E., et al. (2006). CRA-026440: A potent, broad-spectrum, hydroxamic histone deacetylase inhibitor with antiproliferative and antiangiogenic activity in vitro and in vivo. *Molecular Cancer Therapeutics*, *5*, 1693–1701.

Carew, J. S., Giles, F. J., & Nawrocki, S. T. (2008). Histone deacetylase inhibitors: Mechanisms of cell death and promise in combination cancer therapy. *Cancer Letters*, *269*, 7–17.

Carew, J. S., Nawrocki, S. T., Kahue, C. N., Zhang, H., Yang, C., Chung, L., et al. (2007). Targeting autophagy augments the anticancer activity of the histone deacetylase inhibitor SAHA to overcome Bcr-Abl-mediated drug resistance. *Blood*, *110*, 313–322.

Cerveny, L., Svecova, L., Anzenbacherova, E., Vrzal, R., Staud, F., Dvorak, Z., et al. (2007). Valproic acid induces CYP3A4 and MDR1 gene expression by activation of constitutive androstane receptor and pregnane X receptor pathways. *Drug Metabolism and Disposition*, *35*, 1032–1041.

Chang, H. H., Chiang, C. P., Hung, H. C., Lin, C. Y., Deng, Y. T., & Kuo, M. Y. (2009). Histone deacetylase 2 expression predicts poorer prognosis in oral cancer patients. *Oral Oncology*, *45*, 610–614.

Chang, S., McKinsey, T. A., Zhang, C. L., Richardson, J. A., Hill, J. A., & Olson, E. N. (2004). Histone deacetylases 5 and 9 govern responsiveness of the heart to a subset of stress signals and play redundant roles in heart development. *Molecular and Cellular Biology*, *24*, 8467–8476.

Chang, S., Young, B. D., Li, S., Qi, X., Richardson, J. A., & Olson, E. N. (2006). Histone deacetylase 7 maintains vascular integrity by repressing matrix metalloproteinase 10. *Cell*, *126*, 321–334.

Chavez-Blanco, A., Segura-Pacheco, B., Perez-Cardenas, E., Taja-Chayeb, L., Cetina, L., Candelaria, M., et al. (2005). Histone acetylation and histone deacetylase activity of magnesium valproate in tumor and peripheral blood of patients with cervical cancer. A phase I study. *Molecular Cancer*, *4*, 22.

Chen, C. S., Wang, Y. C., Yang, H. C., Huang, P. H., Kulp, S. K., Yang, C. C., et al. (2007). Histone deacetylase inhibitors sensitize prostate cancer cells to agents that produce DNA double-strand breaks by targeting Ku70 acetylation. *Cancer Research*, *67*, 5318–5327.

Chen, L., Fischle, W., Verdin, E., & Greene, W. C. (2001). Duration of nuclear NF-kappaB action regulated by reversible acetylation. *Science*, *293*, 1653–1657.

Chen, Y. D., Jiang, Y. J., Zhou, J. W., Yu, Q. S., & You, Q. D. (2008). Identification of ligand features essential for HDACs inhibitors by pharmacophore modeling. *Journal of Molecular Graphics & Modelling*, *26*, 1160–1168.

Chinnaiyan, P., Chowdhary, S., Potthast, L., Prabhu, A., Tsai, Y. Y., Sarcar, B., et al. (2012). Phase I trial of vorinostat combined with bevacizumab and CPT-11 in recurrent glioblastoma. *Neuro-Oncology*, *14*, 93–100.

Choudhary, C., Kumar, C., Gnad, F., Nielsen, M. L., Rehman, M., Walther, T. C., et al. (2009). Lysine acetylation targets protein complexes and co-regulates major cellular functions. *Science*, *325*, 834–840.

Condorelli, F., Gnemmi, I., Vallario, A., Genazzani, A. A., & Canonico, P. L. (2008). Inhibitors of histone deacetylase (HDAC) restore the p53 pathway in neuroblastoma cells. *British Journal of Pharmacology*, *153*, 657–668.

Conti, C., Leo, E., Eichler, G. S., Sordet, O., Martin, M. M., Fan, A., et al. (2010). Inhibition of histone deacetylase in cancer cells slows down replication forks, activates dormant origins, and induces DNA damage. *Cancer Research*, *70*, 4470–4480.

Corsetti, M. T., Salvi, F., Perticone, S., Baraldi, A., De Paoli, L., Gatto, S., et al. (2011). Hematologic improvement and response in elderly AML/RAEB patients treated with valproic acid and low-dose Ara-C. *Leukemia Research*, *35*, 991–997.

Crump, M., Coiffier, B., Jacobsen, E. D., Sun, L., Ricker, J. L., Xie, H., et al. (2008). Phase II trial of oral vorinostat (suberoylanilide hydroxamic acid) in relapsed diffuse large-B-cell lymphoma. *Annals of Oncology: Official Journal of the ESMO*, *19*, 964–969.

Dai, Y., Rahmani, M., Dent, P., & Grant, S. (2005). Blockade of histone deacetylase inhibitor-induced RelA/p65 acetylation and NF-kappaB activation potentiates apoptosis in leukemia cells through a process mediated by oxidative damage, XIAP downregulation, and c-Jun N-terminal kinase 1 activation. *Molecular and Cellular Biology*, *25*, 5429–5444.

Dannenberg, J. H., & Berns, A. (2010). Drugging drug resistance. *Cell*, *141*, 18–20.

Daud, A. I., Dawson, J., DeConti, R. C., Bicaku, E., Marchion, D., Bastien, S., et al. (2009). Potentiation of a topoisomerase I inhibitor, karenitecin, by the histone deacetylase inhibitor valproic acid in melanoma: Translational and phase I/II clinical trial. *Clinical Cancer Research: An Official Journal of the American Association for Cancer Research*, *15*, 2479–2487.

David, K. A., Mongan, N. P., Smith, C., Gudas, L. J., & Nanus, D. M. (2010). Phase I trial of ATRA-IV and Depakote in patients with advanced solid tumor malignancies. *Cancer Biology & Therapy*, *9*, 678–684.

de Bono, J. S., Kristeleit, R., Tolcher, A., Fong, P., Pacey, S., Karavasilis, V., et al. (2008). Phase I pharmacokinetic and pharmacodynamic study of LAQ824, a hydroxamate histone deacetylase inhibitor with a heat shock protein-90 inhibitory profile, in patients with advanced solid tumors. *Clinical Cancer Research*, *14*, 6663–6673.

Dedes, K. J., Dedes, I., Imesch, P., von Bueren, A. O., Fink, D., & Fedier, A. (2009). Acquired vorinostat resistance shows partial cross-resistance to 'second-generation' HDAC inhibitors and correlates with loss of histone acetylation and apoptosis but not with altered HDAC and HAT activities. *Anti-Cancer Drugs*, *20*, 321–333.

Dickinson, M., Ritchie, D., DeAngelo, D. J., Spencer, A., Ottmann, O. G., Fischer, T., et al. (2009). Preliminary evidence of disease response to the pan deacetylase inhibitor panobinostat (LBH589) in refractory Hodgkin Lymphoma. *British Journal of Haematology*, *147*, 97–101.

Dickson, M. A., Rathkopf, D. E., Carvajal, R. D., Grant, S., Roberts, J. D., Reid, J. M., et al. (2011). A phase I pharmacokinetic study of pulse-dose vorinostat with flavopiridol in solid tumors. *Investigational New Drugs*, *29*, 1004–1012.

Doi, T., Hamaguchi, T., Shirao, K., Chin, K., Hatake, K., Noguchi, K., et al. (2012). Evaluation of safety, pharmacokinetics, and efficacy of vorinostat, a histone deacetylase inhibitor, in the treatment of gastrointestinal (GI) cancer in a phase I clinical trial. *International Journal of Clinical Oncology*, [electronic publishing ahead of print].

Domingo-Domenech, J., Pippa, R., Tapia, M., Gascon, P., Bachs, O., & Bosch, M. (2008). Inactivation of NF-kappaB by proteasome inhibition contributes to increased apoptosis induced by histone deacetylase inhibitors in human breast cancer cells. *Breast Cancer Research and Treatment*, *112*, 53–62.

Duan, H., Heckman, C. A., & Boxer, L. M. (2005). Histone deacetylase inhibitors down-regulate bcl-2 expression and induce apoptosis in t(14;18) lymphomas. *Molecular and Cellular Biology*, *25*, 1608–1619.

Duan, J., Friedman, J., Nottingham, L., Chen, Z., Ara, G., & Van Waes, C. (2007). Nuclear factor-kappaB p65 small interfering RNA or proteasome inhibitor bortezomib sensitizes head and neck squamous cell carcinomas to classic histone deacetylase inhibitors and novel histone deacetylase inhibitor PXD101. *Molecular Cancer Therapeutics*, *6*, 37–50.

Dummer, R., Beyer, M., Hymes, K., Epping, M. T., Bernards, R., Steinhoff, M., et al. (2012). Vorinostat combined with bexarotene for treatment of cutaneous T-cell lymphoma: in vitro and phase I clinical evidence supporting augmentation of retinoic acid receptor/retinoid X receptor activation by histone deacetylase inhibition. *Leukemia & Lymphoma*, *8*, 1501–1508.

Duvic, M., Talpur, R., Ni, X., Zhang, C., Hazarika, P., Kelly, C., et al. (2007). Phase 2 trial of oral vorinostat (suberoylanilide hydroxamic acid, SAHA) for refractory cutaneous T-cell lymphoma (CTCL). *Blood*, *109*, 31–39.

Duvic, M., & Vu, J. (2007). Vorinostat: A new oral histone deacetylase inhibitor approved for cutaneous T-cell lymphoma. *Expert Opinion on Investigational Drugs*, *16*, 1111–1120.

Ellis, L., Bots, M., Lindemann, R. K., Bolden, J. E., Newbold, A., Cluse, L. A., et al. (2009). The histone deacetylase inhibitors LAQ824 and LBH589 do not require death receptor signaling or a functional apoptosome to mediate tumor cell death or therapeutic efficacy. *Blood*, *114*, 380–393.

Ellis, L., Pan, Y., Smyth, G. K., George, D. J., McCormack, C., Williams-Truax, R., et al. (2008). Histone deacetylase inhibitor panobinostat induces clinical responses with associated alterations in gene expression profiles in cutaneous T-cell lymphoma. *Clinical Cancer Research: An Official Journal of the American Association for Cancer Research*, *14*, 4500–4510.

Epping, M. T., Meijer, L. A., Bos, J. L., & Bernards, R. (2009). UNC45A confers resistance to histone deacetylase inhibitors and retinoic acid. *Molecular Cancer Research*, *7*, 1861–1870.

Epping, M. T., Wang, L., Plumb, J. A., Lieb, M., Gronemeyer, H., Brown, R., et al. (2007). A functional genetic screen identifies retinoic acid signaling as a target of histone deacetylase inhibitors. *Proceedings of the National Academy of Sciences of the United States of America*, *104*, 17777–17782.

Fakih, M. G., Fetterly, G., Egorin, M. J., Muindi, J. R., Espinoza-Delgado, I., Zwiebel, J. A., et al. (2010). A phase I, pharmacokinetic, and pharmacodynamic study of two schedules of vorinostat in combination with 5-fluorouracil and leucovorin in patients with refractory solid tumors. *Clinical Cancer Research: An Official Journal of the American Association for Cancer Research*, *16*, 3786–3794.

Fakih, M. G., Groman, A., McMahon, J., Wilding, G., & Muindi, J. R. (2011). A randomized phase II study of two doses of vorinostat in combination with 5-FU/LV in patients with refractory colorectal cancer. *Cancer Chemotherapy and Pharmacology*, *69*, 743–751.

Fakih, M. G., Pendyala, L., Fetterly, G., Toth, K., Zwiebel, J. A., Espinoza-Delgado, I., et al. (2009). A phase I, pharmacokinetic and pharmacodynamic study on vorinostat in combination with 5-fluorouracil, leucovorin, and oxaliplatin in patients with refractory colorectal cancer. *Clinical Cancer Research: An Official Journal of the American Association for Cancer Research*, *15*, 3189–3195.

Fandy, T. E., Herman, J. G., Kerns, P., Jiemjit, A., Sugar, E. A., Choi, S. H., et al. (2009). Early epigenetic changes and DNA damage do not predict clinical response in an overlapping schedule of 5-azacytidine and entinostat in patients with myeloid malignancies. *Blood*, *114*, 2764–2773.

Fantin, V. R., Loboda, A., Paweletz, C. P., Hendrickson, R. C., Pierce, J. W., Roth, J. A., et al. (2008). Constitutive activation of signal transducers and activators of transcription predicts vorinostat resistance in cutaneous T-cell lymphoma. *Cancer Research*, *68*, 3785–3794.

Fiskus, W., Rao, R., Fernandez, P., Herger, B., Yang, Y., Chen, J., et al. (2008). Molecular and biologic characterization and drug sensitivity of pan-histone deacetylase inhibitor-resistant acute myeloid leukemia cells. *Blood*, *112*, 2896–2905.

Fouladi, M., Park, J. R., Stewart, C. F., Gilbertson, R. J., Schaiquevich, P., Sun, J., et al. (2010). Pediatric phase I trial and pharmacokinetic study of vorinostat: A Children's Oncology Group phase I consortium report. *Journal of Clinical Oncology: Official Journal of the American Society of Clinical Oncology*, *28*, 3623–3629.

Friday, B. B., Anderson, S. K., Buckner, J., Yu, C., Giannini, C., Geoffroy, F., et al. (2012). Phase II trial of vorinostat in combination with bortezomib in recurrent glioblastoma: A north central cancer treatment group study. *Neuro-Oncology*, *14*, 215–221.

Frommel, T. O., Coon, J. S., Tsuruo, T., & Roninson, I. B. (1993). Variable effects of sodium butyrate on the expression and function of the MDR1 (P-glycoprotein) gene in colon carcinoma cell lines. *International Journal of Cancer, 55*, 297–302.

Fu, J., Shao, C. J., Chen, F. R., Ng, H. K., & Chen, Z. P. (2010). Autophagy induced by valproic acid is associated with oxidative stress in glioma cell lines. *Neuro-Oncology, 12*, 328–340.

Fujiwara, Y., Yamamoto, N., Yamada, Y., Yamada, K., Otsuki, T., Kanazu, S., et al. (2009). Phase I and pharmacokinetic study of vorinostat (suberoylanilide hydroxamic acid) in Japanese patients with solid tumors. *Cancer Science, 100*, 1728–1734.

Galanis, E., Jaeckle, K. A., Maurer, M. J., Reid, J. M., Ames, M. M., Hardwick, J. S., et al. (2009). Phase II trial of vorinostat in recurrent glioblastoma multiforme: A north central cancer treatment group study. *Journal of Clinical Oncology: Official Journal of the American Society of Clinical Oncology, 27*, 2052–2058.

Galli, M., Salmoiraghi, S., Golay, J., Gozzini, A., Crippa, C., Pescosta, N., et al. (2010). A phase II multiple dose clinical trial of histone deacetylase inhibitor ITF2357 in patients with relapsed or progressive multiple myeloma. *Annals of Hematology, 89*, 185–190.

Garcia-Manero, G., Assouline, S., Cortes, J., Estrov, Z., Kantarjian, H., Yang, H., et al. (2008). Phase 1 study of the oral isotype specific histone deacetylase inhibitor MGCD0103 in leukemia. *Blood, 112*, 981–989.

Garcia-Manero, G., Kantarjian, H. M., Sanchez-Gonzalez, B., Yang, H., Rosner, G., Verstovsek, S., et al. (2006). Phase 1/2 study of the combination of 5-aza-2'-deoxycytidine with valproic acid in patients with leukemia. *Blood, 108*, 3271–3279.

Garcia-Manero, G., Yang, H., Bueso-Ramos, C., Ferrajoli, A., Cortes, J., Wierda, W. G., et al. (2008). Phase 1 study of the histone deacetylase inhibitor vorinostat (suberoylanilide hydroxamic acid [SAHA]) in patients with advanced leukemias and myelodysplastic syndromes. *Blood, 111*, 1060–1066.

Giaccone, G., Rajan, A., Berman, A., Kelly, R. J., Szabo, E., Lopez-Chavez, A., et al. (2011). Phase II study of belinostat in patients with recurrent or refractory advanced thymic epithelial tumors. *Journal of Clinical Oncology: Official Journal of the American Society of Clinical Oncology, 29*, 2052–2059.

Gimsing, P., Hansen, M., Knudsen, L. M., Knoblauch, P., Christensen, I. J., Ooi, C. E., et al. (2008). A phase I clinical trial of the histone deacetylase inhibitor belinostat in patients with advanced hematological neoplasia. *European Journal of Haematology, 81*, 170–176.

Gojo, I., Jiemjit, A., Trepel, J. B., Sparreboom, A., Figg, W. D., Rollins, S., et al. (2007). Phase 1 and pharmacologic study of MS-275, a histone deacetylase inhibitor, in adults with refractory and relapsed acute leukemias. *Blood, 109*, 2781–2790.

Gore, L., Rothenberg, M. L., O'Bryant, C. L., Schultz, M. K., Sandler, A. B., Coffin, D., et al. (2008). A phase I and pharmacokinetic study of the oral histone deacetylase inhibitor, MS-275, in patients with refractory solid tumors and lymphomas. *Clinical Cancer Research, 14*, 4517–4525.

Gottesman, M. M. (2002). Mechanisms of cancer drug resistance. *Annual Review of Medicine, 53*, 615–627.

Grant, C., Rahman, F., Piekarz, R., Peer, C., Frye, R., Robey, R. W., et al. (2010). Romidepsin: A new therapy for cutaneous T-cell lymphoma and a potential therapy for solid tumors. *Expert Review of Anticancer Therapy, 10*, 997–1008.

Gregoretti, I. V., Lee, Y. M., & Goodson, H. V. (2004). Molecular evolution of the histone deacetylase family: Functional implications of phylogenetic analysis. *Journal of Molecular Biology, 338*, 17–31.

Gui, C. Y., Ngo, L., Xu, W. S., Richon, V. M., & Marks, P. A. (2004). Histone deacetylase (HDAC) inhibitor activation of p21WAF1 involves changes in promoter-associated proteins, including HDAC1. *Proceedings of the National Academy of Sciences of the United States of America, 101*, 1241–1246.

Gurvich, N., Tsygankova, O. M., Meinkoth, J. L., & Klein, P. S. (2004). Histone deacetylase is a target of valproic acid-mediated cellular differentiation. *Cancer Research, 64*, 1079–1086.

Haberland, M., Mokalled, M. H., Montgomery, R. L., & Olson, E. N. (2009). Epigenetic control of skull morphogenesis by histone deacetylase 8. *Genes & Development, 23*, 1625–1630.

Hainsworth, J. D., Infante, J. R., Spigel, D. R., Arrowsmith, E. R., Boccia, R. V., & Burris, H. A. (2011). A phase II trial of panobinostat, a histone deacetylase inhibitor, in the treatment of patients with refractory metastatic renal cell carcinoma. *Cancer Investigation, 29*, 451–455.

Hamberg, P., Woo, M. M., Chen, L. C., Verweij, J., Porro, M. G., Zhao, L., et al. (2011). Effect of ketoconazole-mediated CYP3A4 inhibition on clinical pharmacokinetics of panobinostat (LBH589), an orally active histone deacetylase inhibitor. *Cancer Chemotherapy and Pharmacology, 68*, 805–813.

Hauschild, A., Trefzer, U., Garbe, C., Kaehler, K. C., Ugurel, S., Kiecker, F., et al. (2008). Multicenter phase II trial of the histone deacetylase inhibitor pyridylmethyl-N-{4-[(2-aminophenyl)-carbamoyl]-benzyl}-carbamate in pretreated metastatic melanoma. *Melanoma Research, 18*, 274–278.

Hayashi, A., Horiuchi, A., Kikuchi, N., Hayashi, T., Fuseya, C., Suzuki, A., et al. (2010). Type-specific roles of histone deacetylase (HDAC) overexpression in ovarian carcinoma: HDAC1 enhances cell proliferation and HDAC3 stimulates cell migration with down-regulation of E-cadherin. *International Journal of Cancer, 127*, 1332–1346.

Hrzenjak, A., Kremser, M. L., Strohmeier, B., Moinfar, F., Zatloukal, K., & Denk, H. (2008). SAHA induces caspase-independent, autophagic cell death of endometrial stromal sarcoma cells by influencing the mTOR pathway. *The Journal of Pathology, 216*, 495–504.

Hu, E., Dul, E., Sung, C. M., Chen, Z., Kirkpatrick, R., Zhang, G. F., et al. (2003). Identification of novel isoform-selective inhibitors within class I histone deacetylases. *The Journal of Pharmacology and Experimental Therapeutics, 307*, 720–728.

Hu, Y., Lu, W., Chen, G., Zhang, H., Jia, Y., Wei, Y., et al. (2010). Overcoming resistance to histone deacetylase inhibitors in human leukemia with the redox modulating compound beta-phenylethyl isothiocyanate. *Blood, 116*, 2732–2741.

Inoue, S., Walewska, R., Dyer, M. J., & Cohen, G. M. (2008). Downregulation of Mcl-1 potentiates HDACi-mediated apoptosis in leukemic cells. *Leukemia, 22*, 819–825.

Iwamoto, F. M., Lamborn, K. R., Kuhn, J. G., Wen, P. Y., Yung, W. K., Gilbert, M. R., et al. (2011). A phase I/II trial of the histone deacetylase inhibitor romidepsin for adults with recurrent malignant glioma: North American Brain Tumor Consortium Study 03-03. *Neuro-Oncology, 13*, 509–516.

Janku, F., McConkey, D. J., Hong, D. S., & Kurzrock, R. (2011). Autophagy as a target for anticancer therapy. *Nature Reviews. Clinical Oncology, 8*, 528–539.

Jin, S., & Scotto, K. W. (1998). Transcriptional regulation of the MDR1 gene by histone acetyltransferase and deacetylase is mediated by NF-Y. *Molecular and Cellular Biology, 18*, 4377–4384.

Jones, S. F., Bendell, J. C., Infante, J. R., Spigel, D. R., Thompson, D. S., Yardley, D. A., et al. (2011). A phase I study of panobinostat in combination with gemcitabine in the treatment of solid tumors. *Clinical Advances in Hematology & Oncology, 9*, 225–230.

Ju, R., & Muller, M. T. (2003). Histone deacetylase inhibitors activate p21(WAF1) expression via ATM. *Cancer Research, 63*, 2891–2897.

Kachhap, S. K., Rosmus, N., Collis, S. J., Kortenhorst, M. S., Wissing, M. D., Hedayati, M., et al. (2010). Downregulation of homologous recombination DNA repair genes by HDAC inhibition in prostate cancer is mediated through the E2F1 transcription factor. *PLoS One, 5*, e11208.

Kadia, T. M., Yang, H., Ferrajoli, A., Maddipotti, S., Schroeder, C., Madden, T. L., et al. (2010). A phase I study of vorinostat in combination with idarubicin in relapsed or refractory leukaemia. *British Journal of Haematology, 150*, 72–82.

Kahali, S., Sarcar, B., Fang, B., Williams, E. S., Koomen, J. M., Tofilon, P. J., et al. (2010). Activation of the unfolded protein response contributes toward the antitumor activity of vorinostat. *Neoplasia, 12*, 80–86.

Kasperlik-Zaluska, A. A., Zgliczynski, W., Jeske, W., & Zdunowski, P. (2005). ACTH responses to somatostatin, valproic acid and dexamethasone in Nelson's syndrome. *Neuro Endocrinology Letters, 26*, 709–712.

Kelly, W. K., O'Connor, O. A., Krug, L. M., Chiao, J. H., Heaney, M., Curley, T., et al. (2005). Phase I study of an oral histone deacetylase inhibitor, suberoylanilide hydroxamic acid, in patients with advanced cancer. *Journal of Clinical Oncology: Official Journal of the American Society of Clinical Oncology, 23*, 3923–3931.

Khanim, F. L., Bradbury, C. A., Arrazi, J., Hayden, R. E., Rye, A., Basu, S., et al. (2009). Elevated FOSB-expression; a potential marker of valproate sensitivity in AML. *British Journal of Haematology, 144*, 332–341.

Kim, Y. K., Kim, N. H., Hwang, J. W., Song, Y. J., Park, Y. S., Seo, D. W., et al. (2008). Histone deacetylase inhibitor apicidin-mediated drug resistance: Involvement of P-glycoprotein. *Biochemical and Biophysical Research Communications, 368*, 959–964.

Kirschbaum, M., Frankel, P., Popplewell, L., Zain, J., Delioukina, M., Pullarkat, V., et al. (2011). Phase II study of vorinostat for treatment of relapsed or refractory indolent non-Hodgkin's lymphoma and mantle cell lymphoma. *Journal of Clinical Oncology: Official Journal of the American Society of Clinical Oncology, 29*, 1198–1203.

Kirschbaum, M. H., Goldman, B. H., Zain, J. M., Cook, J. R., Rimsza, L. M., Forman, S. J., et al. (2012). A phase 2 study of vorinostat for treatment of relapsed or refractory Hodgkin lymphoma: Southwest Oncology Group Study S0517. *Leukemia & Lymphoma, 53*, 259–262.

Klimek, V. M., Fircanis, S., Maslak, P., Guernah, I., Baum, M., Wu, N., et al. (2008). Tolerability, pharmacodynamics, and pharmacokinetics studies of depsipeptide (romidepsin) in patients with acute myelogenous leukemia or advanced myelodysplastic syndromes. *Clinical Cancer Research: An Official Journal of the American Association for Cancer Research, 14*, 826–832.

Knutson, S. K., Chyla, B. J., Amann, J. M., Bhaskara, S., Huppert, S. S., & Hiebert, S. W. (2008). Liver-specific deletion of histone deacetylase 3 disrupts metabolic transcriptional networks. *The EMBO Journal, 27*, 1017–1028.

Krug, L. M., Curley, T., Schwartz, L., Richardson, S., Marks, P., Chiao, J., et al. (2006). Potential role of histone deacetylase inhibitors in mesothelioma: Clinical experience with suberoylanilide hydroxamic acid. *Clinical Lung Cancer, 7*, 257–261.

Kruh, G. D. (2003). Introduction to resistance to anticancer agents. *Oncogene, 22*, 7262–7264.

Kuendgen, A., Knipp, S., Fox, F., Strupp, C., Hildebrandt, B., Steidl, C., et al. (2005). Results of a phase 2 study of valproic acid alone or in combination with all-trans retinoic acid in 75 patients with myelodysplastic syndrome and relapsed or refractory acute myeloid leukemia. *Annals of Hematology, 84*(Suppl 1), 61–66.

Kuendgen, A., Schmid, M., Schlenk, R., Knipp, S., Hildebrandt, B., Steidl, C., et al. (2006). The histone deacetylase (HDAC) inhibitor valproic acid as monotherapy or in combination with all-trans retinoic acid in patients with acute myeloid leukemia. *Cancer, 106*, 112–119.

Kummar, S., Gutierrez, M., Gardner, E. R., Donovan, E., Hwang, K., Chung, E. J., et al. (2007). Phase I trial of MS-275, a histone deacetylase inhibitor, administered weekly in refractory solid tumors and lymphoid malignancies. *Clinical Cancer Research: An Official Journal of the American Association for Cancer Research, 13*, 5411–5417.

Lagger, G., O'Carroll, D., Rembold, M., Khier, H., Tischler, J., Weitzer, G., et al. (2002). Essential function of histone deacetylase 1 in proliferation control and CDK inhibitor repression. *The EMBO Journal*, *21*, 2672–2681.

Lai, C. J., Bao, R., Tao, X., Wang, J., Atoyan, R., Qu, H., et al. (2010). CUDC-101, a multitargeted inhibitor of histone deacetylase, epidermal growth factor receptor, and human epidermal growth factor receptor 2, exerts potent anticancer activity. *Cancer Research*, *70*, 3647–3656.

Lassen, U., Molife, L. R., Sorensen, M., Engelholm, S. A., Vidal, L., Sinha, R., et al. (2010). A phase I study of the safety and pharmacokinetics of the histone deacetylase inhibitor belinostat administered in combination with carboplatin and/or paclitaxel in patients with solid tumours. *British Journal of Cancer*, *103*, 12–17.

Lee, A. S. (2001). The glucose-regulated proteins: Stress induction and clinical applications. *Trends in Biochemical Sciences*, *26*, 504–510.

Lee, J. H., Choy, M. L., Ngo, L., Foster, S. S., & Marks, P. A. (2010). Histone deacetylase inhibitor induces DNA damage, which normal but not transformed cells can repair. *Proceedings of the National Academy of Sciences of the United States of America*, *107*, 14639–14644.

Lee, J. H., Choy, M. L., Ngo, L., Venta-Perez, G., & Marks, P. A. (2011). Role of checkpoint kinase 1 (Chk1) in the mechanisms of resistance to histone deacetylase inhibitors. *Proceedings of the National Academy of Sciences of the United States of America*, *108*, 19629–19634.

Lee, J. S., Paull, K., Alvarez, M., Hose, C., Monks, A., Grever, M., et al. (1994). Rhodamine efflux patterns predict P-glycoprotein substrates in the National Cancer Institute drug screen. *Molecular Pharmacology*, *46*, 627–638.

Lin, T. Y., Fenger, J., Murahari, S., Bear, M. D., Kulp, S. K., Wang, D., et al. (2010). AR-42, a novel HDAC inhibitor, exhibits biologic activity against malignant mast cell lines via down-regulation of constitutively activated Kit. *Blood*, *115*, 4217–4225.

Lindemann, R. K., Newbold, A., Whitecross, K. F., Cluse, L. A., Frew, A. J., Ellis, L., et al. (2007). Analysis of the apoptotic and therapeutic activities of histone deacetylase inhibitors by using a mouse model of B cell lymphoma. *Proceedings of the National Academy of Sciences of the United States of America*, *104*, 8071–8076.

Lombardi, P. M., Cole, K. E., Dowling, D. P., & Christianson, D. W. (2011). Structure, mechanism, and inhibition of histone deacetylases and related metalloenzymes. *Current Opinion in Structural Biology*, *21*, 735–743.

Lopez, G., Liu, J., Ren, W., Wei, W., Wang, S., Lahat, G., et al. (2009). Combining PCI-24781, a novel histone deacetylase inhibitor, with chemotherapy for the treatment of soft tissue sarcoma. *Clinical Cancer Research: An Official Journal of the American Association for Cancer Research*, *15*, 3472–3483.

Lopez, G., Torres, K., Liu, J., Hernandez, B., Young, E., Belousov, R., et al. (2011). Autophagic survival in resistance to histone deacetylase inhibitors: Novel strategies to treat malignant peripheral nerve sheath tumors. *Cancer Research*, *71*, 185–196.

Lucas, D. M., Alinari, L., West, D. A., Davis, M. E., Edwards, R. B., Johnson, A. J., et al. (2010). The novel deacetylase inhibitor AR-42 demonstrates pre-clinical activity in B-cell malignancies in vitro and in vivo. *PLoS One*, *5*, e10941.

Lucas, D. M., Davis, M. E., Parthun, M. R., Mone, A. P., Kitada, S., Cunningham, K. D., et al. (2004). The histone deacetylase inhibitor MS-275 induces caspase-dependent apoptosis in B-cell chronic lymphocytic leukemia cells. *Leukemia: Official Journal of the Leukemia Society of America, Leukemia Research Fund, U.K*, *18*, 1207–1214.

Lucio-Eterovic, A. K., Cortez, M. A., Valera, E. T., Motta, F. J., Queiroz, R. G., Machado, H. R., et al. (2008). Differential expression of 12 histone deacetylase (HDAC) genes in astrocytomas and normal brain tissue: Class II and IV are hypoexpressed in glioblastomas. *BMC Cancer*, *8*, 243.

Luu, T. H., Morgan, R. J., Leong, L., Lim, D., McNamara, M., Portnow, J., et al. (2008). A phase II trial of vorinostat (suberoylanilide hydroxamic acid) in metastatic breast cancer: A California Cancer Consortium study. *Clinical Cancer Research: An Official Journal of the American Association for Cancer Research, 14*, 7138–7142.

Mackay, H. J., Hirte, H., Colgan, T., Covens, A., MacAlpine, K., Grenci, P., et al. (2010). Phase II trial of the histone deacetylase inhibitor belinostat in women with platinum resistant epithelial ovarian cancer and micropapillary (LMP) ovarian tumours. *European Journal of Cancer, 46*, 1573–1579.

Mann, B. S., Johnson, J. R., He, K., Sridhara, R., Abraham, S., Booth, B. P., et al. (2007). Vorinostat for treatment of cutaneous manifestations of advanced primary cutaneous T-cell lymphoma. *Clinical Cancer Research: An Official Journal of the American Association for Cancer Research, 13*, 2318–2322.

Marks, P. A. (2006). Thioredoxin in cancer—Role of histone deacetylase inhibitors. *Seminars in Cancer Biology, 16*, 436–443.

Marks, P. A. (2010). Histone deacetylase inhibitors: A chemical genetics approach to understanding cellular functions. *Biochimica et Biophysica Acta, 1799*, 717–725.

Marks, P. A., & Breslow, R. (2007). Dimethyl sulfoxide to vorinostat: Development of this histone deacetylase inhibitor as an anticancer drug. *Nature Biotechnology, 25*, 84–90.

Maslak, P., Chanel, S., Camacho, L. H., Soignet, S., Pandolfi, P. P., Guernah, I., et al. (2006). Pilot study of combination transcriptional modulation therapy with sodium phenylbutyrate and 5-azacytidine in patients with acute myeloid leukemia or myelodysplastic syndrome. *Leukemia: Official Journal of the Leukemia Society of America, Leukemia Research Fund, U.K, 20*, 212–217.

Millward, M., Price, T., Townsend, A., Sweeney, C., Spencer, A., Sukumaran, S., et al. (2011). Phase 1 clinical trial of the novel proteasome inhibitor marizomib with the histone deacetylase inhibitor vorinostat in patients with melanoma, pancreatic and lung cancer based on in vitro assessments of the combination. *Investigational New Drugs*, .

Minamiya, Y., Ono, T., Saito, H., Takahashi, N., Ito, M., Mitsui, M., et al. (2011). Expression of histone deacetylase 1 correlates with a poor prognosis in patients with adenocarcinoma of the lung. *Lung Cancer, 74*, 300–304.

Mitsiades, N., Mitsiades, C. S., Richardson, P. G., McMullan, C., Poulaki, V., Fanourakis, G., et al. (2003). Molecular sequelae of histone deacetylase inhibition in human malignant B cells. *Blood, 101*, 4055–4062.

Mizushima, N., Yoshimori, T., & Levine, B. (2010). Methods in mammalian autophagy research. *Cell, 140*, 313–326.

Modesitt, S. C., Sill, M., Hoffman, J. S., & Bender, D. P. (2008). A phase II study of vorinostat in the treatment of persistent or recurrent epithelial ovarian or primary peritoneal carcinoma: A Gynecologic Oncology Group study. *Gynecologic Oncology, 109*, 182–186.

Moffat, D., Patel, S., Day, F., Belfield, A., Donald, A., Rowlands, M., et al. (2010). Discovery of 2-(6-{[(6-fluoroquinolin-2-yl)methyl]amino}bicyclo[3.1.0]hex-3-yl)-N-hydroxypyrim idine-5-carboxamide (CHR-3996), a class I selective orally active histone deacetylase inhibitor. *Journal of Medicinal Chemistry, 53*, 8663–8678.

Mohammed, T. A., Holen, K. D., Jaskula-Sztul, R., Mulkerin, D., Lubner, S. J., Schelman, W. R., et al. (2011). A pilot phase II study of valproic acid for treatment of low-grade neuroendocrine carcinoma. *The Oncologist, 16*, 835–843.

Molife, L. R., Attard, G., Fong, P. C., Karavasilis, V., Reid, A. H., Patterson, S., et al. (2010). Phase II, two-stage, single-arm trial of the histone deacetylase inhibitor (HDACi) romidepsin in metastatic castration-resistant prostate cancer (CRPC). *Annals of Oncology: Official Journal of the ESMO, 21*, 109–113.

Montgomery, R. L., Davis, C. A., Potthoff, M. J., Haberland, M., Fielitz, J., Qi, X., et al. (2007). Histone deacetylases 1 and 2 redundantly regulate cardiac morphogenesis, growth, and contractility. *Genes & Development, 21*, 1790–1802.

Munshi, A., Kurland, J. F., Nishikawa, T., Tanaka, T., Hobbs, M. L., Tucker, S. L., et al. (2005). Histone deacetylase inhibitors radiosensitize human melanoma cells by suppressing DNA repair activity. *Clinical Cancer Research, 11*, 4912–4922.

Munster, P., Marchion, D., Bicaku, E., Lacevic, M., Kim, J., Centeno, B., et al. (2009). Clinical and biological effects of valproic acid as a histone deacetylase inhibitor on tumor and surrogate tissues: Phase I/II trial of valproic acid and epirubicin/FEC. *Clinical Cancer Research: An Official Journal of the American Association for Cancer Research, 15*, 2488–2496.

Munster, P., Marchion, D., Bicaku, E., Schmitt, M., Lee, J. H., DeConti, R., et al. (2007). Phase I trial of histone deacetylase inhibition by valproic acid followed by the topoisomerase II inhibitor epirubicin in advanced solid tumors: A clinical and translational study. *Journal of Clinical Oncology: Official Journal of the American Society of Clinical Oncology, 25*, 1979–1985.

Munster, P. N., Rubin, E. H., Van Belle, S., Friedman, E., Patterson, J. K., Van Dyck, K., et al. (2009). A single supratherapeutic dose of vorinostat does not prolong the QTc interval in patients with advanced cancer. *Clinical Cancer Research: An Official Journal of the American Association for Cancer Research, 15*, 7077–7084.

Munster, P. N., Thurn, K. T., Thomas, S., Raha, P., Lacevic, M., Miller, A., et al. (2011). A phase II study of the histone deacetylase inhibitor vorinostat combined with tamoxifen for the treatment of patients with hormone therapy-resistant breast cancer. *British Journal of Cancer, 104*, 1828–1835.

Niesvizky, R., Ely, S., Mark, T., Aggarwal, S., Gabrilove, J. L., Wright, J. J., et al. (2011). Phase 2 trial of the histone deacetylase inhibitor romidepsin for the treatment of refractory multiple myeloma. *Cancer, 117*, 336–342.

Novotny-Diermayr, V., Sangthongpitag, K., Hu, C. Y., Wu, X., Sausgruber, N., Yeo, P., et al. (2010). SB939, a novel potent and orally active histone deacetylase inhibitor with high tumor exposure and efficacy in mouse models of colorectal cancer. *Molecular Cancer Therapeutics, 9*, 642–652.

O'Connor, O. A., Heaney, M. L., Schwartz, L., Richardson, S., Willim, R., MacGregor-Cortelli, B., et al. (2006). Clinical experience with intravenous and oral formulations of the novel histone deacetylase inhibitor suberoylanilide hydroxamic acid in patients with advanced hematologic malignancies. *Journal of Clinical Oncology: Official Journal of the American Society of Clinical Oncology, 24*, 166–173.

Odenike, O. M., Alkan, S., Sher, D., Godwin, J. E., Huo, D., Brandt, S. J., et al. (2008). Histone deacetylase inhibitor romidepsin has differential activity in core binding factor acute myeloid leukemia. *Clinical Cancer Research: An Official Journal of the American Association for Cancer Research, 14*, 7095–7101.

Oehme, I., Deubzer, H. E., Lodrini, M., Milde, T., & Witt, O. (2009). Targeting of HDAC8 and investigational inhibitors in neuroblastoma. *Expert Opinion on Investigational Drugs, 18*, 1605–1617.

Okada, T., Tanaka, K., Nakatani, F., Sakimura, R., Matsunobu, T., Li, X., et al. (2006). Involvement of P-glycoprotein and MRP1 in resistance to cyclic tetrapeptide subfamily of histone deacetylase inhibitors in the drug-resistant osteosarcoma and Ewing's sarcoma cells. *International Journal of Cancer, 118*, 90–97.

Olsen, E. A., Kim, Y. H., Kuzel, T. M., Pacheco, T. R., Foss, F. M., Parker, S., et al. (2007). Phase IIb multicenter trial of vorinostat in patients with persistent, progressive, or treatment refractory cutaneous T-cell lymphoma. *Journal of Clinical Oncology: Official Journal of the American Society of Clinical Oncology, 25*, 3109–3115.

Osada, H., Tatematsu, Y., Saito, H., Yatabe, Y., Mitsudomi, T., & Takahashi, T. (2004). Reduced expression of class II histone deacetylase genes is associated with poor prognosis in lung cancer patients. *International Journal of Cancer, 112*, 26–32.

Otterson, G. A., Hodgson, L., Pang, H., & Vokes, E. E. (2010). Phase II study of the histone deacetylase inhibitor Romidepsin in relapsed small cell lung cancer (Cancer and Leukemia Group B 30304). *Journal of Thoracic Oncology, 5*, 1644–1648.

Pahl, H. L. (1999). Activators and target genes of Rel/NF-kappaB transcription factors. *Oncogene, 18*, 6853–6866.

Palmieri, D., Lockman, P. R., Thomas, F. C., Hua, E., Herring, J., Hargrave, E., et al. (2009). Vorinostat inhibits brain metastatic colonization in a model of triple-negative breast cancer and induces DNA double-strand breaks. *Clinical Cancer Research: An Official Journal of the American Association for Cancer Research, 15*, 6148–6157.

Parmigiani, R. B., Xu, W. S., Venta-Perez, G., Erdjument-Bromage, H., Yaneva, M., Tempst, P., et al. (2008). HDAC6 is a specific deacetylase of peroxiredoxins and is involved in redox regulation. *Proceedings of the National Academy of Sciences of the United States of America, 105*, 9633–9638.

Peart, M. J., Tainton, K. M., Ruefli, A. A., Dear, A. E., Sedelies, K. A., O'Reilly, L. A., et al. (2003). Novel mechanisms of apoptosis induced by histone deacetylase inhibitors. *Cancer Research, 63*, 4460–4471.

Phuphanich, S., Baker, S. D., Grossman, S. A., Carson, K. A., Gilbert, M. R., Fisher, J. D., et al. (2005). Oral sodium phenylbutyrate in patients with recurrent malignant gliomas: A dose escalation and pharmacologic study. *Neuro-Oncology, 7*, 177–182.

Piekarz, R. L., Frye, R., Prince, H. M., Kirschbaum, M. H., Zain, J., Allen, S. L., et al. (2011). Phase 2 trial of romidepsin in patients with peripheral T-cell lymphoma. *Blood, 117*, 5827–5834.

Piekarz, R. L., Frye, R., Turner, M., Wright, J. J., Allen, S. L., Kirschbaum, M. H., et al. (2009). Phase II multi-institutional trial of the histone deacetylase inhibitor romidepsin as monotherapy for patients with cutaneous T-cell lymphoma. *Journal of Clinical Oncology: Official Journal of the American Society of Clinical Oncology, 27*, 5410–5417.

Piekarz, R. L., Robey, R. W., Zhan, Z., Kayastha, G., Sayah, A., Abdeldaim, A. H., et al. (2004). T-cell lymphoma as a model for the use of histone deacetylase inhibitors in cancer therapy: Impact of depsipeptide on molecular markers, therapeutic targets, and mechanisms of resistance. *Blood, 103*, 4636–4643.

Pilatrino, C., Cilloni, D., Messa, E., Morotti, A., Giugliano, E., Pautasso, M., et al. (2005). Increase in platelet count in older, poor-risk patients with acute myeloid leukemia or myelodysplastic syndrome treated with valproic acid and all-trans retinoic acid. *Cancer, 104*, 101–109.

Qiu, L., Burgess, A., Fairlie, D. P., Leonard, H., Parsons, P. G., & Gabrielli, B. G. (2000). Histone deacetylase inhibitors trigger a G2 checkpoint in normal cells that is defective in tumor cells. *Molecular Biology of the Cell, 11*, 2069–2083.

Quint, K., Agaimy, A., Di Fazio, P., Montalbano, R., Steindorf, C., Jung, R., et al. (2011). Clinical significance of histone deacetylases 1, 2, 3, and 7: HDAC2 is an independent predictor of survival in HCC. *Virchows Archiv, 459*, 129–139.

Raffoux, E., Cras, A., Recher, C., Boelle, P. Y., de Labarthe, A., Turlure, P., et al. (2010). Phase 2 clinical trial of 5-azacitidine, valproic acid, and all-trans retinoic acid in patients with high-risk acute myeloid leukemia or myelodysplastic syndrome. *Oncotarget, 1*, 34–42.

Ramalingam, S. S., Belani, C. P., Ruel, C., Frankel, P., Gitlitz, B., Koczywas, M., et al. (2009). Phase II study of belinostat (PXD101), a histone deacetylase inhibitor, for second line therapy of advanced malignant pleural mesothelioma. *Journal of Thoracic Oncology, 4*, 97–101.

Ramalingam, S. S., Kummar, S., Sarantopoulos, J., Shibata, S., LoRusso, P., Yerk, M., et al. (2010). Phase I study of vorinostat in patients with advanced solid tumors and hepatic dysfunction: A National Cancer Institute Organ Dysfunction Working Group study.

Journal of Clinical Oncology: Official Journal of the American Society of Clinical Oncology, 28, 4507–4512.

Ramalingam, S. S., Parise, R. A., Ramanathan, R. K., Lagattuta, T. F., Musguire, L. A., Stoller, R. G., et al. (2007). Phase I and pharmacokinetic study of vorinostat, a histone deacetylase inhibitor, in combination with carboplatin and paclitaxel for advanced solid malignancies. *Clinical Cancer Research: An Official Journal of the American Association for Cancer Research, 13*, 3605–3610.

Ramaswamy, B., Fiskus, W., Cohen, B., Pellegrino, C., Hershman, D. L., Chuang, E., et al. (2011). Phase I-II study of vorinostat plus paclitaxel and bevacizumab in metastatic breast cancer: Evidence for vorinostat-induced tubulin acetylation and Hsp90 inhibition in vivo. *Breast Cancer Research and Treatment, 132*, 1063–1072.

Rambaldi, A., Dellacasa, C. M., Finazzi, G., Carobbio, A., Ferrari, M. L., Guglielmelli, P., et al. (2010). A pilot study of the Histone-Deacetylase inhibitor Givinostat in patients with JAK2V617F positive chronic myeloproliferative neoplasms. *British Journal of Haematology, 150*, 446–455.

Rao, R., Nalluri, S., Fiskus, W., Savoie, A., Buckley, K. M., Ha, K., et al. (2010). Role of CAAT/enhancer binding protein homologous protein in panobinostat-mediated potentiation of bortezomib-induced lethal endoplasmic reticulum stress in mantle cell lymphoma cells. *Clinical Cancer Research: An Official Journal of the American Association for Cancer Research, 16*, 4742–4754.

Rao, R., Nalluri, S., Kolhe, R., Yang, Y., Fiskus, W., Chen, J., et al. (2010). Treatment with panobinostat induces glucose-regulated protein 78 acetylation and endoplasmic reticulum stress in breast cancer cells. *Molecular Cancer Therapeutics, 9*, 942–952.

Rascle, A., Johnston, J. A., & Amati, B. (2003). Deacetylase activity is required for recruitment of the basal transcription machinery and transactivation by STAT5. *Molecular and Cellular Biology, 23*, 4162–4173.

Rathkopf, D., Wong, B. Y., Ross, R. W., Anand, A., Tanaka, E., Woo, M. M., et al. (2010). A phase I study of oral panobinostat alone and in combination with docetaxel in patients with castration-resistant prostate cancer. *Cancer Chemotherapy and Pharmacology, 66*, 181–189.

Razak, A. R., Hotte, S. J., Siu, L. L., Chen, E. X., Hirte, H. W., Powers, J., et al. (2011). Phase I clinical, pharmacokinetic and pharmacodynamic study of SB939, an oral histone deacetylase (HDAC) inhibitor, in patients with advanced solid tumours. *British Journal of Cancer, 104*, 756–762.

Ree, A. H., Dueland, S., Folkvord, S., Hole, K. H., Seierstad, T., Johansen, M., et al. (2010). Vorinostat, a histone deacetylase inhibitor, combined with pelvic palliative radiotherapy for gastrointestinal carcinoma: The Pelvic Radiation and Vorinostat (PRAVO) phase 1 study. *The Lancet Oncology, 11*, 459–464.

Reid, T., Valone, F., Lipera, W., Irwin, D., Paroly, W., Natale, R., et al. (2004). Phase II trial of the histone deacetylase inhibitor pivaloyloxymethyl butyrate (Pivanex, AN-9) in advanced non-small cell lung cancer. *Lung Cancer, 45*, 381–386.

Richards, D. A., Boehm, K. A., Waterhouse, D. M., Wagener, D. J., Krishnamurthi, S. S., Rosemurgy, A., et al. (2006). Gemcitabine plus CI-994 offers no advantage over gemcitabine alone in the treatment of patients with advanced pancreatic cancer: Results of a phase II randomized, double-blind, placebo-controlled, multicenter study. *Annals of Oncology: Official Journal of the ESMO, 17*, 1096–1102.

Richardson, P., Mitsiades, C., Colson, K., Reilly, E., McBride, L., Chiao, J., et al. (2008). Phase I trial of oral vorinostat (suberoylanilide hydroxamic acid, SAHA) in patients with advanced multiple myeloma. *Leukemia & Lymphoma, 49*, 502–507.

Richon, V. M., Sandhoff, T. W., Rifkind, R. A., & Marks, P. A. (2000). Histone deacetylase inhibitor selectively induces p21WAF1 expression and gene-associated histone acetylation. *Proceedings of the National Academy of Sciences of the United States of America, 97*, 10014–10019.

Robert, T., Vanoli, F., Chiolo, I., Shubassi, G., Bernstein, K. A., Rothstein, R., et al. (2011). HDACs link the DNA damage response, processing of double-strand breaks and autophagy. *Nature, 471*, 74–79.

Robey, R. W., Zhan, Z., Piekarz, R. L., Kayastha, G. L., Fojo, T., & Bates, S. E. (2006). Increased MDR1 expression in normal and malignant peripheral blood mononuclear cells obtained from patients receiving depsipeptide (FR901228, FK228, NSC630176). *Clinical Cancer Research, 12*, 1547–1555.

Rocca, A., Minucci, S., Tosti, G., Croci, D., Contegno, F., Ballarini, M., et al. (2009). A phase I-II study of the histone deacetylase inhibitor valproic acid plus chemo-immunotherapy in patients with advanced melanoma. *British Journal of Cancer, 100*, 28–36.

Romanski, A., Bacic, B., Bug, G., Pfeifer, H., Gul, H., Remiszewski, S., et al. (2004). Use of a novel histone deacetylase inhibitor to induce apoptosis in cell lines of acute lymphoblastic leukemia. *Haematologica, 89*, 419–426.

Ron, D., & Walter, P. (2007). Signal integration in the endoplasmic reticulum unfolded protein response. *Nature Reviews. Molecular Cell Biology, 8*, 519–529.

Rosato, R. R., Almenara, J. A., Maggio, S. C., Coe, S., Atadja, P., Dent, P., et al. (2008). Role of histone deacetylase inhibitor-induced reactive oxygen species and DNA damage in LAQ-824/fludarabine antileukemic interactions. *Molecular Cancer Therapeutics, 7*, 3285–3297.

Rosato, R. R., Kolla, S. S., Hock, S. K., Almenara, J. A., Patel, A., Amin, S., et al. (2010). Histone deacetylase inhibitors activate NF-kappaB in human leukemia cells through an ATM/NEMO-related pathway. *The Journal of Biological Chemistry, 285*, 10064–10077.

Rubin, E. H., Agrawal, N. G., Friedman, E. J., Scott, P., Mazina, K. E., Sun, L., et al. (2006). A study to determine the effects of food and multiple dosing on the pharmacokinetics of vorinostat given orally to patients with advanced cancer. *Clinical Cancer Research: An Official Journal of the American Association for Cancer Research, 12*, 7039–7045.

Rundall, B. K., Denlinger, C. E., & Jones, D. R. (2004). Combined histone deacetylase and NF-kappaB inhibition sensitizes non-small cell lung cancer to cell death. *Surgery, 136*, 416–425.

Ryan, Q. C., Headlee, D., Acharya, M., Sparreboom, A., Trepel, J. B., Ye, J., et al. (2005). Phase I and pharmacokinetic study of MS-275, a histone deacetylase inhibitor, in patients with advanced and refractory solid tumors or lymphoma. *Journal of Clinical Oncology: Official Journal of the American Society of Clinical Oncology, 23*, 3912–3922.

Sandor, V., Senderowicz, A., Mertins, S., Sackett, D., Sausville, E., Blagosklonny, M. V., et al. (2000). P21-dependent g(1)arrest with downregulation of cyclin D1 and upregulation of cyclin E by the histone deacetylase inhibitor FR901228. *British Journal of Cancer, 83*, 817–825.

Santo, L., Hideshima, T., Kung, A. L., Tseng, J. C., Tamang, D., Yang, M., et al. (2012). Preclinical activity, pharmacodynamic and pharmacokinetic properties of a selective HDAC6 inhibitor, ACY-1215, in combination with bortezomib in multiple myeloma. *Blood, 119*, 2579–2589.

Schaefer, E. W., Loaiza-Bonilla, A., Juckett, M., DiPersio, J. F., Roy, V., Slack, J., et al. (2009). A phase 2 study of vorinostat in acute myeloid leukemia. *Haematologica, 94*, 1375–1382.

Scherpereel, A., Berghmans, T., Lafitte, J. J., Colinet, B., Richez, M., Bonduelle, Y., et al. (2011). Valproate-doxorubicin: Promising therapy for progressing mesothelioma. A phase II study. *The European Respiratory Journal, 37*, 129–135.

Schneider, B. J., Kalemkerian, G. P., Bradley, D., Smith, D. C., Egorin, M. J., Daignault, S., et al. (2012). Phase I study of vorinostat (suberoylanilide hydroxamic acid, NSC 701852) in combination with docetaxel in patients with advanced and relapsed solid malignancies. *Investigational New Drugs, 30*, 249–257.

Schrump, D. S., Fischette, M. R., Nguyen, D. M., Zhao, M., Li, X., Kunst, T. F., et al. (2008). Clinical and molecular responses in lung cancer patients receiving Romidepsin. *Clinical Cancer Research: An Official Journal of the American Association for Cancer Research*, *14*, 188–198.

Schuetz, A., Min, J., Allali-Hassani, A., Schapira, M., Shuen, M., Loppnau, P., et al. (2008). Human HDAC7 harbors a class IIa histone deacetylase-specific zinc binding motif and cryptic deacetylase activity. *The Journal of Biological Chemistry*, *283*, 11355–11363.

Shao, W., Growney, J. D., Feng, Y., O'Connor, G., Pu, M., Zhu, W., et al. (2010). Activity of deacetylase inhibitor panobinostat (LBH589) in cutaneous T-cell lymphoma models: Defining molecular mechanisms of resistance. *International Journal of Cancer*, *127*, 2199–2208.

Shao, Y., Gao, Z., Marks, P. A., & Jiang, X. (2004). Apoptotic and autophagic cell death induced by histone deacetylase inhibitors. *Proceedings of the National Academy of Sciences of the United States of America*, *101*, 18030–18035.

Sharma, S. V., Lee, D. Y., Li, B., Quinlan, M. P., Takahashi, F., Maheswaran, S., et al. (2010). A chromatin-mediated reversible drug-tolerant state in cancer cell subpopulations. *Cell*, *141*, 69–80.

Shi, Y., Gerritsma, D., Bowes, A. J., Capretta, A., & Werstuck, G. H. (2007). Induction of GRP78 by valproic acid is dependent upon histone deacetylase inhibition. *Bioorganic & Medicinal Chemistry Letters*, *17*, 4491–4494.

Siu, L. L., Pili, R., Duran, I., Messersmith, W. A., Chen, E. X., Sullivan, R., et al. (2008). Phase I study of MGCD0103 given as a three-times-per-week oral dose in patients with advanced solid tumors. *Journal of Clinical Oncology*, *26*, 1940–1947.

Somoza, J. R., Skene, R. J., Katz, B. A., Mol, C., Ho, J. D., Jennings, A. J., et al. (2004). Structural snapshots of human HDAC8 provide insights into the class I histone deacetylases. *Structure*, *12*, 1325–1334.

Soriano, A. O., Yang, H., Faderl, S., Estrov, Z., Giles, F., Ravandi, F., et al. (2007). Safety and clinical activity of the combination of 5-azacytidine, valproic acid, and all-trans retinoic acid in acute myeloid leukemia and myelodysplastic syndrome. *Blood*, *110*, 2302–2308.

Stadler, W. M., Margolin, K., Ferber, S., McCulloch, W., & Thompson, J. A. (2006). A phase II study of depsipeptide in refractory metastatic renal cell cancer. *Clinical Genitourinary Cancer*, *5*, 57–60.

Stathis, A., Hotte, S. J., Chen, E. X., Hirte, H. W., Oza, A. M., Moretto, P., et al. (2011). Phase I study of decitabine in combination with vorinostat in patients with advanced solid tumors and non-Hodgkin's lymphomas. *Clinical Cancer Research: An Official Journal of the American Association for Cancer Research*, *17*, 1582–1590.

Steele, N. L., Plumb, J. A., Vidal, L., Tjornelund, J., Knoblauch, P., Rasmussen, A., et al. (2008). A phase 1 pharmacokinetic and pharmacodynamic study of the histone deacetylase inhibitor belinostat in patients with advanced solid tumors. *Clinical Cancer Research: An Official Journal of the American Association for Cancer Research*, *14*, 804–810.

Stimson, L., & La Thangue, N. B. (2009). Biomarkers for predicting clinical responses to HDAC inhibitors. *Cancer Letters*, *280*, 177–183.

Su, J. M., Li, X. N., Thompson, P., Ou, C. N., Ingle, A. M., Russell, H., et al. (2011). Phase 1 study of valproic acid in pediatric patients with refractory solid or CNS tumors: A children's oncology group report. *Clinical Cancer Research: An Official Journal of the American Association for Cancer Research*, *17*, 589–597.

Sung, M. W., & Waxman, S. (2007). Combination of cytotoxic-differentiation therapy with 5-fluorouracil and phenylbutyrate in patients with advanced colorectal cancer. *Anticancer Research*, *27*, 995–1001.

Traynor, A. M., Dubey, S., Eickhoff, J. C., Kolesar, J. M., Schell, K., Huie, M. S., et al. (2009). Vorinostat (NSC# 701852) in patients with relapsed non-small cell lung cancer: A Wisconsin Oncology Network phase II study. *Journal of Thoracic Oncology*, *4*, 522–526.
Trivedi, C. M., Luo, Y., Yin, Z., Zhang, M., Zhu, W., Wang, T., et al. (2007). Hdac2 regulates the cardiac hypertrophic response by modulating Gsk3 beta activity. *Nature Medicine*, *13*, 324–331.
Ungerstedt, J. S., Sowa, Y., Xu, W. S., Shao, Y., Dokmanovic, M., Perez, G., et al. (2005). Role of thioredoxin in the response of normal and transformed cells to histone deacetylase inhibitors. *Proceedings of the National Academy of Sciences of the United States of America*, *102*, 673–678.
Vannini, A., Volpari, C., Filocamo, G., Casavola, E. C., Brunetti, M., Renzoni, D., et al. (2004). Crystal structure of a eukaryotic zinc-dependent histone deacetylase, human HDAC8, complexed with a hydroxamic acid inhibitor. *Proceedings of the National Academy of Sciences of the United States of America*, *101*, 15064–15069.
Vansteenkiste, J., Van Cutsem, E., Dumez, H., Chen, C., Ricker, J. L., Randolph, S. S., et al. (2008). Early phase II trial of oral vorinostat in relapsed or refractory breast, colorectal, or non-small cell lung cancer. *Investigational New Drugs*, *26*, 483–488.
Vega, R. B., Matsuda, K., Oh, J., Barbosa, A. C., Yang, X., Meadows, E., et al. (2004). Histone deacetylase 4 controls chondrocyte hypertrophy during skeletogenesis. *Cell*, *119*, 555–566.
Wada, C. K., Frey, R. R., Ji, Z., Curtin, M. L., Garland, R. B., Holms, J. H., et al. (2003). Alpha-keto amides as inhibitors of histone deacetylase. *Bioorganic & Medicinal Chemistry Letters*, *13*, 3331–3335.
Wang, J. F., Bown, C., & Young, L. T. (1999). Differential display PCR reveals novel targets for the mood-stabilizing drug valproate including the molecular chaperone GRP78. *Molecular Pharmacology*, *55*, 521–527.
Wang, Y., Fiskus, W., Chong, D. G., Buckley, K. M., Natarajan, K., Rao, R., et al. (2009). Cotreatment with panobinostat and JAK2 inhibitor TG101209 attenuates JAK2V617F levels and signaling and exerts synergistic cytotoxic effects against human myeloproliferative neoplastic cells. *Blood*, *114*, 5024–5033.
Warrell, R. P., Jr., He, L. Z., Richon, V., Calleja, E., & Pandolfi, P. P. (1998). Therapeutic targeting of transcription in acute promyelocytic leukemia by use of an inhibitor of histone deacetylase. *Journal of the National Cancer Institute*, *90*, 1621–1625.
Wegener, D., Deubzer, H. E., Oehme, I., Milde, T., Hildmann, C., Schwienhorst, A., et al. (2008). HKI 46F08, a novel potent histone deacetylase inhibitor, exhibits antitumoral activity against embryonic childhood cancer cells. *Anticancer drugs*, *19*, 849–857.
Weichert, W., Denkert, C., Noske, A., Darb-Esfahani, S., Dietel, M., Kalloger, S. E., et al. (2008). Expression of class I histone deacetylases indicates poor prognosis in endometrioid subtypes of ovarian and endometrial carcinomas. *Neoplasia*, *10*, 1021–1027.
Whitehead, R. P., Rankin, C., Hoff, P. M., Gold, P. J., Billingsley, K. G., Chapman, R. A., et al. (2009). Phase II trial of romidepsin (NSC-630176) in previously treated colorectal cancer patients with advanced disease: A Southwest Oncology Group study (S0336). *Investigational New Drugs*, *27*, 469–475.
Whittaker, S. J., Demierre, M. F., Kim, E. J., Rook, A. H., Lerner, A., Duvic, M., et al. (2010). Final results from a multicenter, international, pivotal study of romidepsin in refractory cutaneous T-cell lymphoma. *Journal of Clinical Oncology: Official Journal of the American Society of Clinical Oncology*, *28*, 4485–4491.
Wilson, P. M., El-Khoueiry, A., Iqbal, S., Fazzone, W., LaBonte, M. J., Groshen, S., et al. (2010). A phase I/II trial of vorinostat in combination with 5-fluorouracil in patients with metastatic colorectal cancer who previously failed 5-FU-based chemotherapy. *Cancer Chemotherapy and Pharmacology*, *65*, 979–988.

Witt, O., Deubzer, H. E., Milde, T., & Oehme, I. (2009). HDAC family: What are the cancer relevant targets? *Cancer Letters, 277*, 8–21.

Wolff, J. E., Kramm, C., Kortmann, R. D., Pietsch, T., Rutkowski, S., Jorch, N., et al. (2008). Valproic acid was well tolerated in heavily pretreated pediatric patients with high-grade glioma. *Journal of Neuro-Oncology, 90*, 309–314.

Xiao, J. J., Huang, Y., Dai, Z., Sadee, W., Chen, J., Liu, S., et al. (2005). Chemoresistance to depsipeptide FK228 [(E)-(1S,4S,10S,21R)-7-[(Z)-ethylidene]-4,21-diisopropyl-2-oxa-12, 13-dithi a-5,8,20,23-tetraazabicyclo[8,7,6]-tricos-16-ene-3,6,9,22-pentanone] is mediated by reversible MDR1 induction in human cancer cell lines. *The Journal of Pharmacology and Experimental Therapeutics, 314*, 467–475.

Yamada, H., Arakawa, Y., Saito, S., Agawa, M., Kano, Y., & Horiguchi-Yamada, J. (2006). Depsipeptide-resistant KU812 cells show reversible P-glycoprotein expression, hyperacetylated histones, and modulated gene expression profile. *Leukemia Research, 30*, 723–734.

Yaneva, M., Li, H., Marple, T., & Hasty, P. (2005). Non-homologous end joining, but not homologous recombination, enables survival for cells exposed to a histone deacetylase inhibitor. *Nucleic Acids Research, 33*, 5320–5330.

Yong, W. P., Goh, B. C., Soo, R. A., Toh, H. C., Ethirajulu, K., Wood, J., et al. (2011). Phase I and pharmacodynamic study of an orally administered novel inhibitor of histone deacetylases, SB939, in patients with refractory solid malignancies. *Annals of Oncology: Official Journal of the ESMO, 22*, 2516–2522.

Younes, A., Oki, Y., Bociek, R. G., Kuruvilla, J., Fanale, M., Neelapu, S., et al. (2011). Mocetinostat for relapsed classical Hodgkin's lymphoma: An open-label, single-arm, phase 2 trial. *The Lancet Oncology, 12*, 1222–1228.

Yu, H., & Jove, R. (2004). The STATs of cancer—New molecular targets come of age. *Nature Reviews. Cancer, 4*, 97–105.

Zhang, C., Richon, V., Ni, X., Talpur, R., & Duvic, M. (2005). Selective induction of apoptosis by histone deacetylase inhibitor SAHA in cutaneous T-cell lymphoma cells: Relevance to mechanism of therapeutic action. *The Journal of Investigative Dermatology, 125*, 1045–1052.

Zhang, C. L., McKinsey, T. A., Chang, S., Antos, C. L., Hill, J. A., & Olson, E. N. (2002). Class II histone deacetylases act as signal-responsive repressors of cardiac hypertrophy. *Cell, 110*, 479–488.

Zhang, Y., Adachi, M., Kawamura, R., & Imai, K. (2006). Bmf is a possible mediator in histone deacetylase inhibitors FK228 and CBHA-induced apoptosis. *Cell Death and Differentiation, 13*, 129–140.

Zhang, Y., Carr, T., Dimtchev, A., Zaer, N., Dritschilo, A., & Jung, M. (2007). Attenuated DNA damage repair by trichostatin A through BRCA1 suppression. *Radiation Research, 168*, 115–124.

Zhang, Y., Kwon, S., Yamaguchi, T., Cubizolles, F., Rousseaux, S., Kneissel, M., et al. (2008). Mice lacking histone deacetylase 6 have hyperacetylated tubulin but are viable and develop normally. *Molecular and Cellular Biology, 28*, 1688–1701.

Zhou, J., Bi, C., Chng, W. J., Cheong, L. L., Liu, S. C., Mahara, S., et al. (2011). PRL-3, a metastasis associated tyrosine phosphatase, is involved in FLT3-ITD signaling and implicated in anti-AML therapy. *PLoS One, 6*, e19798.

CHAPTER THREE

HDAC Inhibitors: Roles of DNA Damage and Repair

Carine Robert, Feyruz V. Rassool[1]

Department of Radiation Oncology and Greenebaum Cancer Center, University of Maryland School of Medicine, Baltimore, Maryland, USA

[1]Corresponding author: e-mail address: frassool@som.umaryland.edu

Contents

Abstract

Histone deacetylase inhibitors (HDACis) increase gene expression through induction of histone acetylation. However, it remains unclear whether specific gene expression changes determine the apoptotic response following HDACis administration. Herein, we discuss evidence that HDACis trigger in cancer and leukemia cells not only

Advances in Cancer Research, Volume 116
ISSN 0065-230X
http://dx.doi.org/10.1016/B978-0-12-394387-3.00003-3

widespread histone acetylation but also actual increases in reactive oxygen species (ROS) and DNA damage that are further increased following treatment with DNA-damaging chemotherapies. While the origins of ROS production are not completely understood, mechanisms, including inflammation and altered antioxidant signaling, have been reported. While the generation of ROS is an explanation, at least in part, for the source of DNA damage observed with HDACi treatment, DNA damage can also be independently induced by changes in the DNA repair activity and chromatin remodeling factors. Recent development of sirtuin inhibitors (SIRTis) has shown that, similar to HDACis, these drugs induce increases in ROS and DNA damage used singly, or in combination with HDACis and other drugs. Thus, induction of apoptosis by HDACis/SIRTis may result through oxidative stress and DNA damage mechanisms in addition to direct activation of apoptosis-inducing genes. Nevertheless, while DNA damage and stress responses could be of interest as markers for clinical responses, they have yet to be validated as markers for responses to HDACi treatment in clinical trials, alone, and in combination.

1. INTRODUCTION

Histone deacetylases (HDACs) catalyze the removal of acetyl groups from the amino-terminal lysine residues of histone proteins, while histone acetyltransferases (HATs) promote their addition, leading to local remodeling of chromatin, critical steps for the access of regulatory proteins to DNA (Selvi & Kundu, 2009). In addition to their activity toward histone proteins, HDACs also target nonhistone proteins such as transcription factors, chaperone proteins, signaling mediators, hormone receptors, and proteins involved in the DNA damage response (DDR, Choudhary et al., 2009; Xu, Parmigiani, & Marks, 2007); therefore, HDACs are now also called lysine DACs or KDACs (Choudhary et al., 2009).

Eighteen HDACs have been discovered in humans and classified into four classes based on their homology to yeast HDACs, their cellular localization, and enzymes activities. Class I HDACs are constituted by HDAC1, 2, 3, and 8; are similar to the yeast Rpd3 deacetylase; and are localized to the nucleus. Class II HDACs are constituted by HDAC4, 5, 7, and 9 and have similarity to Hda1 in yeast. These HDACs are localized in the nucleus and the cytoplasm. Class IV HDACs have only one member, HDAC11 that shares similarities with both class I and II HDACs (Table 3.1). Class III HDACs are constituted by sirtuin (SIRT)1–7 and are related to Sir2 in yeast. Their function is nicotinamide adenine dinucleotide (NAD+) dependent, and they are localized to the nucleus, mitochondria, and cytoplasm (Table 3.1; Bolden, Peart, & Johnstone, 2006; Rajendran et al., 2011).

Table 3.1 KDACs family: localization and cancer involvement

Class	HDAC	Localization	HDAC cancer involvement
I	HDAC1	Nucleus	↑ Prostate, gastric, colorectal, pancreas, and eosophage cancer
	HDAC2	Nucleus	↑ Colorectal, gastric, cervical dysplasia, and invasive carcinoma
	HDAC3	Nucleus	↑ Lung, prostate, and colon cancer
	HDAC8	Nucleus	↑ Poor outcome in pediatric neuroblastoma
IIa	HDAC4	Nucleus/cytoplasm	↑ Breast cancer
	HDAC5	Nucleus/cytoplasm	↓ Colon and AML, and poor outcome in lung cancer
	HDAC7	Nucleus/cytoplasm	↑ Colon cancer
	HDAC9	Nucleus/cytoplasm	↑ Medulloblastoma and astrocytoma
IIb	HDAC6	Cytoplasm	↑ Ovarian and acute myeloid leukemia
	HDAC10	Cytoplasm	↑ Hepatocellular carcinoma
IV	HDAC11	Nucleus/cytoplasm	↑ Breast cancer
III	SIRT1	Nucleus	↑ AML, colon, prostate, skin cancer, and BCLL ↓ Glioma, prostate, ovarian, and bladder
	SIRT2	Cytoplasm	↓ Glioma
	SIRT3	Nucleus/mitochondria	↓ Breast cancer, glioblastoma, prostate, and head and neck
	SIRT4	Mitochondria	↑ Breast cancer
	SIRT5	Mitochondria	↑ Pancreas and breast cancer
	SIRT6	Nucleus	↑ Colon and breast cancer
	SIRT7	Nucleus	↑ Breast cancer

↑, increase expression or overexpression; ↓, decrease expression. AML, acute myeloid leukemia; BCLL, B-cell chronic lymphocytic leukemia; HDAC; histone deacetylase; SIRT, sirtuin.
For review: Carafa, Nebbioso, and Altucci (2012), Khan and La Thangue (2012), McGuinness, McGuinness, McCaul, and Shiels (2011), and Rajendran, Garva, Krstic-Demonacos, and Demonacos (2011).

Altered function of all classes of HDACs has been mechanistically linked to development of diseases such as cancer and leukemia (Bolden et al., 2006; Rajendran et al., 2011): (1) Changes in the global pattern of acetylation are characteristic features of cancer cells (Fraga et al., 2005); (2) HDACs have been shown to be upregulated in lymphomas (Adams, Fritzsche, Dirnhofer, Kristiansen, & Tzankov, 2010; Marquard et al., 2009) and cancers such as breast (Zhang et al., 2005), ovarian (Jin et al., 2008), lung (Nakagawa et al., 2007), and prostate cancer (Weichert et al., 2008); (3) They are aberrantly recruited to the promoter of oncogene's target gene inducing a repression of their transcription and a blockage in differentiation as described in acute promyelocytic leukemia (APL; Mercurio, Minucci, & Pelicci, 2010); (4) Sirtuins have also been shown to be upregulated in acute myeloid leukemia (AML), colon cancer, and prostate cancer (McGuinness et al., 2011; Rajendran et al., 2011); and (5) HDAC upregulation is usually associated with an oncogenic function, while deregulation of SIRTs expression in cancer has been associated with both oncogenic and tumor suppressor activities at least for SIRT1 (Table 3.1; Carafa et al., 2012; Khan & La Thangue, 2012; McGuinness et al., 2011; Rajendran et al., 2011).

The regulation of HDAC function with the use of small molecule inhibitors (HDACis) is one of the most promising drug developments for cancer and other diseases (Bradbury et al., 2005; Minucci & Pelicci, 2006). HDACis can be purified from natural and synthetic sources and can be classified in six groups: (i) hydroxamate, such as trichostatin A (TSA), suberoylanilide hydroxamic acid (SAHA also called Vorinostat), LAQ824, LBH589 (Panabinostat), or PXD101 (Belinostat); (ii) aliphatic acid, such as sodium butyrate (NaB), valproic acid (VPA), or AN-9 (Pivanex); (iii) benzamides, such as MS-275 (Entinostat); (iv) cyclic tetrapeptides, such as apicidin or depsipeptide (Romidepsin); (v) electrophilic ketone; and (vi) sirtuins inhibitors (SIRTis), such as the pan inhibitor nicotinamide or the specific SIRT1, 2 inhibitors sirtinol, cambinol, or EX-527 (Drummond et al., 2005; Lawson et al., 2010). The specificity and efficacy of some of these inhibitors are presented in Table 3.2 (Drummond et al., 2005; Lawson et al., 2010), and many of these compounds are in clinical trials alone or in combination with other agents for the treatment of variety of leukemias or solid tumor malignancies, such as ovarian, breast, lung, AML, and lymphoma (Carafa et al., 2012; Khan & La Thangue, 2012; McGuinness et al., 2011; Rajendran et al., 2011). Moreover, two of HDACis, Vorinostat and Romidepsin, have already been approved by the Food and Drug

Table 3.2 KDACis specificity, potency, and clinical trials

	HDACi	**Target**	**Potency**	**Clinical relevance**
Hydroxamates	SAHA (Vorinostat)	HDAC1-11	nM	FDA approved for CTCL
	LBH-589 (Panabinostat)	HDAC1-11	nM	Phase II and III
	PXD101 (Belinostat)	HDAC1-11	nM	Phase I and II
	TSA	HDAC1-11	nM–μM	High toxicity
Aliphatic acids	VPA	HDAC1-9	mM	Phase I–III
	NaB	HDAC1-9	mM	Low penetration
	AN-9 (Pivanex)	HDAC1-9	mM	Phase I and II
Benzamides	MS-275 (Entinostat)	HDAC1-3	μM	Phase II
Cyclic tetrapeptides	Depsipeptide (Romidepsin)	HDAC1-2	nM	FDA approved for CTCL
–	EX-527	SIRT1	nM	Phase I

CTCL, cutaneous T cell lymphoma; HDAC, histone deacetylase; NaB, sodium butyrate; TSA, trichostatin A; SAHA, suberoylanilide hydroxamic acid; SIRT, sirtuin; VPA, valproic acid.
For review: Carafa, Nebbioso, and Altucci (2012), Drummond et al. (2005), and Lawson et al. (2010).

administration (FDA) for the treatment of cutaneous T cell lymphoma (CTCL) in 2006 and 2009, respectively, showing that indeed HDACis have achieved some of the anticipated clinical-translational promise.

However, while HDACis show anticancer activity *in vitro* and *in vivo* by inducing growth arrest, differentiation, apoptosis, autophagy, mitotic cell death, angiogenesis, the mechanisms underlying those effects are still not clear. In particular, the response to HDACi treatment appears dependent on: (1) the type of disease; (2) the molecular defects present in individual cancer cells; and (3) the dose and specificity of the HDACis used (Bolden et al., 2006; Xu et al., 2007). Importantly, the mechanism(s) underlying one of the key features for measuring clinical responses to HDACi treatment, apoptosis of cancer cells *in vivo*, are still quite controversial. Some authors have reported that HDACis induce apoptosis by upregulation of both death receptors and their ligands in leukemia cells (Insinga et al., 2005; Nebbioso et al., 2005), or by activation of the mitochondrial/intrinsic apoptotic pathway by selective activation of Bid and induction of reactive oxygen species (ROS; Ruefli et al., 2001). We and others have also shown that HDACis can cause DNA damage that is significantly increased in solid tumor (Munshi et al., 2005) and leukemia cells (Gaymes et al., 2006) compared with normal cells, and this effect could underlie at least in part, apoptotic effects. In addition, HDACis have been shown to affect the expression of genes involved in DNA repair such as ataxia telangiectasia mutated and rad3-related (ATR), BLM, BRCA1 and 2, or NBS1 (Xu et al., 2007). HDACis also show increases in acetylation that regulate the activity of numerous DNA repair proteins such as P53 (Yang & Seto, 2008), ku70 (Cohen, Lavu, et al., 2004; Cohen, Miller, et al., 2004), ataxia telangiectasia mutated (ATM; Sun, Jiang, Chen, Fernandes, & Price, 2005), or DNA-dependent protein kinase (DNA PK; Jiang, Sun, Chen, Roy, & Price, 2006), which sense DNA damage and induce an appropriate DDR. In addition, chromatin remodeling factors and acetylation of key histone marks are important in inducing DDR (Rossetto, Truman, Kron, & Cote, 2010).

HDACis also generate ROS in many cancer settings (Marks, 2006). While the mechanisms for ROS production are still not well understood, mitochondrial membrane disruption (Ruefli et al., 2001) and decreased oxidative repair and inflammation (Rosato et al., 2008, 2010) have been implicated in increased oxidative stress following treatment with HDACis. Notably, increased ROS levels could at least in part explain increased levels of DNA damage observed in HDACi-treated cells. Recently, SIRTis have been developed showing similar effects to HDACis in cancers and leukemias (McGuinness et al., 2011; Rajendran et al., 2011). Together, those results strongly suggest that the increase in damage, including ROS induction and altered repair, could explain some of the apoptotic effects of HDACis alone and in combination with SIRTis and/or other agents.

This review will focus on the pathways by which HDACis and SIRTis induce DNA damage, and how DNA repair may be altered by these agents (Eot-Houllier, Fulcrand, Magnaghi-Jaulin, & Jaulin, 2009; Rajendran et al., 2011). We will next describe the mechanisms by which HDACis and SIRTis generate oxidative stress, and whether this may be the source of DNA damage. Finally, we will discuss how understanding ROS production and the DDR may reveal insights into anticancer effects of these inhibitors and suggest therapeutic strategies to enhance anticancer responses *in vivo*.

2. HDACis AND DNA DAMAGE

An earlier study by Bakkenist and Kastan (2003) suggested that chromatin changes induced by HDACis can directly activate DDR without the requirement for DNA double-strand breaks (DSBs) (Bakkenist & Kastan, 2003). In their study, a key initiating event in cellular responses to ionizing radiation (IR), activation of ATM kinase, was not dependent on direct binding to DNA strand breaks but resulted from changes in the structure of chromatin (Bakkenist & Kastan, 2003). In contrast, other studies revealed that HDACis may induce actual DNA damage, such as DSBs, as evidenced by a sustained increase in variant histone H2AX phosphorylation on serine 139, γH2AX, an established marker of DSBs (Gaymes et al., 2006; Munshi et al., 2005) and/or indirectly through increase in oxidative stress via production of ROS (Ruefli et al., 2001). Moreover, recent studies have revealed that the pathways for chromatin remodeling and DNA damage signaling are inextricably linked, in that DNA damage involves changes in chromatin and vice versa, that is, chromatin changes can also lead to DNA damage (Fig. 3.1; Lukas, Lukas, & Bartek, 2011; Misteli & Soutoglou, 2009).

2.1. Induction of actual DNA damage

We previously reported that HDACi treatment of normal and leukemic cells resulted in histone acetylation and rapid induction of γH2AX in leukemia cells and to a lesser extent in normal cells (Gaymes et al., 2006). The AML cell line, HL60, and normal IL-2-stimulated peripheral blood lymphocytes (PBL) were treated with varying concentrations of TSA, NaB, or apicidin for 24 h. Administration of TSA (300 nM) in HL60 cells resulted in phosphorylation of H2AX that appeared by 3 min (0.05 h), peaked at 30 min (0.5 h), and was still evident at 24 h, while resolved several hours after treatment in normal PBL. The timing of this change corresponded to concomitant increases in the acetylation of histone H4 that could also be demonstrated on chromatin fibers in both HL60 and PBL. Importantly, the same results were observed in primary cells from AML patients and other myeloid leukemia cell lines such as K562 and NB4. In accordance with the work of Bakkenist and Kastan (2003), we showed increased phosphorylation of ATM in HL60 cells treated with TSA. This phosphorylation peaked at 10 min after drug administration and disappeared over 8 h (Gaymes et al., 2006).

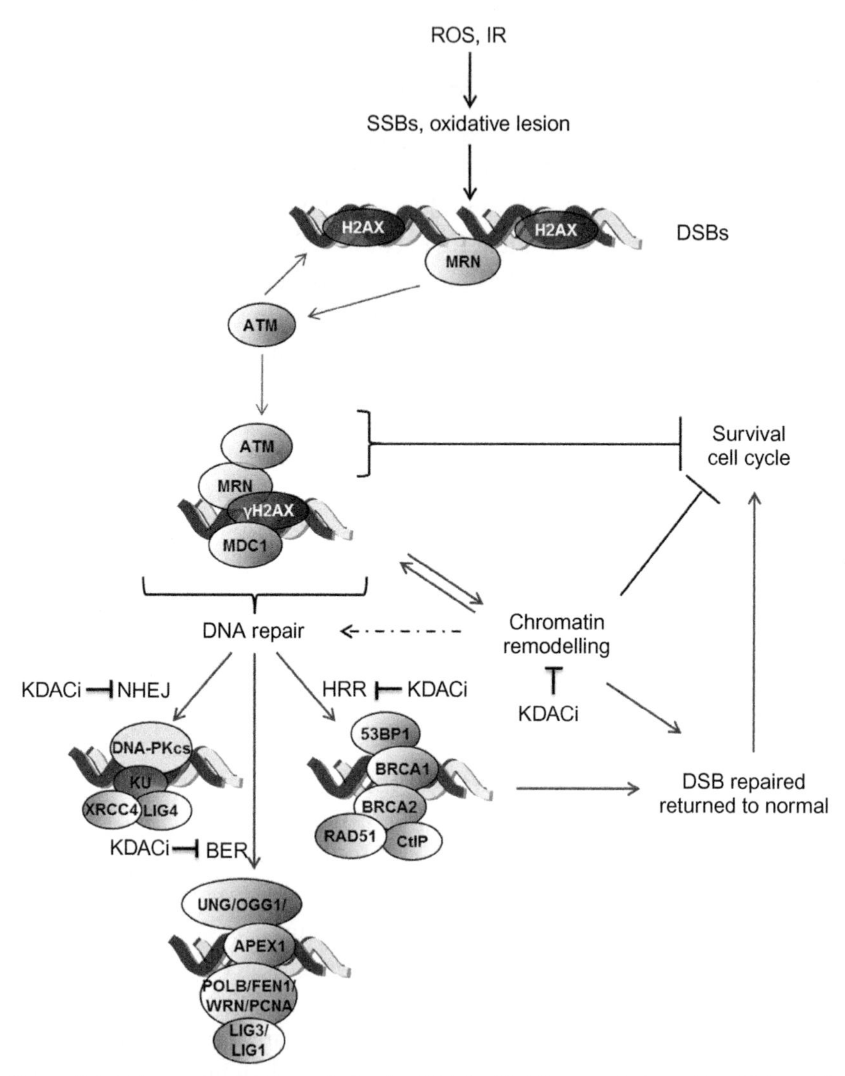

Figure 3.1 Schematic of the regulation by KDACi of DNA repair induced by ROS and IR. Reactive oxygen species (ROS) and ionizing radiation (IR) induce oxidative lesions and single-strand breaks (SSBs) that could result in double-strand breaks (DSBs) during replication. These lesions recruit the Mre11–RAD50–NBS1 (MRN) complex to the site of damage and induce the phosphorylation of H2AX at Ser139 (γH2AX). Mediator of DNA damage checkpoint protein 1 (MDC1) is then recruited to the DNA damage site anchoring the ATM/MRN complex to the DNA, ensuring the recruitment of non-homologous end-joining (NHEJ) and homology-directed repair (HRR) proteins to repair DSBs, and base excision repair (BER) and nucleotide excision repair (NER) to repair SSBs and oxidative lesions. For accurate repair, the chromatin needs to be remodeled in order to facilitate access of repair factors to the site of the DNA lesion, while replication and transcription are shut down. After repair has occurred, the chromatin is again

HDACis have been shown to induce apoptosis in leukemic cells, and it is now well documented that apoptosis induction is linked to its antitumor effects, as all the HDACis have been reported to induce apoptotic mechanisms (Carew, Giles, & Nawrocki, 2008; Johnstone, 2002; Marks, Richon, & Rifkind, 2000). We showed that the onset of apoptosis, as marked by the cleavage of caspase-3 and poly(ADP-ribose)polymerase1 (PARP1), began as early as 4 h after HDACi administration in HL60 cells and well after the appearance of increased histone acetylation, activation of γH2AX, and ATM phosphorylation. Importantly, pretreatment of HL60 and NB4 cells with the caspase inhibitor zVAD-fmk abrogated endogenous caspase activity following administration of HDACis, but still induced early and sustainable DNA damage as measured by γH2AX. This suggests that HDACi-dependent DNA damage induction does not result from apoptosis but likely precedes it. In addition, we showed reduced cell death in cells pretreated with the caspase inhibitor followed by HDACi treatment, as determined by trypan blue exclusion. All of these data suggest that in neoplastic cells, HDACis specifically cause DNA damage and activate the DDR in association with the acetylation of histone proteins, followed by apoptotic events (Gaymes et al., 2006).

To determine whether the increased DNA damage we observed *in vitro* also occurs *in vivo*, we used an APL transplantable mouse model treated with HDACis (Brown et al., 1997). Spleen cells from APL mice exhibited increased DNA damage compared with cells from control mice. We were also able to demonstrate selective apoptosis of APL mice spleen cells with TSA, whereas TSA had little effect on control mouse spleen cells as measured by caspase cleavage assays. Thus, the amount of DNA damage and apoptosis were dramatically increased in leukemic versus normal cells with TSA treatment *in vivo* as had been demonstrated *in vitro* (Gaymes et al., 2006). We, therefore, proposed a model for the action of HDACis resulting in chromatin changes that give rise to DNA damage: leukemic cells sustain more DNA damage than normal cells, and this may be due to the production of increased ROS in the former (Rahmani et al., 2005; Ungerstedt et al., 2005).

---

remodeled, allowing cells to reenter the cell cycle without damage. HDACis and SIRTis (KDACi) inhibit the DNA repair by downregulation or inhibition of the activity of DNA repair proteins or other proteins involved in chromatin remodeling by acetylation, inducing cell death specifically in cancer cells. Arrows represent activation, barred lines represent inhibition, and dotted arrows when both inhibition and activation could occur. (See Page 4 in Color Section at the back of the book.)

This induction of excessive DNA damage and subsequent apoptosis may be the key elements underlying the antitumor effects seen *in vitro* and in patients treated with HDACis. Since then, several HDACis (i.e., TSA, NaB, SAHA, MS275, LAQ824, LBH589, and FK228) have been shown to induce DNA damage, as measured by the induction of γH2AX in leukemia (Dasmahapatra et al., 2011; Gaymes et al., 2006; Petruccelli et al., 2011; Rosato et al., 2008, 2010) and cancer such as prostate (Chen et al., 2007; Lee, Choy, Ngo, Foster, & Marks, 2010), breast (Namdar, Perez, Ngo, & Marks, 2010), lung (Chen et al., 2008; Lee et al., 2010), and osteosarcoma (Miller et al., 2010; Table 3.3).

Table 3.3 Effects of histone deacetylases inhibitors (HDACis) on DNA damage response (DDR)-related proteins

Class	HDAC	DDR-related target	Inhibition-related effect
I	HDAC1	H3K56, H4K16, P53, Ku70, ATM, ATR, APE1/Ref1, CtIP	Chromatin remodeling, apoptosis, NHEJ, and HRR downregulation
	HDAC2	H3K56, H4K16, Ku70	Chromatin remodeling and apoptosis
	HDAC3	H3K9 and K14, H4K5 and K12	Chromatin remodeling
IIa	HDAC4	53BP1	NHEJ and HRR downregulation
	HDAC9	?	HRR downregulation
IIb	HDAC6	Ku70 GADD153	Apoptosis
	HDAC10	?	HRR downregulation
III	SIRT1	H3K9, H4K16, P53, Ku70, FOXO, WRN, NBS1, XPA, XPC, APE1/Ref1, RelA/P65, TIP60, DNMT1, P300	Chromatin remodeling, apoptosis, NER, BER, NHEJ, and HRR
	SIRT3	H4K16, Idh2	Chromatin remodeling and oxidative stress
	SIRT6	H3K9, H3K56, CtIP, XPA, DNA PK	Chromatin remodeling, BER, NHEJ, and HRR

APE1, apurinic/apyrimidinic endonuclease-1; ATM, ataxia telangiectasia mutated; ATR, ATM and Rad3-related protein; BER, base excision repair; CtIP, C-terminal binding protein interacting protein; DNA PK, DNA-dependent protein kinase; DNMT, DNA methyltransferase; HRR, homology-directed repair; NER, nucleotide excision repair, NHEJ, nonhomologous end-joining, Tip60, TAT-interacting protein 60, XPA, xeroderma pigmentosum A.

2.2. Increased DNA damage: Synergy with chemotherapeutic agents

Several lines of evidence suggest that HDACis can synergize with other therapeutic agents to kill tumor cells. As early as 1985, it was shown that NaB could sensitize colon cancer cells to radiotherapy *in vitro* (Arundel, Glicksman, & Leith, 1985). Since then, numerous studies have confirmed that HDACis (SAHA, NaB, TSA, and MS275) could sensitize tumor cells, such as melanoma (Munshi et al., 2005, 2006), prostate, gliomas (Camphausen et al., 2004), or osteosarcoma (Miller et al., 2010) to IR (also reviewed in Karagiannis, Kn, & El-Osta, 2006). These studies showed that cells treated with HDACis followed by IR demonstrated sustained γH2AX expression and histone hyperacetylation, compared to HDACi treatment alone. In addition, Camphausen et al. (2004) showed that histone acetylation and γH2AX expression could be correlated to HDACi-induced radiosensitivity *in vitro* (Camphausen et al., 2004). Notably, some of the studies showing increased DSBs following HDACi treatment revealed an important feature of the effect of HDACis on normal versus cancer cells. In normal cells, γH2AX was slightly induced by HDACis but its phosphorylation was downregulated to baseline levels within 2 h. In contrast, in cancer cells, HDACi-induced γH2AX persisted longer following HDACi treatment (Gaymes et al., 2006; Lee et al., 2010; Miller et al., 2010) or in combination with DNA-damaging agents such as IR (Karagiannis et al., 2006). Indeed, the sensitizing effects of HDACis have been demonstrated in combination with a variety of DNA-damaging agents, such as topoisomerase inhibitors, alkylating agents, and chemotherapy. Thus, DNA damage resulting from HDACi treatment could be due to (i) profound changes in the chromatin structure exposing sections of the DNA normally protected from damage to DNA-damaging agents; (ii) inhibition of the DNA repair machinery that could lead to accumulation of unrepaired damage; and/or (iii) modulation in expression of DNA damage and cell survival genes (Fig. 3.1).

2.3. DSB repair responses

The next section will review DSB repair pathways and the altered response of DSB repair components to treatment with HDACis, including their effects on transcriptional regulation and differential acetylation.

DSBs are the most lethal form of DNA damage in mammalian cells (Khanna & Jackson, 2001). The cytotoxicity of DSBs presumably reflects the difficulty of repairing these lesions because, unlike almost other types of DNA damage that have an intact undamaged template strand to guide the repair, the integrity of both strands of the duplex is lost. ATM protein that is defective in the cancer-prone radiation-sensitive human disease, ataxia telangiectasia, is a central player in the activation of cell cycle checkpoints by DSBs. ATM is recruited to DSBs by the MRN complex (MRE11/NBS1/RAD50; Falck, Coates, & Jackson, 2005). This results in ATM autophosphorylation and its conversion from an inactive dimer to an active monomer (Bakkenist & Kastan, 2003). Once activated, ATM phosphorylates MRN and downstream effector proteins to initiate cell cycle checkpoints at the G1/S, intra-S, and G2/M boundaries (O'Driscoll & Jeggo, 2006). The activation of these checkpoints allows increased time for the repair of DNA damage before it is replicated and/or passed on to daughter cells, thereby increasing cell survival and preserving genomic integrity. In addition to its role in DSB-activated signal transduction pathways, there is emerging evidence that ATM is involved in the repair of a subset of DSBs in both G0 and G2 cells (Beucher et al., 2009; Riballo et al., 2004).

The repair of DSBs occurs via two mechanistically distinct groups of pathways: homology-directed pathways (HRR) and nonhomologous end-joining (NHEJ) pathways (Fig. 3.1).

2.3.1 HRR

This pathway repairs replication-associated DSBs and is characterized by the invasion of single-stranded DNA into a homologous duplex (Hartlerode & Scully, 2009; Khanna & Jackson, 2001). This repair pathway, which is active in late S phase and in the G2 phase of the cell cycle, utilizes the undamaged sister chromatid as the template for repair and so is usually error free. The first step in HR is resection of the DSB in a 5′-3′ manner that involves the human MRN complex and CtIP (Huertas & Jackson, 2009; Sartori et al., 2007). Next, hRad51 is recruited and assembled into a nucleoprotein filament, assisted by several accessory proteins including hRad52, XRCC2, XRCC3, and BRCA2 (Liu et al., 1998; Petalcorin, Sandall, Wigley, & Boulton, 2006; Sonoda et al., 2007). The recruitment and assembly of hRad51 nucleoprotein filaments results in generating two identical sister chromatids (Mimitou & Symington, 2009).

2.3.2 NHEJ

The repair of DSBs by NHEJ is initiated by the Ku70/Ku86 heterodimer, a ring-shaped complex that binds to and encircles DNA ends (Walker, Corpina, & Goldberg, 2001). This serves to protect the DNA ends from degradation and to recruit the catalytic subunit of DNA PK (Falzon, Fewell, & Kuff, 1993; Gottlieb & Jackson, 1993; Mimori & Hardin, 1986) to form the activated DNA PK (Calsou et al., 1999; Singleton, Torres-Arzayus, Rottinghaus, Taccioli, & Jeggo, 1999). The key step in NHEJ is the physical juxtaposition of DNA ends. When juxtaposed ends can be directly ligated, the repair reaction is completed by DNA ligase IV/XRCC4, which is recruited to DSBs by interactions with DNA PK (Lobrich & Jeggo, 2005). The joining of DSBs by DNA PK-dependent NHEJ often results in the loss or addition of a few nucleotides and the presence of short complementary sequences, microhomologies, at the break site that presumably contribute to the alignment of the DNA ends (Roth, Porter, & Wilson, 1985; Roth & Wilson, 1986). An alternative (Alt) version of NHEJ that results in larger deletions and chromosomal translocations (Iliakis, 2009; Nussenzweig & Nussenzweig, 2007) also exists. The hallmark features of the Alt NHEJ pathway are larger deletions and insertions, longer tracts of microhomology, and a much higher frequency of chromosomal translocations compared with DNA PK-dependent NHEJ (Nussenzweig & Nussenzweig, 2007). A number of DNA repair proteins, including PARP 1, MRN, WRN, and DNA ligase IIIα/XRCC1 (Audebert, Salles, & Calsou, 2008; Deriano, Stracker, Baker, Petrini, & Roth, 2009; Dinkelmann et al., 2009; Rass et al., 2009; Sallmyr, Tomkinson, & Rassool, 2008; Wang et al., 2005, 2006; Xie, Kwok, & Scully, 2009), have been implicated in Alt NHEJ and shown to be increased in cancer cells, but the mechanisms and regulation of this repair pathway are poorly defined.

2.4. HDACis and transcriptional downregulation of DSB repair

In addition to increasing DNA damage, other studies have implicated HDACis in the transcriptional downregulation of DNA repair factors as a possible mechanism for sustained γH2AX and, sensitivity of cancer cells to DNA-damaging agents (Table 3.3 and Fig. 3.1; Kachhap et al., 2010; Lee et al., 2010; Munshi et al., 2005, 2006; Rosato et al., 2008). SAHA has been shown to upregulate and sustain γH2AX expression by downregulating the expression of RAD50 and MRE11, two proteins involved in HRR and the sensing of damage/NHEJ repair pathway in cancer cells (prostate and lung) but not normal cells (Lee et al., 2010). SAHA has also been shown to downregulate the expression of NHEJ proteins Ku70, Ku80, and HRR factor RAD50 in melanoma cells (Munshi et al., 2006), while NaB induced downregulation of NHEJ components, Ku70, Ku80, and DNA PK (Munshi et al., 2005). Moreover, HDACis LAQ824 and MS275 downregulate BRCA1, RAD50, Ku80, EXO1, and CHK2 in AML cells (Rosato et al., 2008). Recently, Kachhap et al. (2010) described a possible mechanism to explain the downregulation of all these DSB repair proteins. First, they analyzed the DNA repair genes that were downregulated after SAHA and VPA treatment in prostate cancer cells and found that RAD51, BRCA1, CHK1, and BLM were actually downregulated by treatment with both HDACis. By measuring DSB repair efficiency, RAD51 and BRCA1 foci by immunofluorescence, these authors determined that not only there was a decrease in expression levels of these proteins but also their recruitment to the repair foci was impaired. This correlated with a decrease in DSB repair efficiency and an increase in levels of DNA damage, as measured by the comet assay and γH2AX foci formation (Kachhap et al., 2010). Moreover, HDACis were shown to decrease expression levels of the E2F1 transcription factor, a known activator of RAD51, CHK1, and BRCA1 gene expression (Bindra & Glazer, 2007; Carrassa, Broggini, Vikhanskaya, & Damia, 2003). Using chromatin immunoprecipitation assays, these authors confirmed that in the presence of HDACis, there is decreased recruitment of E2F1 to the promoters of RAD51, BRCA1, and CHK1 (Kachhap et al., 2010).

2.5. HDACis acetylation/deacetylation of DSB repair proteins

Lysine acetylation has emerged as a major posttranslational protein modification that regulates not only gene expression (i.e. when occuring on histone proteins as discussed in the next paragraph) but many other physiological processes, such as DDR. (Peserico & Simone, 2011). By immunoaffinity purification with an antiacetyl lysine antibody followed by enrichment with isoelectric focusing and high-resolution mass spectrophotometry, Choudhary et al. (2009) evaluated the acetylome changes after inhibition of HDACs with SAHA and MS275 (Choudhary et al., 2009) in three different cancer cell lines (leukemia, MV411 and Jurkat, and lung cancer, A549).

Their study identified 3600 acetylation sites in 1750 human proteins, most of which were new targets involved in all major nuclear processes such as cell cycle, chromatin remodeling, DNA replication, transcription, nuclear transport, histone modification, and DNA repair (Choudhary et al., 2009). Most of the targets of both SAHA and MS275 were histone related with a special emphasis on a variant histone H2AZ which is a mark for DNA damage sites. Interestingly, the use of SAHA increased acetylation of H2AZ by 20-fold while MS275 moderately increased it by 2-fold reflecting the different HDACs specificity of these drugs. However, comparing the acetylated profile between the three cell lines, Choudhary et al. (2009) confirmed acetylation of DNA repair proteins, such as P53, Ku70, flap endonuclease (FEN1), WRN, or ATM, that had previously been described by other studies. In addition, they identified new targets, such as MDC1 and DNA PK, that are differentially acetylated following HDACi treatment but the regulatory role of acetylation of those proteins by HDACis still needs to be determined. (Choudhary et al., 2009).

One of the main HDACis targets is the DDR protein P53 which has been shown to be stabilized by acetylation, leading to transcription of its target genes, including p21 that culminates in cell cycle arrest or cell death (Ito et al., 2002; Luo, Su, Chen, Shiloh, & Gu, 2000). HDACis such as TSA, SAHA, MS275, and OSU-HDAC42 have also been shown to induce Ku70 acetylation in prostate cancer cells (Chen et al., 2007) and colon cancer cells (Kerr et al., 2012). Notably, Ku70 acetylation reduces its binding to DNA and sensitizes cells to DNA-damaging agents, such as topoisomerase II inhibitors (VP16 and doxorubicin; Chen et al., 2007). In addition to its role in DNA repair, Ku70 has been shown to play a role in regulation of apoptosis. Normally existing in a complex with Bax in the cytoplasm, HDACi-mediated acetylation of Ku70 results in the release of Bax and its transport to the mitochondria, leading to cell death through the intrinsic apoptotic pathway (Chen et al., 2007). Apparently, Ku70 also complexes with antiapoptotic protein FLIP that inhibits the extrinsic cell death pathways (Kerr et al., 2012). Acetylation of Ku70 disrupts the Ku70–FLIP complex, inducing inhibition of the extrinsic apoptotic pathway (Kerr et al., 2012). Together, these results suggest that acetylation of DDR proteins and especially DSB repair is a crucial step for the radiosensitization that is observed upon HDACi treatment in tumor cells (Table 3.3 and Fig. 3.1).

2.6. HDACs, histone marks, and DSB repair

Most HDACs and HATs are present in the same complex as chromatin modifiers such as DNA methyltransferases (DNMTs), transcription factors, and DNA repair proteins, in order to regulate the balance between repair of damage and transcription of the gene. As an example, early DDR factor, ATM was shown to form a complex with HDAC1 and the binding of these proteins is increased following IR (Kim et al., 1999). In contrast, the interaction between ATM and HDAC1 was not detected by immunoprecipitation analysis in cell lines which were derived from AT patients. Also, cells expressing a mutated form of ATM demonstrate decreased HDAC1–ATM complex formation, preventing the histone deacetylation apparent in normal cells after IR. Disruption of this complex could explain the radiosensitivity of cancer cells (Kim et al., 1999). Further evidence for the role of HDACs in DSB repair came from a study by Miller et al. (2010) who demonstrated the presence of HDAC 1 and 2 at site of DSBs in osteosarcoma cells and that they were responsible for deacetylation of histone marks H3K56 and H4K16 (Miller et al., 2010). Inhibition or depletion of HDAC1 and 2 rendered cells hypersensitive to IR, with increased levels of γH2AX and induction of P53, CHK1 and 2 phosphorylation, compared to control cells (Miller et al., 2010). In addition, they showed that depletion of HDAC1 and 2 following IR diminished DSB repair capacity as demonstrated by the comet assay. Notably, NHEJ appeared more impaired than HRR, as demonstrated by the persistence of Ku70 at DSBs induced by laser microirradiation (Miller et al., 2010). These results suggest that HDAC1 and 2 could promote DSB repair by local condensation of the chromatin. The authors proposed that deacetylation of these histones (i) condenses chromatin and downregulates transcription so as not to interfere with DSB repair and/or (ii) prevents Ku70 from sliding off naked DNA so the repair can occur (Miller et al., 2010). Indeed, deacetylation could be important for the chromatin compaction that signals both the beginning and the end of the repair process (Lukas et al., 2011; Miller et al., 2010).

HDAC3 has also recently been implicated in the regulation of DSB repair (Bhaskara et al., 2008, 2010). Indeed, HDAC3 was shown to deacetylate histone marks H3K9 and K14, and H4K5 and K12 during the late S phase. Notably, mouse embryonic fibroblasts (MEFs) from mice knocked out for HDAC3(−/−) showed increased apoptosis, formation of DSBs, and an increased sensitivity to DSB-inducing agents (i.e., doxorubicin and cisplatin, topoisomerase II inhibitors). This appeared to result from a decrease in NHEJ and HRR activity, as measured by plasmid-based repair assays (Bhaskara et al., 2008, 2010). This diminution of DSB repair was not due to transcriptional downregulation of DSB repair proteins, but due to maintenance of acetylation on lysine residues of histones H3K9Ac and H3K14Ac and concomitant downregulation of methylation marks on H3K9me3 (Bhaskara et al., 2010) which are crucial for the recruitment of chromatin remodeling protein TIP60 (discussed in the next section) and various DNA repair proteins to the site of γH2AX (Sun et al., 2009). Together, these results show that chromatin conformation is a regulator of DSB repair and that not only HATs but also HDAC that deacetylate and condensed chromatin are involved in regulation of DSB repair.

HDAC4 has also been shown to be important in the DDR. HDAC4 colocalizes with HRR protein 53BP1 under stress and can, when silenced, switch off the DDR. HDAC4 in Hela cells abolishes the G2 checkpoint and renders these cells radiosensitive (Basile, Mantovani, & Imbriano, 2006; Kao et al., 2003). HDAC9 and 10 have also been implicated into the regulation of HRR, as depletion of HDAC9 and 10 increases sensitivity to mytomycin C. However, specific HRR targets have yet to be identified (Kotian, Liyanarachchi, Zelent, & Parvin, 2011).

Thus, modulation of chromatin architecture through acetylation but also other modifications not discussed here, such as phosphorylation, poly(ADP) ribosylation, ubiquitinylation, and sumoylation are essential for DDR by allowing the repair factor to access the damaged DNA (Table 3.3; Figs. 3.1 and 3.2; Lukas et al., 2011; Rossetto et al., 2010).

2.7. Chromatin remodeling and DSB repair

Several lines of evidence suggest that repair of DNA damage involves remodeling of chromatin (Rossetto et al., 2010). For example, induction of DSBs induces activation of γH2AX that accumulates at the sites of DSBs, forming nuclear foci that may represent a chromatin region exceeding 1 Mbp around each DSB site (Rogakou, Pilch, Orr, Ivanova, & Bonner, 1998). γH2AX induces the recruitment of mediator of DNA damage checkpoint protein 1 (MDC1) to DSBs (Polo, Kaidi, Baskcomb, Galanty, & Jackson, 2010) (Fig. 3.1). This anchors ATM to the chromatin through activated NBS1 (Lukas et al., 2011). The MRN complex promotes the preparation of DNA damage site for repair, while phosphorylation of MDC1 by CK2 orchestrates all the downstream chromatin modifications (Fig. 3.1; Ciccia & Elledge, 2010; Lukas et al., 2011). MDC1 coordinates the recruitment of the chromatin remodeling complex NuA4, that contains the HAT HIV-1,

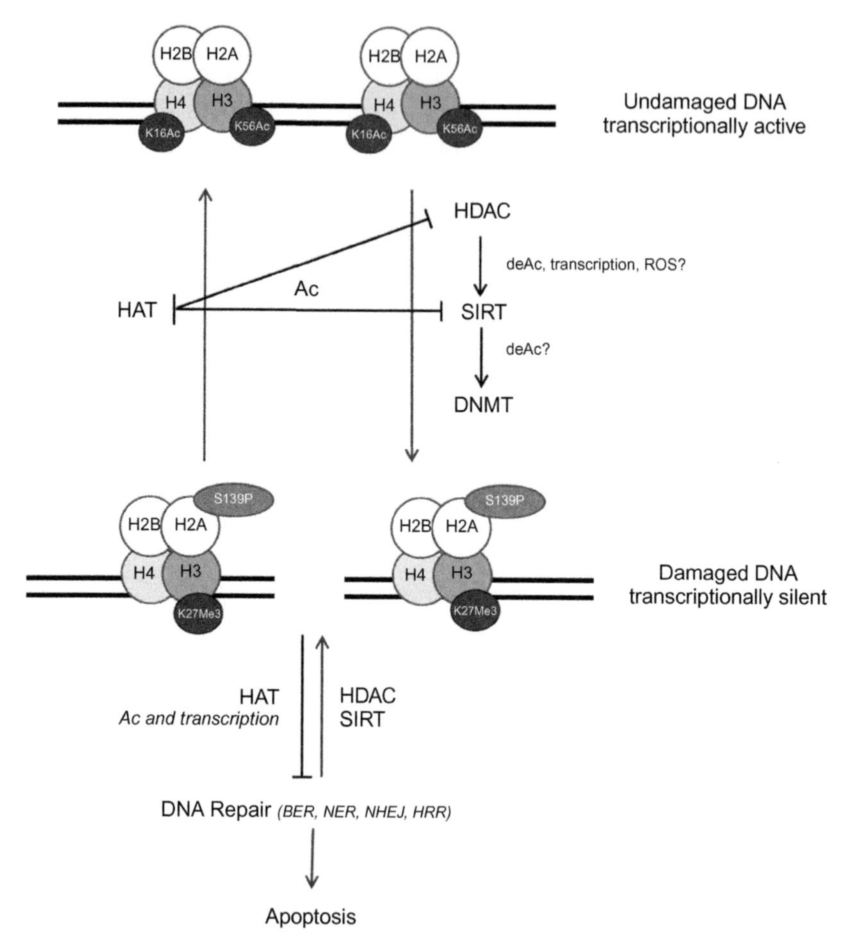

Figure 3.2 Schematic for the role of HDACs and HATs in modifying lysine residues on histone proteins at DNA damage sites. Histone acetyl transferase (HAT)s regulate in part the compaction of the chromatin and the DNA repair by acetylation of specific lysine residues on histones, such as H4K16 and H3K56, while histone deacetylase (HDAC) and sirtuins (SIRT) deacetylate these residues, permitting the trimethylation of H3K27 leading to local decompaction of the chromatin, silencing of the transcription and the recruitment of DNA repair proteins at the DNA damage site. HAT activity can be inhibited by the deacetylation activity of KDACs and vice versa. HDACs could also activate SIRTs that in turn activate DNMTs. Arrows represent activation and barred lines represent inhibition. (See Page 5 in Color Section at the back of the book.)

Tat interacting protein 60 kDa (TIP60-p400) to the site of γH2AX (Sun et al., 2009). This chromatin remodeler TIP60 is a HAT that acetylates histones, H2A, and H4K16 (Sun, Jiang, & Price, 2010), relaxing the chromatin so that repair factors can access the DNA lesions (Murr et al., 2006). Indeed, inactivation of TIP60 leads to increased radiosensitivity (Sun et al., 2009) and upregulation of TIP60 has been detected in many cancers such as prostate, breast, and colon. In addition, acetylation of TIP60 at lysine residue 3016 positively regulates ATM, inducing a feedback loop that boosts the DDR (Sun, Xu, Roy, & Price, 2007). "Males absent on the first" (MOF), is another HAT that is responsible for the majority of acetylation on histone mark H4K16 in the cell, that plays a critical role in DSB repair (Sharma et al., 2010). Depletion of MOF in irradiated breast cancer cells delays γH2AX foci formation at a stably transfected ISCE-inducible DSB. This results in inactivation of ATM and downregulation of MDC1, 53BP1, RAD51, and DNA PK at ISCE1 site, altering DSB repair by both HRR and NHEJ (Sharma et al., 2010). Interestingly, TIP60 and MOF appear to play different roles in the regulation of DDR as measured by gH2AX foci kinetics. By double depletion of TIP60 and MOF Sharma et al. (2010) showed that appearance of IR-induced gH2Ax is dependent on MOF activity while its disappearance is controlled by TIP60 (Sharma et al., 2010).

Acetylation of histone H3K9 by P300/CBP also appears critical for DSB repair, as silencing or inhibition of P300/CBP by the HAT inhibitor anacardic acid in lung cancer cells challenged by IR suppresses recruitment of NHEJ factors Ku70 and Ku80 to the site of an inducible DSB (Ogiwara et al., 2011). In addition, the SWI/SNF chromatin remodeling complex known to enhance NHEJ by local relaxation of the chromatin is not recruited following HAT inhibitor treatment (Park et al., 2006), emphasizing the necessary role of chromatin relaxation in the DNA repair response.

A simplified schematic of the different effects of HDACis on the DDR and chromatin remodeling is presented in Figs. 3.1 and 3.2.

3. HDACis AND OXIDATIVE STRESS

Generation of oxidative stress has been proposed as a mechanism by which HDACis (i.e., SAHA, CS055, MS275, NaB, LAQ824, LBH589, NCH-51, and PCI-24781) exert their lethal effects in leukemia (Bhalla et al., 2009; Dai, Rahmani, Dent, & Grant, 2005; Gong, Xie, Yi, & Li, 2012; Petruccelli et al., 2011; Rosato et al., 2008, 2010, 2006; Ruefli et al., 2001) and cancer cells (Butler et al., 2002; Ungerstedt et al., 2005; Xu, Ngo, Perez, Dokmanovic, & Marks, 2006). If oxidative damage is not repaired, this can lead to single-strand breaks (SSBs) and DSBs (Caldecott, 2007; Lee & Pervaiz, 2011).

3.1. ROS production and elimination

Oxidative damage such as 8-oxodeoxyguanosine (8-oxodG) lesions in DNA is generated by endogenous by-products of cellular metabolism such as ROS. Superoxide (O_2^-), hydrogen peroxide (H_2O_2), hydroxyl radical, and singlet oxygen comprise the main ROS that cause the peroxidation of nucleic acids, lipids, amino acids, and carbohydrates (Li, Yang, Ming, & Liu, 2011). Excessive amounts of ROS could lead to cellular senescence, apoptosis, or carcinogenesis (Li et al., 2011), but redox levels are also important for the differentiation and survival of normal cells (Chen & Pervaiz, 2007; Clement & Stamenkovic, 1996; Sattler et al., 1999). ROS mainly develops in the mitochondria (Balaban, Nemoto, & Finkel, 2005), but cells can also actively generate ROS through NADPH oxidase complexes that use NADPH to generate NADP+ and O_2^- (Bedard & Krause, 2007). O_2^- is converted into H_2O_2 by Mn-containing mitochondrial superoxide dismutase (MnSOD, SOD2) in the mitochondria or by SOD1 and SOD3 in the cytosol (Pervaiz, Taneja, & Ghaffari, 2009). H_2O_2 is then converted into OH^- which will cause lipid, protein oxidization, and DNA damage such as 8-oxodG and SSBs (Kobayashi & Suda, 2012). A number of studies have linked ROS directly to HDACi-induced cell death: (1) the use of antioxidant *N*-acetylcysteine (NAC) treatment or other ROS-mimicking scavenger proteins prior to administration of HDACis, reduces or abolishes HDACi-dependent apoptosis (Dasmahapatra et al., 2011; Gong et al., 2012; Petruccelli et al., 2011; Rosato, Almenara, & Grant, 2003; Rosato et al., 2008; Ruefli et al., 2001; Sanda et al., 2007) and (2) ROS levels are further increased in the presence of HDACis in combination with DNA-damaging agents or chemo/radiotherapies, such as proteasome inhibitors (Bortezomib; Bhalla et al., 2009) or fludarabine (Rosato et al., 2008).

3.2. HDACis and antioxidant pathways

One possible mechanism by which HDACis induce ROS is through downregulation of the antioxidant pathways (Butler et al., 2002; Ungerstedt et al., 2005). To cope with excessive ROS, cells upregulate scavenger enzymes such as glutathione (GSH)/GSSG and thioredoxin (TRX)/TRX reductase (TRXR) system. H_2O_2 is neutralized by several enzymes such as glutathione peroxidase (GPX), peroxiredoxin (Hofmann, Hecht, & Flohe, 2002), or catalase (CAT; Michiels, Raes, Toussaint, & Remacle, 1994). The first proof that HDACis may affect antioxidant pathways came from a study of prostate, bladder, and breast carcinoma cells treated with SAHA (Butler et al., 2002). These authors demonstrated that TRX binding protein-2 (TBP-2), that regulates the activity of the antioxidant protein TRX, was overexpressed in normal tissue, compared to cancer cells, and that SAHA treatment upregulated TBP-2 in cancer cells while TRX was decreased (Butler et al., 2002). TRX appeared to act by directly modulating the DNA-binding activity of transcription factors, such as nuclear factor kappa-light-chain-enhancer of activated B cells (NF-κB) and apurinic/apyrimidinic endonuclease-1 (APE-1/Ref-1; Hayashi, Ueno, & Okamoto, 1993; Hirota et al., 1997). In addition, reduction of this powerful antioxidant TRX leads to increased levels of ROS (Nakamura et al., 1994). TRX is also an inhibitor of apoptosis signal-regulating kinase1 (ASK1) that regulates apoptosis through activation of SEK1-JNK and p38/MAPK pathways also called stress-associated protein kinase (Saitoh et al., 1998). A later study demonstrated that SAHA and MS275 could induce ROS in lung and breast cancer but not in normal cells, while the other canonical markers of HDACi treatment, such as acetylation of histone H3 and H4, increased levels of P21 and G1 arrest could be seen in both normal and cancer cells (Ungerstedt et al., 2005). Like other reports using DNA damage as a marker following treatment with HDACis, ROS levels persisted in cancer cells leading to caspase-dependent apoptosis. Louis et al. (2004) demonstrated that NaB treatment of breast cancer cell lines could induce downregulation of another scavenger protein GPX, leading to apoptosis of cancer cells (Louis et al., 2004). Furthermore, apoptosis may occur through an ASK1/JNK/P38-dependent mechanism. Chen et al. (2008) also showed that treatment of lung cancer cells with Romidepsin could induce the upregulation of TRXR that correlates with HDACi resistance while downregulation of TRXR was observed in sensitive cells (Chen et al., 2008). Together, these studies show that HDACis increase ROS and induce cell death in cancer cells by downregulation of antioxidant proteins, while normal cells are protected from HDACi-induced ROS and apoptosis because of increased levels of antioxidant proteins (Ungerstedt et al., 2005).

3.3. HDACis and repair of oxidative DNA damage

Other studies suggest that increased ROS levels may be generated through downregulation of the pathways for repair of oxidative DNA damage, base excision repair (BER), and nucleotide excision repair (NER; Butler et al., 2002; Chen et al., 2008; Dai et al., 2005; Dasmahapatra et al., 2011; Gong et al., 2012; Louis et al., 2004; Ruefli et al., 2001; Ungerstedt et al., 2005).

3.3.1 BER

BER corrects base modifications in DNA, including oxidative damage of nucleotide bases, which is the most common form of damage found in cells. BER is initiated by the recognition of the base damage by DNA glycosylases, such as 8-oxoguanine glycosylase, which liberates the damaged base, thus generating an abasic site (Lindahl & Wood, 1999; Scharer & Jiricny, 2001; Wyatt & Pittman, 2006). Upon the removal of the damaged base, an apurinated (AP) site is formed in the DNA strand. 5′-APE1 recognizes the AP sites and creates a nick in the DNA to the 5′-end of the damaged base by hydrolyzing the corresponding phosphodiester bond resulting in a 3′-OH group and a 5′-deoxyribose phosphate (dRP) group. Completion of the repair can then take place by one of the two alternate routes. In the majority (~80–90%) of the AP sites, repair occurs by short-patch BER, which requires both 5′-lyase activity and polymerase activity of DNA polymerase beta (pol β) to remove dRP and fill the gap with correct nucleotide, followed by sealing of nicked DNA by DNA ligase I alone or Ligase III/XRCC1 complex. Alternatively, cells can utilize the long-patch repair, in which pol β mediates the replacement of 2–10 nucleotides by strand displacement DNA synthesis utilizing FEN1 in the process (Memisoglu & Samson, 2000). Once the SSBs are detected, they undergo end processing where 3′ and 5′ termini of SSBs are converted to 3′OH and 5′P for gap filling and ligation by pol β, polynucleotide kinase, and the nuclease aprataxin and indirectly with DNA ligase IIIα and tyrosyl-DNA phosphodiesterase 1 (Fig. 3.1; Maynard, Schurman, Harboe, de Souza-Pinto, & Bohr, 2009).

3.3.2 NER

NER is used for some forms of ROS-oxidative damages (Hansen & Kelley, 2000; Sancar, Lindsey-Boltz, Unsal-Kacmaz, & Linn, 2004). Defects in NER proteins result in the development of human disorders such as xeroderma pigmentosum (XP), Cockayne's syndrome, and trichothiodystrophy. NER is activated by such proteins as XPC–RAD23B complexes. XPC is essential for the recruitment of downstream NER factors including TFIIH, which comprises the XPB and XPD subunits, followed by XPA, replication protein A, and the incision enzymes XPF-ERCC1 and XPG. Following the incision, the resulting gap is filled by DNA polymerases ε or δ and the nick is sealed by DNA ligase (Fig. 3.1).

3.3.3 HDACis and downregulation of oxidative repair activity

As reported earlier, HDACis could induce DNA repair proteins down-regulation and/or acetylation. Rosato et al. (2008) reported that HDACis could induce downregulation of not only DSB repair proteins BRCA1 and RAD51 but also other DDR proteins, including BER and NER proteins EXO1, FEN1, and XPA. Simultaneously, Ku70 was acetylated resulting in abrogation of its binding to DNA. The use of NAC did not rescue any of these effects showing that altered gene expression is independent of HDACi-induced ROS generation and is a parallel effect induced by HDACi treatment (Rosato et al., 2008). Thus, HDACis could down-regulate not only NHEJ, HRR as described earlier, but also BER and NER that could be responsible of the maintenance of oxidative damage (Table 3.3 and Fig. 3.1).

3.4. HDACis, ROS, and mitochondrial disruption

Other studies suggest that HDACis may generate ROS through mitochondrial disruption that is part of the intrinsic apoptotic pathway. Ruefli et al. (2001) demonstrated that in T cell leukemia SAHA could induce the transcription of Bid, a BH3-only proapoptotic protein, as actinomycin D, an inhibitor of transcription, inhibits cell death. Next, fixation of a truncated form of Bid, t-Bid into the mitochondrial membrane, induces changes in mitochondrial potential, release of ROS, and cytochrome *c* that activate the downstream proapoptotic factors, caspases-3 and -7 (Ruefli et al., 2001). This study was the first to report mitochondrial injury as a mechanism for the antitumor effect of HDACis, and since then, others have reported similar HDACi-dependent mechanisms (Ruefli et al., 2001). Depolarization of the mitochondrial membrane has also been shown to induce ROS (Magne et al., 2006) and could be triggered by HDACis (Rosato et al., 2008, 2010).

3.5. HDACis, ROS, and inflammatory responses

As demonstrated earlier, HDACis have also been shown to induce a ROS-dependent activation of the NF-κB pathway through activation of ATM/NEMO (Rosato et al., 2010). Panobinostat (LBH589) in leukemic cell lines (U937) increases the antioxidant SOD2 protein while simultaneously upregulating TBP-2 (Rosato et al., 2008, 2010) through the activation of the cytoprotective pathway NF-κB. Interestingly, LBH589 triggers acetylation of NF-kB complex component p65/RelA (Rosato et al., 2010), NF-κB-dependent induction of antioxidant SOD2, activation of ATM, and increased association of ATM with NF-κB essential modulator (NEMO) activating the NF-κB survival pathway (Miyamoto, 2011; Rosato et al., 2008, 2010). NF-κB has been shown to have proinflammatory properties through induction of its target gene such as the proinflammatory cytokines TNFα, IL1, IL2, and IL6. This inflammatory response dependent on NF-κB signaling induced by HDACi treatment, such as LBH589, LAQ824, Vorinostat, or NaB (Rosato et al., 2008, 2010), could be a mechanism for induction of ROS (Kim, Lee, Park, & Yoo, 2010; Rahman, Marwick, & Kirkham, 2004). Those results suggest that HDACis by regulation of NF-κB can play a dual role in cancer therapy: (i) inducing upregulation of protective pathway that could help regulating the redox status of cancer cells and probably allowing cells time to repair oxidative lesions and (ii) inducing cancer cell death. These results suggest that inhibiting the NF-κB pathway can enhance the antileukemic effect of HDACis. However, this response to HDACis is dependent on the nature of HDACis, concentration and/or time of exposure, and the cell type (Montero & Jassem, 2011).

3.6. ROS as a source of DSBs

ROS can induce mutagenesis by base oxidation resulting in mutagenic 8-oxodG. If these oxidative lesions are not repaired, the resultant SSBs will be converted to DSBs during replication when these lesions encounter a replication fork or during transcription (Caldecott, 2007). Several lines of evidence suggest that the generation of ROS may be the source of the increased DSBs that are observed with HDACis treatment. In particular, the study of Rosato et al. (2008) demonstrated that treatment of leukemia cell lines HL60, K562, Jurkat with HDACis, such as LAQ824 and MS275, followed by administration of the chemotherapeutic drug fludarabine, used in the treatment of hematological malignancies, such as AML, CML, and T cell lymphoma, exert a synergistic effect on apoptosis or clonogenicity (Rosato et al., 2008). In contrast, simultaneous treatment with the two drugs or treatment of fludarabine followed by HDACis was found to be minimally toxic (Rosato et al., 2008). Looking at the mechanism of apoptosis, they found that LAQ824/fludarabine induced a loss of mitochondrial potential accompanied by release of proapoptotic cytochrome *c*, activation of t-Bid, and activation of the caspase cascade (Rosato et al., 2008). Moreover, they showed that a transient and early induction of ROS was dependent on HDACis treatment. Notably, treatment of leukemia cells with antioxidant NAC abrogated HDACi-induced ROS production and protected the cells from apoptosis (Rosato et al., 2008). Concomitantly, they observed an increase in antioxidant TBP-2 and MnSOD2. Furthermore, overexpression of MnSOD2 totally abrogated ROS induction and apoptosis (Rosato et al., 2008). Next, these authors examined DNA damage and demonstrated that LAQ824 induced an increase in γH2AX and ATM phosphorylation that is further increased by the fludarabine treatment. The use of NAC prior to combined treatment reduced ROS, γH2AX, and ATM phosphorylation (Rosato et al., 2008). Using a pan caspase inhibitor, they also showed that HDACi-mediated ROS induction and DNA damage was a cause rather than a consequence of cell death, similar to the earlier study of Gaymes et al. (2006). As described earlier, HDACi treatment induced a down-regulation of DNA repair proteins such as BRCA1, CHK1, EXO1, FEN1, RAD51, and XPA, and acetylation and impaired DNA binding of Ku70 (Rosato et al., 2008). Notably, the use of NAC did not rescue any of these effects showing that altered gene expression is independent of HDACi-induced ROS generation. Furthermore, addition of fludarabine triggers a dramatic increase in DNA damage, either by inducing damage (Huang & Plunkett, 1995) or by disrupting repair mechanisms (Moufarij, Sampath, Keating, & Plunkett, 2006), that is responsible for the lethal HDACi-induced effect (Rosato et al., 2008). Recent studies confirmed these results by showing that SAHA and LBH-589 could induce not only DSBs but also oxidative stress, as measured by 8oxodG (SAHA) and SSBs (LBH-589) in AML cells (Petruccelli et al., 2011; Rosato et al., 2010) and patient-derived leukemic blasts (Petruccelli et al., 2011). These studies showed (i) the mechanistic link between HDACi-dependent ROS induction and DNA damage and (ii) the synergistic effect of fludarabine and HDACis, leading to cell death (Fig. 3.3).

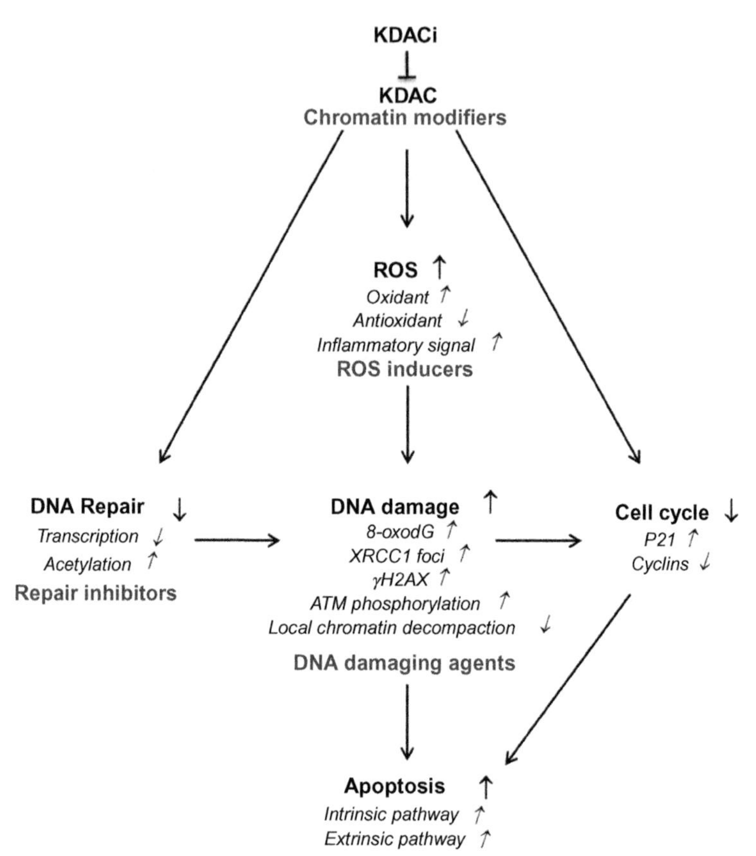

Figure 3.3 Model for the action of KDACis alone and in combination with other agents in cancers cells. HDACis and SIRTis induce cell cycle arrest, ROS production, and decrease in DNA repair activity independently (bold type). ROS also increases the DNA damage load in cells that have a KDACi-induced deficit in DNA repair activity. These combined activities lead to apoptosis. Combined treatment with chromatin modifiers, such as DNMTis, ROS inducers (such as proteasome inhibitors), DNA repair inhibitors (such as PARP inhibitors), or DNA-damaging agents (such as IR), would, in this context, amplify the ROS and DNA damage signal, leading to apoptosis. (For color version of this figure, the reader is referred to the online version of this chapter.)

3.7. HDACis: From bench to bedside

While mechanisms have been proposed to explain the anticancer activities of HDACis and suggesting promise in cancer therapy, the mechanisms by which they induce apoptosis are still under debate (Khan & La Thangue, 2012; Miller, Singh, Rivera-Del Valle, Manton, & Chandra, 2011; Nalabothula & Carrier, 2011; Wagner, Hackanson, Lubbert, & Jung, 2010). Moreover, their effects *in vivo*, when used as single agents in leukemia cells and solid tumors, are modest and vary with HDACis type (Chang et al., 2012), concentration (Gong et al., 2012), and course of treatment (Chang et al., 2012; Lee et al., 2010; Rosato et al., 2010). Based on these observations, studying the mechanism of action of HDACis in the context of specific cancer cell types versus normal cells will give new insights and markers that predict a specific combination of treatment that is patient and disease specific. In this regard, Vorinostat (SAHA) and Romidepsin (FK228, Depsipeptide) have been FDA approved in 2006 and 2009, respectively, for the treatment of CTCL (Table 3.2).

While ROS and DNA damage have been demonstrated as potential markers for clinical responses to these drugs, and several clinical trials combining HDACis with others agents are ongoing in cancer or leukemia (www.clinicaltrials.com), the clinical relevance of ROS and DNA damage has yet to be determined. In fact, a clinical trial in myelodysplastic syndromes (MDS) that combined HDACi MS275 with demethylating agent 5-azacytidine showed no correlation of γH2AX, chromatin methylation, or DNA acetylation with clinical responses to these drugs (Gore et al., 2006). Furthermore, correlative studies measuring ROS levels as markers for response in clinical trials using HDACis have yet to be performed.

4. SIRTs: DNA DAMAGE AND OXIDATIVE STRESS RESPONSES

The class III HDACs, the sirtuins, also play important roles in DDR and the response to oxidative stress (Table 3.3 and Fig. 3.1). While the first studies showing involvement of SIRTs in the DDR and stress responses came from knockout mice, inhibitors have recently been developed for potential clinical use (Table 3.2).

4.1. Studies in knockout mice

Evidence from knockout mouse studies demonstrate that SIRT1 −/− and SIRT6 −/− mice exhibit increased radiosensitivity, chromosomal aberrations, and DNA repair defects with an increased rate of tumorigenesis (Mostoslavsky et al., 2006; Wang et al., 2008), showing that SIRTs could act as tumor suppressors probably through regulating DNA repair. Notably, depletion of SIRT1 and 6 resembles the effects of HDACis or depletion of HDACs *in vitro* and *in vivo* in cancer cells. MEFs from SIRT1 −/− mice show an increase in acetylation of histones H3K9 and

H4K16, that correlates with a defect in the heterochromatin compaction compared to wild-type (WT) cells (Wang et al., 2008), recapitulating the phenotype obtained with TSA or MS275 in osteosarcoma cells in Miller et al. study (Miller et al., 2010). Following IR, SIRT1 −/− MEFs also demonstrate a twofold increase of DNA damage, as measured by comet assay, showing that the radiosensitivity of the SIRT1 −/− MEFs is probably due to a defect in DNA repair. Indeed, in the presence of IR or UV, SIRT1 −/− MEFs show defective repair of DSBs and SSBs, as measured by plasmid reactivation assays and decreased recruitment of RAD51, BRCA1, and NBS1 to the sites of DSBs. Nevertheless, the level of these proteins was not altered between SIRT1 −/− and WT MEFs (Wang et al., 2008), suggesting that protein acetylation altered specific activities involved in recruitment of these factors to DSBs.

SIRT6 −/− embryonic stem cells are also more sensitive to DNA-damaging agents such as H_2O_2, MMS, and IR, compared to WT MEFs. Surprisingly, while DSBs repair pathways appear intact, BER is defective. Moreover, overexpression of the Pol β dRP lyase domain rescued hypersensitivity of SIRT6-deficient cells to DNA-damaging agents, showing that BER is specifically regulated by SIRT6. Even though BER is impaired, defects in recruitment to sites of oxidative damage or decreased levels of BER proteins were not identified. Thus, these authors proposed that SIRT6 may promote BER by creating accessibility for BER factors via modification of histones or other chromatin-related factors. Interestingly, MEFs from knockout of another SIRT, SIRT3 −/−, demonstrate abnormal mitochondrial function, increased stress-induced ROS, and genomic instability (Kim et al., 2010). These authors demonstrated that SIRT3 −/− MEFs show an increase in superoxide radicals that is further increased when challenged by IR or an inhibitor of the mitochondrial electron transport chain, suggesting that SIRT3 could be involved in redox maintenance. In addition, this effect was concomitant with higher chromosomal instability induced by genotoxic stress (Kim, Patel, et al., 2010). In order to assess the possible antitumor role of SIRT3, these authors infected SIRT3 −/− cells with the c-Myc oncogene, alone, or in combination with the Ras oncogene and studied the transformation and clonogenicity of these cells compared to similarly infected WT cells. SIRT3 −/− cells became immortalized if infected with only one of the oncogenes and their clonogenicity increased even further if both c-Myc and Ras are expressed, compared to WT SIRT3 +/+ cells that require infection with both oncogenes or untransfected SIRT3 −/− cells (Kim, Patel, et al., 2010). Importantly, levels of superoxide were higher, and mitochondrial

ATP levels were decreased in SIRT3−/− cells infected with oncogenes compared to controls. This result suggested that SIRT3−/− cells are more metabolically active either by producing more ROS or by decreasing ROS scavenging systems (Kim, Patel, et al., 2010). Indeed, overexpressing MnSOD in SIRT3−/− cells infected with one or two oncogene increased the proliferation rate, prevented immortalization induced by oncogene expression, and reverted clonogenicity and genomic instability to levels in WT cells. This showed that increased levels of superoxide in SIRT3−/− may play a central role in promoting cellular transformation. Interestingly, MnSOD is down-regulated in SIRT3−/− cells compared to WT cells, and binding of transcription factors that regulate MnSOD expression, FOXO3a, and NF-κB, is also decreased at the MnSOD promoter. One of the mechanisms by which SIRTs could regulate FOXO3a activity is through acetylation/deacetylation (Tao et al., 2010). SIRT3 deacetylates FOXO3a, increasing its presence and activity in the nucleus (Tao et al., 2010). Furthermore, expressing FOXO3a prevented immortalization of oncogene-infected SIRT3−/− cells, rescued MnSOD expression, and restored mitochondrial metabolism (Kim, Patel, et al., 2010). Together, these results show that SIRT3 is a major HDAC, with tumor suppressor activity that plays a central role in the regulation of stress-induced superoxide levels and genomic stability. Thus, sirtuins, in particular, SIRT1, 3, and 6, recapitulate the characteristics of HDACs and make them relevant targets for inhibition in the treatment of cancers.

4.2. SIRTs and chromatin remodeling

In vitro and *in vivo*, SIRT1 has been shown to regulate the deacetylation of lysine residues on histones H3K9 and H4K16 (Liu, Liu, & Marshall, 2009), and SIRT6 deacetylates lysine residues on histones H3K9 and H3K56. While SIRT3 has been shown to deacetylate H3K9 and H4K16 *in vitro*, results *in vivo* are controversial (Scher, Vaquero, & Reinberg, 2007). Importantly, and contrary to the action of HDACs, SIRTs induce acetylation of histones targets that is dependent on DNA damage induction (Miller et al., 2011). SIRTs also facilitate histone methylation thus enhancing global transcriptional repression that is believed to facilitate DNA repair (Alcain & Villalba, 2009; Miller et al., 2010). SIRT1 has been implicated in the transcriptional repression that is necessary to promote the DNA repair (Espada & Esteller, 2007; O'Hagan, Mohammad, & Baylin, 2008). After a DSB occurs, H2AX is phosphorylated and histone H4K16 is acetylated causing the chromatin to open, allowing repair factors access to the break and a stimulation of DNA damage signaling. Next, SIRT1 and DNA methyl transferases (DNMT1 and DNMT3B) are recruited to the area around the break site, resulting in a decrease in acetylation of histone H4K16Ac and an increase in methylation of histone H3K27me3 (Espada & Esteller, 2007; O'Hagan et al., 2008). In addition, SIRT1 was recently shown to deacetylate DNMT1 on lysines 1349 and 1415 which increases DNMT1 methyltransferase activity, while deacetylation on residues 1111, 1113, 1115, and 1117 reduces its methyltransferase-independent transcription repression function (Peng et al., 2011). Thus, compaction of chromatin around the DNA break site and inhibition of transcription which is permissive for DNA repair are likely under the control of SIRT1 (Peng et al., 2011). In fact, chromatin remodeling factors, such as TIP60 (Wang & Chen, 2010), are involved in regulating deacetylation/acetylation of a particular histone residue in concert with SIRT1 (Hajji et al., 2010). Hajji et al. (2010) demonstrated that sensitivity of multidrug-resistant cell lines to etoposide could be reverted by overexpression of the remodeling factor MOF which induced acetylation of H4K16 and chromatin relaxation (Hajji et al., 2010). Interestingly, overexpression of SIRT1 that deacetylates H4K16 and leads to chromatin compaction renders the cells resistant to etoposide treatment (Hajji et al., 2010). Thus, the ratio of SIRT1 to MOF may be a biomarker for the sensitivity of HDACi-treated lung cancer cells to DNA-damaging agents (Table 3.3 and Fig. 3.2; Hajji et al., 2010).

4.3. SIRT and modulation of DNA repair

Several recent reports have shown the importance of SIRT1 in enhancing DNA repair. SIRT1 regulates Ku70 acetylation and the presence of BAX at mitochondrial membrane as previously described for some HDACs (Jeong et al., 2007). Following DSB induction, NBS1 is also a target for deacetylation by SIRT1 which promotes NBS1 activation and consequent DNA repair (Yuan, Zhang, Sengupta, Lane, & Seto, 2007). SIRT1 has also been shown to regulate the deacetylation of WRN, stabilizing it at DSBs and regulating its helicase and exonuclease activities (Li et al., 2008), promoting HRR repair (Uhl et al., 2010). SIRT1 also regulates other DNA repair pathways, such as BER and NER, through increased transcription of XPA, XPC group proteins (Fan & Luo, 2010; Ming et al., 2010). Furthermore, SIRT1 directly binds to, deacetylates, activates APE1, increasing BER (Yamamori et al., 2010). Interestingly, SIRT1 regulation of FOXO1 and 4 leads to increased oxidative stress through regulation of MnSOD, CAT, and DNA damage inducible factor GADD45α (Kobayashi et al., 2005; van der Horst et al., 2004). SIRT6 is also involved in the deacetylation of HRR factor CtIP that promotes end-resection of DSBs, as discussed earlier for HDACs, showing that HDACs and SIRTs share common targets (Kaidi, Weinert, Choudhary, & Jackson, 2010). In addition, McCord et al. (2009) showed that SIRT6 could promote NHEJ repair by forming a complex with DNA PK (McCord et al., 2009). SIRT6 partners with DNA repair factor PARP1 under conditions of oxidative stress, enhancing NHEJ and BER (Mao et al., 2011). Together, these studies demonstrate that SIRTs are involved in the DDR by regulating NHEJ, BER, HRR, and redox status of the cells (Table 3.3).

4.4. SIRT inhibitors for preclinical and clinical use

Recent studies have evaluated the role of SIRT inhibitors for therapeutics (Carafa et al., 2012; Cea et al., 2011; Chauhan et al., 2011; Kalle et al., 2010; Yuan et al., 2012; Table 3.2). Cea et al. (2011) demonstrated that combinations of HDACis (i.e., SAHA, VPA, and NaB) and SIRT inhibitors (i.e., Sirtinol, Cambinol, and EX527) or NAD+ biosynthesis inhibitors (i.e., FK866) can potentiate the antitumor effect of HDACis in leukemia cells, but not in normal leukocytes and hematopoietic CD34+ progenitors (Cea et al., 2011). This effect was not dependent on the level of SIRT expression but was correlated with increased mitochondrial membrane depolarization and an increase in the level of BAX (Cea et al., 2011). This study showed that SIRTs and HDACs cooperate in leukemia cells to avoid apoptosis and that targeting SIRT and HDAC inhibition could be of real interest in the treatment of cancers as neither normal CD34+ nor PBMCs were targeted by the treatment (Cea et al., 2011). Moreover, HDACis and now SIRTis are anticancer and antileukemia drugs that could also overcome resistance to therapeutic agents, as demonstrated by the recent work of Nguyen et al. (2011). These authors demonstrated that treatment of imatinib-resistant BCR-ABL positive cell lines with SAHA, Entinostat, or Apicidin resensitizes cells to tyrosine kinase inhibitors (TKIs), causing lethality of both TKI-resistant and -sensitive cells *in vitro* and *in vivo* in murine xenografts bearing imatinib-resistant ALL cells (Nguyen et al., 2011). This effect is also only seen in CD34+ from CML while absent from normal CD34+ population (Nguyen et al., 2011). More interestingly, these authors demonstrated that the lethality of cells correlated with apoptosis as measured by Tunel and caspases activation, G2/M arrest, DNA damage induction as measured by γH2AX which occurred prior to caspase activation, and ROS levels. Treatment with a ROS scavenger and SOD2 mimetic Mn(III) meso-tetrakis (4-benzoic acid) porphyrin (Mn-TBAP) totally abrogates ROS, DSB induction, and caspase activation without inactivating BCR-ABL targets (Nguyen et al., 2011). Indeed, BCR-ABL has also been shown to activate SIRT1 (Nguyen et al., 2011; Yuan et al., 2012). Thus, SIRT1 inhibitors could be attractive therapies for CML patients that are resistant to TKIs.

A recent paper also provided an insight into the mechanisms of action of an epigenetic inhibitor, UVI5008 that inhibits HDACs, SIRTs, and DNMTs (Nebbioso et al., 2011). As suggested earlier, HDACs could regulate SIRT1 activity which has been shown to regulate DNMT activity (Hajji et al., 2010; Peng et al., 2011), so intuitively, targeting all three factors could be of potential therapeutic interest (Fig. 3.2). As a matter of fact, Nebbioso et al. (2011) demonstrated that UVI5008 could induce apoptosis in a variety of cancer cells, including leukemias, breast, osteosarcoma, prostate, colon, carcinoma, melanomas *in vitro*, and tumor xenografts of human breast cancer (Nebbioso et al., 2011). These authors demonstrated that UVI5008 induced its anticancer activities through a P53-independent activation of the death receptors and an increase in ROS production (Nebbioso et al., 2011). Together, these studies suggest that increasing ROS or oxidative damage induction represent one mechanism of targeting cancer cells and that regulation of chromatin through inhibition of DNMTs, HDACs, and SIRTs may yield additional therapeutic effects (Fig. 3.3).

5. CONCLUDING REMARKS AND FUTURE DIRECTIONS

HDACis and SIRTis act through direct acetylation ofDNA repair proteins and modification of chromatin to increase ROS and DNA damage, leading to apoptosis of cancer cells. These effects are increased even further in combination with DNA-damaging agents (Fig. 3.3). Several lines of evidence suggest that drugs that induce oxidative stress and DNA damage yield additional therapeutic effects in combination with HDACis and/or SIRTis. Thus, using a prooxidant approach, agents that are ROS inducers, such as proteasome inhibitors, or others that downregulate antioxidant pathways, such as NF-κB inhibitor, parthenolide (Gopal, Arora, & Van Dyke, 2007), are attractive agents in future combination therapies (Fig. 3.3). Given the success in combining demethylating agents with HDACis in treatment of leukemias, including MDS and AML (Gore et al., 2006), combining DNA methyl transferase inhibitors DNMTi with HDACis and SIRTis would be an attractive therapeutic strategy in leukemias and other cancers. Indeed, treatment of cancer cells with UVI5008 that simultaneously inhibits DNMTs, HDACs, and SIRTs has been shown to increase ROS and oxidative DNA damage, leading to apoptosis (Nebbioso et al., 2011). However, while the oxidant and DNA damage approaches hold promise as markers for apoptotic effects of HDACis and SIRTis, this has yet to be validated in correlative studies in clinical trials. In fact, Gore et al. (2006) showed no correlation of DNA damage markers with response to HDACis combined with DNMTis in clinical trials (Gore et al., 2006). In addition, while high doses of these drugs induce rapid DNA damage and cytotoxicity, this does not explain the prolonged time to the response observed in leukemia patients treated with these drugs. There is some evidence that low doses of HDACis and demethylating agents may exert their effects not through DNA damage, but reprogramming of cells (Tsai et al., 2012). In this regard, we have recently showed that transient exposure of cultured and primary leukemic and epithelial tumor cells to clinically relevant nanomolar doses of DNMTis, without causing increased DSBs and cytotoxicity, produces an antitumor "memory" response, including inhibition of subpopulations of cancer stem-like cells. These effects are accompanied by sustained decreases in genome-wide promoter DNA methylation, gene reexpression, and antitumor changes in key cellular regulatory pathways (Tsai et al., 2012). Notably, this approach appears to work in certain subtypes of cancer and leukemia (Tsai et al., 2012). Therefore, whether cytotoxic effects of HDACis and SIRTis may be harnessed to benefit specific cancers, while reprogramming with low doses of these agents may be used in others, has yet to be determined.

ACKNOWLEDGMENTS

This work was supported by the Cigarette Restitution Funds of Maryland and the V foundation.

REFERENCES

Adams, H., Fritzsche, F. R., Dirnhofer, S., Kristiansen, G., & Tzankov, A. (2010). Class I histone deacetylases 1, 2 and 3 are highly expressed in classical Hodgkin's lymphoma. *Expert Opinion on Therapeutic Targets*, *14*, 577–584.

Alcain, F. J., & Villalba, J. M. (2009). Sirtuin inhibitors. *Expert Opinion on Therapeutic Patents*, *19*, 283–294.

Arundel, C. M., Glicksman, A. S., & Leith, J. T. (1985). Enhancement of radiation injury in human colon tumor cells by the maturational agent sodium butyrate (NaB). *Radiation Research*, *104*, 443–448.

Audebert, M., Salles, B., & Calsou, P. (2008). Effect of double-strand break DNA sequence on the PARP-1 NHEJ pathway. *Biochemical and Biophysical Research Communications*, *369*, 982–988.

Bakkenist, C. J., & Kastan, M. B. (2003). DNA damage activates ATM through intermolecular autophosphorylation and dimer dissociation. *Nature*, *421*, 499–506.

Balaban, R. S., Nemoto, S., & Finkel, T. (2005). Mitochondria, oxidants, and aging. *Cell*, *120*, 483–495.

Basile, V., Mantovani, R., & Imbriano, C. (2006). DNA damage promotes histone deacetylase 4 nuclear localization and repression of G2/M promoters, via p53 C-terminal lysines. *The Journal of Biological Chemistry*, *281*, 2347–2357.

Bedard, K., & Krause, K. H. (2007). The NOX family of ROS-generating NADPH oxidases: Physiology and pathophysiology. *Physiological Reviews*, *87*, 245–313.

Beucher, A., Birraux, J., Tchouandong, L., Barton, O., Shibata, A., Conrad, S., et al. (2009). ATM and Artemis promote homologous recombination of radiation-induced DNA double-strand breaks in G2. *The EMBO Journal*, *28*, 3413–3427.

Bhalla, S., Balasubramanian, S., David, K., Sirisawad, M., Buggy, J., Mauro, L., et al. (2009). PCI-24781 induces caspase and reactive oxygen species-dependent apoptosis through NF-kappaB mechanisms and is synergistic with bortezomib in lymphoma cells. *Clinical Cancer Research*, *15*, 3354–3365.

Bhaskara, S., Chyla, B. J., Amann, J. M., Knutson, S. K., Cortez, D., Sun, Z. W., et al. (2008). Deletion of histone deacetylase 3 reveals critical roles in S phase progression and DNA damage control. *Molecular Cell*, *30*, 61–72.

Bhaskara, S., Knutson, S. K., Jiang, G., Chandrasekharan, M. B., Wilson, A. J., Zheng, S., et al. (2010). Hdac3 is essential for the maintenance of chromatin structure and genome stability. *Cancer Cell*, *18*, 436–447.

Bindra, R. S., & Glazer, P. M. (2007). Repression of RAD51 gene expression by E2F4/p130 complexes in hypoxia. *Oncogene*, *26*, 2048–2057.

Bolden, J. E., Peart, M. J., & Johnstone, R. W. (2006). Anticancer activities of histone deacetylase inhibitors. *Nature Reviews. Drug Discovery*, *5*, 769–784.

Bradbury, C. A., Khanim, F. L., Hayden, R., Bunce, C. M., White, D. A., Drayson, M. T., et al. (2005). Histone deacetylases in acute myeloid leukaemia show a distinctive pattern of expression that changes selectively in response to deacetylase inhibitors. *Leukemia*, *19*, 1751–1759.

Brown, D., Kogan, S., Lagasse, E., Weissman, I., Alcalay, M., Pelicci, P. G., et al. (1997). A PMLRARalpha transgene initiates murine acute promyelocytic leukemia. *Proceedings of the National Academy of Sciences of the United States of America*, *94*, 2551–2556.

Butler, L. M., Zhou, X., Xu, W. S., Scher, H. I., Rifkind, R. A., Marks, P. A., et al. (2002). The histone deacetylase inhibitor SAHA arrests cancer cell growth, up-regulates thioredoxin-binding protein-2, and down-regulates thioredoxin. *Proceedings of the National Academy of Sciences of the United States of America*, *99*, 11700–11705.

Caldecott, K. W. (2007). Mammalian single-strand break repair: Mechanisms and links with chromatin. *DNA Repair*, *6*, 443–453.

Calsou, P., Frit, P., Humbert, O., Muller, C., Chen, D. J., & Salles, B. (1999). The DNA-dependent protein kinase catalytic activity regulates DNA end processing by means of Ku entry into DNA. *The Journal of Biological Chemistry*, *274*, 7848–7856.

Camphausen, K., Burgan, W., Cerra, M., Oswald, K. A., Trepel, J. B., Lee, M. J., et al. (2004). Enhanced radiation-induced cell killing and prolongation of gH2AX foci expression by the histone deacetylase inhibitor MS-275. *Cancer Research*, *64*, 316–321.

Carafa, V., Nebbioso, A., & Altucci, L. (2012). Sirtuins and disease: The road ahead. *Frontiers in Pharmacology*, *3*, 4.

Carew, J. S., Giles, F. J., & Nawrocki, S. T. (2008). Histone deacetylase inhibitors: Mechanisms of cell death and promise in combination cancer therapy. *Cancer Letters, 269*, 7–17.

Carrassa, L., Broggini, M., Vikhanskaya, F., & Damia, G. (2003). Characterization of the 5′flanking region of the human Chk1 gene: Identification of E2F1 functional sites. *Cell Cycle, 2*, 604–609.

Cea, M., Soncini, D., Fruscione, F., Raffaghello, L., Garuti, A., Emionite, L., et al. (2011). Synergistic interactions between HDAC and sirtuin inhibitors in human leukemia cells. *PLoS One, 6*, e22739.

Chang, J., Varghese, D. S., Gillam, M. C., Peyton, M., Modi, B., Schiltz, R. L., et al. (2012). Differential response of cancer cells to HDAC inhibitors trichostatin A and depsipeptide. *British Journal of Cancer, 106*, 116–125.

Chauhan, D., Bandi, M., Singh, A. V., Ray, A., Raje, N., Richardson, P., et al. (2011). Preclinical evaluation of a novel SIRT1 modulator SRT1720 in multiple myeloma cells. *British Journal of Haematology, 155*, 588–598.

Chen, G., Li, A., Zhao, M., Gao, Y., Zhou, T., Xu, Y., et al. (2008). Proteomic analysis identifies protein targets responsible for depsipeptide sensitivity in tumor cells. *Journal of Proteome Research*, 7, 2733–2742.

Chen, Z. X., & Pervaiz, S. (2007). Bcl-2 induces pro-oxidant state by engaging mitochondrial respiration in tumor cells. *Cell Death and Differentiation, 14*, 1617–1627.

Chen, C. S., Wang, Y. C., Yang, H. C., Huang, P. H., Kulp, S. K., Yang, C. C., et al. (2007). Histone deacetylase inhibitors sensitize prostate cancer cells to agents that produce DNA double-strand breaks by targeting Ku70 acetylation. *Cancer Research, 67*, 5318–5327.

Choudhary, C., Kumar, C., Gnad, F., Nielsen, M. L., Rehman, M., Walther, T. C., et al. (2009). Lysine acetylation targets protein complexes and co-regulates major cellular functions. *Science, 325*, 834–840.

Ciccia, A., & Elledge, S. J. (2010). The DNA damage response: Making it safe to play with knives. *Molecular Cell, 40*, 179–204.

Clement, M. V., & Stamenkovic, I. (1996). Superoxide anion is a natural inhibitor of FAS-mediated cell death. *The EMBO Journal, 15*, 216–225.

Cohen, H. Y., Lavu, S., Bitterman, K. J., Hekking, B., Imahiyerobo, T. A., Miller, C., et al. (2004). Acetylation of the C terminus of Ku70 by CBP and PCAF controls Bax-mediated apoptosis. *Molecular Cell, 13*, 627–638.

Cohen, H. Y., Miller, C., Bitterman, K. J., Wall, N. R., Hekking, B., Kessler, B., et al. (2004). Calorie restriction promotes mammalian cell survival by inducing the SIRT1 deacetylase. *Science, 305*, 390–392.

Dai, Y., Rahmani, M., Dent, P., & Grant, S. (2005). Blockade of histone deacetylase inhibitor-induced RelA/p65 acetylation and NF-kappaB activation potentiates apoptosis in leukemia cells through a process mediated by oxidative damage, XIAP downregulation, and c-Jun N-terminal kinase 1 activation. *Molecular and Cellular Biology, 25*, 5429–5444.

Dasmahapatra, G., Lembersky, D., Son, M. P., Attkisson, E., Dent, P., Fisher, R. I., et al. (2011). Carfilzomib interacts synergistically with histone deacetylase inhibitors in mantle cell lymphoma cells in vitro and in vivo. *Molecular Cancer Therapeutics, 10*, 1686–1697.

Deriano, L., Stracker, T. H., Baker, A., Petrini, J. H., & Roth, D. B. (2009). Roles for NBS1 in alternative nonhomologous end-joining of V(D)J recombination intermediates. *Molecular Cell, 34*, 13–25.

Dinkelmann, M., Spehalski, E., Stoneham, T., Buis, J., Wu, Y., Sekiguchi, J. M., et al. (2009). Multiple functions of MRN in end-joining pathways during isotype class switching. *Nature Structural & Molecular Biology, 16*, 808–813.

Drummond, D. C., Marx, C., Guo, Z., Scott, G., Noble, C., Wang, D., et al. (2005). Enhanced pharmacodynamic and antitumor properties of a histone deacetylase inhibitor encapsulated in liposomes or ErbB2-targeted immunoliposomes. *Clinical Cancer Research, 11*, 3392–3401.

Eot-Houllier, G., Fulcrand, G., Magnaghi-Jaulin, L., & Jaulin, C. (2009). Histone deacetylase inhibitors and genomic instability. *Cancer Letters, 274*, 169–176.

Espada, J., & Esteller, M. (2007). Epigenetic control of nuclear architecture. *Cellular and Molecular Life Sciences, 64*, 449–457.

Falck, J., Coates, J., & Jackson, S. P. (2005). Conserved modes of recruitment of ATM, ATR and DNA-PKcs to sites of DNA damage. *Nature, 434*, 605–611.

Falzon, M., Fewell, J. W., & Kuff, E. L. (1993). EBP-80, a transcription factor closely resembling the human autoantigen Ku, recognizes single- to double-strand transitions in DNA. *The Journal of Biological Chemistry, 268*, 10546–10552.

Fan, W., & Luo, J. (2010). SIRT1 regulates UV-induced DNA repair through deacetylating XPA. *Molecular Cell, 39*, 247–258.

Fraga, M. F., Ballestar, E., Villar-Garea, A., Boix-Chornet, M., Espada, J., Schotta, G., et al. (2005). Loss of acetylation at Lys16 and trimethylation at Lys20 of histone H4 is a common hallmark of human cancer. *Nature Genetics, 37*, 391–400.

Gaymes, T. J., Padua, R. A., Pla, M., Orr, S., Omidvar, N., Chomienne, C., et al. (2006). Histone deacetylase inhibitors (HDI) cause DNA damage in leukemia cells: A mechanism for leukemia-specific HDI-dependent apoptosis? *Molecular Cancer Research, 4*, 563–573.

Gong, K., Xie, J., Yi, H., & Li, W. (2012). CS055 (Chidamide/HBI-8000), a novel histone deacetylase inhibitor, induces G1-arrest, ROS-dependent apoptosis and differentiation in human leukemia cells. *The Biochemical Journal, 443*(3), 735–746.

Gopal, Y. N., Arora, T. S., & Van Dyke, M. W. (2007). Parthenolide specifically depletes histone deacetylase 1 protein and induces cell death through ataxia telangiectasia mutated. *Chemistry & Biology, 14*, 813–823.

Gore, S. D., Baylin, S., Sugar, E., Carraway, H., Miller, C. B., Carducci, M., et al. (2006). Combined DNA methyltransferase and histone deacetylase inhibition in the treatment of myeloid neoplasms. *Cancer Research, 66*, 6361–6369.

Gottlieb, T. M., & Jackson, S. P. (1993). The DNA-dependent protein kinase: Requirement for DNA ends and association with Ku antigen. *Cell, 72*, 131–142.

Hajji, N., Wallenborg, K., Vlachos, P., Fullgrabe, J., Hermanson, O., & Joseph, B. (2010). Opposing effects of hMOF and SIRT1 on H4K16 acetylation and the sensitivity to the topoisomerase II inhibitor etoposide. *Oncogene, 29*, 2192–2204.

Hansen, W. K., & Kelley, M. R. (2000). Review of mammalian DNA repair and translational implications. *The Journal of Pharmacology and Experimental Therapeutics, 295*, 1–9.

Hartlerode, A. J., & Scully, R. (2009). Mechanisms of double-strand break repair in somatic mammalian cells. *The Biochemical Journal, 423*, 157–168.

Hayashi, T., Ueno, Y., & Okamoto, T. (1993). Oxidoreductive regulation of nuclear factor kappa B. Involvement of a cellular reducing catalyst thioredoxin. *The Journal of Biological Chemistry, 268*, 11380–11388.

Hirota, K., Matsui, M., Iwata, S., Nishiyama, A., Mori, K., & Yodoi, J. (1997). AP-1 transcriptional activity is regulated by a direct association between thioredoxin and Ref-1. *Proceedings of the National Academy of Sciences of the United States of America, 94*, 3633–3638.

Hofmann, B., Hecht, H. J., & Flohe, L. (2002). Peroxiredoxins. *Biological Chemistry, 383*, 347–364.

Huang, P., & Plunkett, W. (1995). Fludarabine- and gemcitabine-induced apoptosis: Incorporation of analogs into DNA is a critical event. *Cancer Chemotherapy and Pharmacology, 36*, 181–188.

Huertas, P., & Jackson, S. P. (2009). Human CtIP mediates cell cycle control of DNA end resection and double strand break repair. *The Journal of Biological Chemistry, 284*, 9558–9565.

Iliakis, G. (2009). Backup pathways of NHEJ in cells of higher eukaryotes: Cell cycle dependence. *Radiotherapy and Oncology, 92*, 310–315.
Insinga, A., Monestiroli, S., Ronzoni, S., Gelmetti, V., Marchesi, F., Viale, A., et al. (2005). Inhibitors of histone deacetylases induce tumor-selective apoptosis through activation of the death receptor pathway. *Nature Medicine, 11*, 71–76.
Ito, A., Kawaguchi, Y., Lai, C. H., Kovacs, J. J., Higashimoto, Y., Appella, E., et al. (2002). MDM2-HDAC1-mediated deacetylation of p53 is required for its degradation. *The EMBO Journal, 21*, 6236–6245.
Jeong, J., Juhn, K., Lee, H., Kim, S. H., Min, B. H., Lee, K. M., et al. (2007). SIRT1 promotes DNA repair activity and deacetylation of Ku70. *Experimental & Molecular Medicine, 39*, 8–13.
Jiang, X., Sun, Y., Chen, S., Roy, K., & Price, B. D. (2006). The FATC domains of PIKK proteins are functionally equivalent and participate in the Tip60-dependent activation of DNA-PKcs and ATM. *The Journal of Biological Chemistry, 281*, 15741–15746.
Jin, K. L., Pak, J. H., Park, J. Y., Choi, W. H., Lee, J. Y., Kim, J. H., et al. (2008). Expression profile of histone deacetylases 1, 2 and 3 in ovarian cancer tissues. *Journal of Gynecologic Oncology, 19*, 185–190.
Johnstone, R. W. (2002). Histone-deacetylase inhibitors: Novel drugs for the treatment of cancer. *Nature Reviews. Drug Discovery, 1*, 287–299.
Kachhap, S. K., Rosmus, N., Collis, S. J., Kortenhorst, M. S., Wissing, M. D., Hedayati, M., et al. (2010). Downregulation of homologous recombination DNA repair genes by HDAC inhibition in prostate cancer is mediated through the E2F1 transcription factor. *PLoS One, 5*, e11208.
Kaidi, A., Weinert, B. T., Choudhary, C., & Jackson, S. P. (2010). Human SIRT6 promotes DNA end resection through CtIP deacetylation. *Science, 329*, 1348–1353.
Kalle, A. M., Mallika, A., Badiger, J., Alinakhi, Talukdar, P., & Sachchidanand, (2010). Inhibition of SIRT1 by a small molecule induces apoptosis in breast cancer cells. *Biochemical and Biophysical Research Communications, 401*, 13–19.
Kao, G. D., McKenna, W. G., Guenther, M. G., Muschel, R. J., Lazar, M. A., & Yen, T. J. (2003). Histone deacetylase 4 interacts with 53BP1 to mediate the DNA damage response. *The Journal of Cell Biology, 160*, 1017–1027.
Karagiannis, T. C., Kn, H., & El-Osta, A. (2006). The epigenetic modifier, valproic acid, enhances radiation sensitivity. *Epigenetics, 1*, 131–137.
Kerr, E., Holohan, C., McLaughlin, K. M., Majkut, J., Dolan, S., Redmond, K., et al. (2012). Identification of an acetylation-dependant Ku70/FLIP complex that regulates FLIP expression and HDAC inhibitor-induced apoptosis. *Cell Death and Differentiation, 19*(8), 1317–1327.
Khan, O., & La Thangue, N. B. (2012). HDAC inhibitors in cancer biology: Emerging mechanisms and clinical applications. *Immunology and Cell Biology, 90*, 85–94.
Khanna, K. K., & Jackson, S. P. (2001). DNA double-strand breaks: Signaling, repair and the cancer connection. *Nature Genetics, 27*, 247–254.
Kim, G. D., Choi, Y. H., Dimtchev, A., Jeong, S. J., Dritschilo, A., & Jung, M. (1999). Sensing of ionizing radiation-induced DNA damage by ATM through interaction with histone deacetylase. *The Journal of Biological Chemistry, 274*, 31127–31130.
Kim, J. J., Lee, S. B., Park, J. K., & Yoo, Y. D. (2010). TNF-alpha-induced ROS production triggering apoptosis is directly linked to Romo1 and Bcl-X(L). *Cell Death and Differentiation, 17*, 1420–1434.
Kim, H. S., Patel, K., Muldoon-Jacobs, K., Bisht, K. S., Aykin-Burns, N., Pennington, J. D., et al. (2010). SIRT3 is a mitochondria-localized tumor suppressor required for maintenance of mitochondrial integrity and metabolism during stress. *Cancer Cell, 17*, 41–52.

Kobayashi, Y., Furukawa-Hibi, Y., Chen, C., Horio, Y., Isobe, K., Ikeda, K., et al. (2005). SIRT1 is critical regulator of FOXO-mediated transcription in response to oxidative stress. *International Journal of Molecular Medicine, 16*, 237–243.

Kobayashi, C. I., & Suda, T. (2012). Regulation of reactive oxygen species in stem cells and cancer stem cells. *Journal of Cellular Physiology, 227*, 421–430.

Kotian, S., Liyanarachchi, S., Zelent, A., & Parvin, J. D. (2011). Histone deacetylases 9 and 10 are required for homologous recombination. *The Journal of Biological Chemistry, 286*, 7722–7726.

Lawson, M., Uciechowska, U., Schemies, J., Rumpf, T., Jung, M., & Sippl, W. (2010). Inhibitors to understand molecular mechanisms of NAD(+)-dependent deacetylases (sirtuins). *Biochimica et Biophysica Acta, 1799*, 726–739.

Lee, J. H., Choy, M. L., Ngo, L., Foster, S. S., & Marks, P. A. (2010). Histone deacetylase inhibitor induces DNA damage, which normal but not transformed cells can repair. *Proceedings of the National Academy of Sciences of the United States of America, 107*, 14639–14644.

Lee, S. F., & Pervaiz, S. (2011). Assessment of oxidative stress-induced DNA damage by immunoflourescent analysis of 8-oxodG. *Methods in Cell Biology, 103*, 99–113.

Li, K., Casta, A., Wang, R., Lozada, E., Fan, W., Kane, S., et al. (2008). Regulation of WRN protein cellular localization and enzymatic activities by SIRT1-mediated deacetylation. *The Journal of Biological Chemistry, 283*, 7590–7598.

Li, Z. Y., Yang, Y., Ming, M., & Liu, B. (2011). Mitochondrial ROS generation for regulation of autophagic pathways in cancer. *Biochemical and Biophysical Research Communications, 414*, 5–8.

Lindahl, T., & Wood, R. D. (1999). Quality control by DNA repair. *Science, 286*, 1897–1905.

Liu, N., Lamerdin, J. E., Tebbs, R. S., Schild, D., Tucker, J. D., Shen, M. R., et al. (1998). XRCC2 and XRCC3, new human Rad51-family members, promote chromosome stability and protect against DNA cross-links and other damages. *Molecular Cell, 1*, 783–793.

Liu, T., Liu, P. Y., & Marshall, G. M. (2009). The critical role of the class III histone deacetylase SIRT1 in cancer. *Cancer Research, 69*, 1702–1705.

Lobrich, M., & Jeggo, P. A. (2005). The two edges of the ATM sword: Co-operation between repair and checkpoint functions. *Radiotherapy and Oncology, 76*, 112–118.

Louis, M., Rosato, R. R., Brault, L., Osbild, S., Battaglia, E., Yang, X. H., et al. (2004). The histone deacetylase inhibitor sodium butyrate induces breast cancer cell apoptosis through diverse cytotoxic actions including glutathione depletion and oxidative stress. *International Journal of Oncology, 25*, 1701–1711.

Lukas, J., Lukas, C., & Bartek, J. (2011). More than just a focus: The chromatin response to DNA damage and its role in genome integrity maintenance. *Nature Cell Biology, 13*, 1161–1169.

Luo, J., Su, F., Chen, D., Shiloh, A., & Gu, W. (2000). Deacetylation of p53 modulates its effect on cell growth and apoptosis. *Nature, 408*, 377–381.

Magne, N., Toillon, R. A., Bottero, V., Didelot, C., Houtte, P. V., Gerard, J. P., et al. (2006). NF-kappaB modulation and ionizing radiation: Mechanisms and future directions for cancer treatment. *Cancer Letters, 231*, 158–168.

Mao, Z., Hine, C., Tian, X., Van Meter, M., Au, M., Vaidya, A., et al. (2011). SIRT6 promotes DNA repair under stress by activating PARP1. *Science, 332*, 1443–1446.

Marks, P. A. (2006). Thioredoxin in cancer—Role of histone deacetylase inhibitors. *Seminars in Cancer Biology, 16*, 436–443.

Marks, P. A., Richon, V. M., & Rifkind, R. A. (2000). Histone deacetylase inhibitors: Inducers of differentiation or apoptosis of transformed cells. *Journal of the National Cancer Institute, 92*, 1210–1216.

Marquard, L., Poulsen, C. B., Gjerdrum, L. M., de Nully Brown, P., Christensen, I. J., Jensen, P. B., et al. (2009). Histone deacetylase 1, 2, 6 and acetylated histone H4 in B- and T-cell lymphomas. *Histopathology, 54*, 688–698.

Maynard, S., Schurman, S. H., Harboe, C., de Souza-Pinto, N. C., & Bohr, V. A. (2009). Base excision repair of oxidative DNA damage and association with cancer and aging. *Carcinogenesis*, *30*, 2–10.
McCord, R. A., Michishita, E., Hong, T., Berber, E., Boxer, L. D., Kusumoto, R., et al. (2009). SIRT6 stabilizes DNA-dependent protein kinase at chromatin for DNA double-strand break repair. *Aging (Albany, NY)*, *1*, 109–121.
McGuinness, D., McGuinness, D. H., McCaul, J. A., & Shiels, P. G. (2011). Sirtuins, bioageing, and cancer. *Journal of Aging Research*, *2011*, 235754.
Memisoglu, A., & Samson, L. (2000). Base excision repair in yeast and mammals. *Mutation Research*, *451*, 39–51.
Mercurio, C., Minucci, S., & Pelicci, P. G. (2010). Histone deacetylases and epigenetic therapies of hematological malignancies. *Pharmacological Research*, *62*, 18–34.
Michiels, C., Raes, M., Toussaint, O., & Remacle, J. (1994). Importance of Se-glutathione peroxidase, catalase, and Cu/Zn-SOD for cell survival against oxidative stress. *Free Radical Biology & Medicine*, *17*, 235–248.
Miller, C. P., Singh, M. M., Rivera-Del Valle, N., Manton, C. A., & Chandra, J. (2011). Therapeutic strategies to enhance the anticancer efficacy of histone deacetylase inhibitors. *Journal of Biomedicine and Biotechnology*, *2011*, 514261.
Miller, K. M., Tjeertes, J. V., Coates, J., Legube, G., Polo, S. E., Britton, S., et al. (2010). Human HDAC1 and HDAC2 function in the DNA-damage response to promote DNA nonhomologous end-joining. *Nature Structural & Molecular Biology*, *17*, 1144–1151.
Mimitou, E. P., & Symington, L. S. (2009). Nucleases and helicases take center stage in homologous recombination. *Trends in Biochemical Sciences*, *34*, 264–272.
Mimori, T., & Hardin, J. A. (1986). Mechanism of interaction between Ku protein and DNA. *The Journal of Biological Chemistry*, *261*, 10375–10379.
Ming, M., Shea, C. R., Guo, X., Li, X., Soltani, K., Han, W., et al. (2010). Regulation of global genome nucleotide excision repair by SIRT1 through xeroderma pigmentosum C. *Proceedings of the National Academy of Sciences of the United States of America*, *107*, 22623–22628.
Minucci, S., & Pelicci, P. G. (2006). Histone deacetylase inhibitors and the promise of epigenetic (and more) treatments for cancer. *Nature Reviews. Cancer*, *6*, 38–51.
Misteli, T., & Soutoglou, E. (2009). The emerging role of nuclear architecture in DNA repair and genome maintenance. *Nature Reviews. Molecular Cell Biology*, *10*, 243–254.
Miyamoto, S. (2011). Nuclear initiated NF-kappaB signaling: NEMO and ATM take center stage. *Cell Research*, *21*, 116–130.
Montero, A. J., & Jassem, J. (2011). Cellular redox pathways as a therapeutic target in the treatment of cancer. *Drugs*, *71*, 1385–1396.
Mostoslavsky, R., Chua, K. F., Lombard, D. B., Pang, W. W., Fischer, M. R., Gellon, L., et al. (2006). Genomic instability and aging-like phenotype in the absence of mammalian SIRT6. *Cell*, *124*, 315–329.
Moufarij, M. A., Sampath, D., Keating, M. J., & Plunkett, W. (2006). Fludarabine increases oxaliplatin cytotoxicity in normal and chronic lymphocytic leukemia lymphocytes by suppressing interstrand DNA crosslink removal. *Blood*, *108*, 4187–4193.
Munshi, A., Kurland, J. F., Nishikawa, T., Tanaka, T., Hobbs, M. L., Tucker, S. L., et al. (2005). Histone deacetylase inhibitors radiosensitize human melanoma cells by suppressing DNA repair activity. *Clinical Cancer Research*, *11*, 4912–4922.
Munshi, A., Tanaka, T., Hobbs, M. L., Tucker, S. L., Richon, V. M., & Meyn, R. E. (2006). Vorinostat, a histone deacetylase inhibitor, enhances the response of human tumor cells to ionizing radiation through prolongation of g-H2AX foci. *Molecular Cancer Therapeutics*, *5*, 1967–1974.
Murr, R., Loizou, J. I., Yang, Y. G., Cuenin, C., Li, H., Wang, Z. Q., et al. (2006). Histone acetylation by Trrap-Tip60 modulates loading of repair proteins and repair of DNA double-strand breaks. *Nature Cell Biology*, *8*, 91–99.

Nakagawa, M., Oda, Y., Eguchi, T., Aishima, S., Yao, T., Hosoi, F., et al. (2007). Expression profile of class I histone deacetylases in human cancer tissues. *Oncology Reports, 18*, 769–774.

Nakamura, H., Matsuda, M., Furuke, K., Kitaoka, Y., Iwata, S., Toda, K., et al. (1994). Adult T cell leukemia-derived factor/human thioredoxin protects endothelial F-2 cell injury caused by activated neutrophils or hydrogen peroxide. *Immunology Letters, 42*, 75–80.

Nalabothula, N., & Carrier, F. (2011). Cancer cells' epigenetic composition and predisposition to histone deacetylase inhibitor sensitization. *Epigenomics, 3*, 145–155.

Namdar, M., Perez, G., Ngo, L., & Marks, P. A. (2010). Selective inhibition of histone deacetylase 6 (HDAC6) induces DNA damage and sensitizes transformed cells to anticancer agents. *Proceedings of the National Academy of Sciences of the United States of America, 107*, 20003–20008.

Nebbioso, A., Clarke, N., Voltz, E., Germain, E., Ambrosino, C., Bontempo, P., et al. (2005). Tumor-selective action of HDAC inhibitors involves TRAIL induction in acute myeloid leukemia cells. *Nature Medicine, 11*, 77–84.

Nebbioso, A., Pereira, R., Khanwalkar, H., Matarese, F., Garcia-Rodriguez, J., Miceli, M., et al. (2011). Death receptor pathway activation and increase of ROS production by the triple epigenetic inhibitor UVI5008. *Molecular Cancer Therapeutics, 10*, 2394–2404.

Nguyen, T., Dai, Y., Attkisson, E., Kramer, L., Jordan, N., Nguyen, N., et al. (2011). HDAC inhibitors potentiate the activity of the BCR/ABL kinase inhibitor KW-2449 in imatinib-sensitive or -resistant BCR/ABL+ leukemia cells in vitro and in vivo. *Clinical Cancer Research, 17*, 3219–3232.

Nussenzweig, A., & Nussenzweig, M. C. (2007). A backup DNA repair pathway moves to the forefront. *Cell, 131*, 223–225.

O'Driscoll, M., & Jeggo, P. A. (2006). The role of double-strand break repair—Insights from human genetics. *Nature Reviews. Genetics, 7*, 45–54.

O'Hagan, H. M., Mohammad, H. P., & Baylin, S. B. (2008). Double strand breaks can initiate gene silencing and SIRT1-dependent onset of DNA methylation in an exogenous promoter CpG island. *PLoS Genetics, 4*, e1000155.

Ogiwara, H., Ui, A., Otsuka, A., Satoh, H., Yokomi, I., Nakajima, S., et al. (2011). Histone acetylation by CBP and p300 at double-strand break sites facilitates SWI/SNF chromatin remodeling and the recruitment of non-homologous end joining factors. *Oncogene, 30*, 2135–2146.

Park, J. H., Park, E. J., Lee, H. S., Kim, S. J., Hur, S. K., Imbalzano, A. N., et al. (2006). Mammalian SWI/SNF complexes facilitate DNA double-strand break repair by promoting gamma-H2AX induction. *The EMBO Journal, 25*, 3986–3997.

Peng, L., Yuan, Z., Ling, H., Fukasawa, K., Robertson, K., Olashaw, N., et al. (2011). SIRT1 deacetylates the DNA methyltransferase 1 (DNMT1) protein and alters its activities. *Molecular and Cellular Biology, 31*, 4720–4734.

Pervaiz, S., Taneja, R., & Ghaffari, S. (2009). Oxidative stress regulation of stem and progenitor cells. *Antioxidants & Redox Signaling, 11*, 2777–2789.

Peserico, A., & Simone, C. (2011). Physical and functional HAT/HDAC interplay regulates protein acetylation balance. *Journal of Biomedicine and Biotechnology, 2011*, 371832.

Petalcorin, M. I., Sandall, J., Wigley, D. B., & Boulton, S. J. (2006). CeBRC-2 stimulates D-loop formation by RAD-51 and promotes DNA single-strand annealing. *Journal of Molecular Biology, 361*, 231–242.

Petruccelli, L. A., Dupere-Richer, D., Pettersson, F., Retrouvey, H., Skoulikas, S., & Miller, W. H., Jr. (2011). Vorinostat induces reactive oxygen species and DNA damage in acute myeloid leukemia cells. *PLoS One, 6*, e20987.

Polo, S. E., Kaidi, A., Baskcomb, L., Galanty, Y., & Jackson, S. P. (2010). Regulation of DNA-damage responses and cell-cycle progression by the chromatin remodelling factor CHD4. *The EMBO Journal, 29*, 3130–3139.

Rahman, I., Marwick, J., & Kirkham, P. (2004). Redox modulation of chromatin remodeling: Impact on histone acetylation and deacetylation, NF-kappaB and pro-inflammatory gene expression. *Biochemical Pharmacology, 68*, 1255–1267.

Rahmani, M., Reese, E., Dai, Y., Bauer, C., Payne, S. G., Dent, P., et al. (2005). Coadministration of histone deacetylase inhibitors and perifosine synergistically induces apoptosis in human leukemia cells through Akt and ERK1/2 inactivation and the generation of ceramide and reactive oxygen species. *Cancer Research, 65*, 2422–2432.

Rajendran, R., Garva, R., Krstic-Demonacos, M., & Demonacos, C. (2011). Sirtuins: Molecular traffic lights in the crossroad of oxidative stress, chromatin remodeling, and transcription. *Journal of Biomedicine and Biotechnology, 2011*, 368276.

Rass, E., Grabarz, A., Plo, I., Gautier, J., Bertrand, P., & Lopez, B. S. (2009). Role of Mre11 in chromosomal nonhomologous end joining in mammalian cells. *Nature Structural & Molecular Biology, 16*, 819–824.

Riballo, E., Kuhne, M., Rief, N., Doherty, A., Smith, G. C., Recio, M. J., et al. (2004). A pathway of double-strand break rejoining dependent upon ATM, Artemis, and proteins locating to gamma-H2AX foci. *Molecular Cell, 16*, 715–724.

Rogakou, E. P., Pilch, D. R., Orr, A. H., Ivanova, V. S., & Bonner, W. M. (1998). DNA double-stranded breaks induce histone H2AX phosphorylation on serine 139. *The Journal of Biological Chemistry, 273*, 5858–5868.

Rosato, R. R., Almenara, J. A., & Grant, S. (2003). The histone deacetylase inhibitor MS-275 promotes differentiation or apoptosis in human leukemia cells through a process regulated by generation of reactive oxygen species and induction of p21CIP1/WAF1 1. *Cancer Research, 63*, 3637–3645.

Rosato, R. R., Almenara, J. A., Maggio, S. C., Coe, S., Atadja, P., Dent, P., et al. (2008). Role of histone deacetylase inhibitor-induced reactive oxygen species and DNA damage in LAQ-824/fludarabine antileukemic interactions. *Molecular Cancer Therapeutics, 7*, 3285–3297.

Rosato, R. R., Kolla, S. S., Hock, S. K., Almenara, J. A., Patel, A., Amin, S., et al. (2010). Histone deacetylase inhibitors activate NF-kappaB in human leukemia cells through an ATM/NEMO-related pathway. *The Journal of Biological Chemistry, 285*, 10064–10077.

Rosato, R. R., Maggio, S. C., Almenara, J. A., Payne, S. G., Atadja, P., Spiegel, S., et al. (2006). The histone deacetylase inhibitor LAQ824 induces human leukemia cell death through a process involving XIAP down-regulation, oxidative injury, and the acid sphingomyelinase-dependent generation of ceramide. *Molecular Pharmacology, 69*, 216–225.

Rossetto, D., Truman, A. W., Kron, S. J., & Cote, J. (2010). Epigenetic modifications in double-strand break DNA damage signaling and repair. *Clinical Cancer Research, 16*, 4543–4552.

Roth, D. B., Porter, T. N., & Wilson, J. H. (1985). Mechanisms of nonhomologous recombination in mammalian cells. *Molecular and Cellular Biology, 5*, 2599–2607.

Roth, D. B., & Wilson, J. H. (1986). Nonhomologous recombination in mammalian cells: Role for short sequence homologies in the joining reaction. *Molecular and Cellular Biology, 6*, 4295–4304.

Ruefli, A. A., Ausserlechner, M. J., Bernhard, D., Sutton, V. R., Tainton, K. M., Kofler, R., et al. (2001). The histone deacetylase inhibitor and chemotherapeutic agent suberoylanilide hydroxamic acid (SAHA) induces a cell-death pathway characterized by cleavage of Bid and production of reactive oxygen species. *Proceedings of the National Academy of Sciences of the United States of America, 98*, 10833–10838.

Saitoh, M., Nishitoh, H., Fujii, M., Takeda, K., Tobiume, K., Sawada, Y., et al. (1998). Mammalian thioredoxin is a direct inhibitor of apoptosis signal-regulating kinase (ASK) 1. *The EMBO Journal, 17*, 2596–2606.

Sallmyr, A., Tomkinson, A. E., & Rassool, F. V. (2008). Up-regulation of WRN and DNA ligase IIIalpha in chronic myeloid leukemia: Consequences for the repair of DNA double-strand breaks. *Blood*, *112*, 1413–1423.

Sancar, A., Lindsey-Boltz, L. A., Unsal-Kacmaz, K., & Linn, S. (2004). Molecular mechanisms of mammalian DNA repair and the DNA damage checkpoints. *Annual Review of Biochemistry*, *73*, 39–85.

Sanda, T., Okamoto, T., Uchida, Y., Nakagawa, H., Iida, S., Kayukawa, S., et al. (2007). Proteome analyses of the growth inhibitory effects of NCH-51, a novel histone deacetylase inhibitor, on lymphoid malignant cells. *Leukemia*, *21*, 2344–2353.

Sartori, A. A., Lukas, C., Coates, J., Mistrik, M., Fu, S., Bartek, J., et al. (2007). Human CtIP promotes DNA end resection. *Nature*, *450*, 509–514.

Sattler, M., Winkler, T., Verma, S., Byrne, C. H., Shrikhande, G., Salgia, R., et al. (1999). Hematopoietic growth factors signal through the formation of reactive oxygen species. *Blood*, *93*, 2928–2935.

Scharer, O. D., & Jiricny, J. (2001). Recent progress in the biology, chemistry and structural biology of DNA glycosylases. *BioEssays*, *23*, 270–281.

Scher, M. B., Vaquero, A., & Reinberg, D. (2007). SirT3 is a nuclear NAD+-dependent histone deacetylase that translocates to the mitochondria upon cellular stress. *Genes & Development*, *21*, 920–928.

Selvi, R. B., & Kundu, T. K. (2009). Reversible acetylation of chromatin: Implication in regulation of gene expression, disease and therapeutics. *Biotechnology Journal*, *4*, 375–390.

Sharma, G. G., So, S., Gupta, A., Kumar, R., Cayrou, C., Avvakumov, N., et al. (2010). MOF and histone H4 acetylation at lysine 16 are critical for DNA damage response and double-strand break repair. *Molecular and Cellular Biology*, *30*, 3582–3595.

Singleton, B. K., Torres-Arzayus, M. I., Rottinghaus, S. T., Taccioli, G. E., & Jeggo, P. A. (1999). The C terminus of Ku80 activates the DNA-dependent protein kinase catalytic subunit. *Molecular and Cellular Biology*, *19*, 3267–3277.

Sonoda, E., Zhao, G. Y., Kohzaki, M., Dhar, P. K., Kikuchi, K., Redon, C., et al. (2007). Collaborative roles of gammaH2AX and the Rad51 paralog Xrcc3 in homologous recombinational repair. *DNA Repair*, *6*, 280–292.

Sun, Y., Jiang, X., Chen, S., Fernandes, N., & Price, B. D. (2005). A role for the Tip60 histone acetyltransferase in the acetylation and activation of ATM. *Proceedings of the National Academy of Sciences of the United States of America*, *102*, 13182–13187.

Sun, Y., Jiang, X., & Price, B. D. (2010). Tip60: Connecting chromatin to DNA damage signaling. *Cell Cycle*, *9*, 930–936.

Sun, Y., Jiang, X., Xu, Y., Ayrapetov, M. K., Moreau, L. A., Whetstine, J. R., et al. (2009). Histone H3 methylation links DNA damage detection to activation of the tumour suppressor Tip60. *Nature Cell Biology*, *11*, 1376–1382.

Sun, Y., Xu, Y., Roy, K., & Price, B. D. (2007). DNA damage-induced acetylation of lysine 3016 of ATM activates ATM kinase activity. *Molecular and Cellular Biology*, *27*, 8502–8509.

Tao, R., Coleman, M. C., Pennington, J. D., Ozden, O., Park, S. H., Jiang, H., et al. (2010). Sirt3-mediated deacetylation of evolutionarily conserved lysine 122 regulates MnSOD activity in response to stress. *Molecular Cell*, *40*, 893–904.

Tsai, H. C., Li, H., Van Neste, L., Cai, Y., Robert, C., Rassool, F. V., et al. (2012). Transient low doses of DNA-demethylating agents exert durable antitumor effects on hematological and epithelial tumor cells. *Cancer Cell*, *21*, 430–446.

Uhl, M., Csernok, A., Aydin, S., Kreienberg, R., Wiesmuller, L., & Gatz, S. A. (2010). Role of SIRT1 in homologous recombination. *DNA Repair*, *9*, 383–393.

Ungerstedt, J. S., Sowa, Y., Xu, W. S., Shao, Y., Dokmanovic, M., Perez, G., et al. (2005). Role of thioredoxin in the response of normal and transformed cells to histone

deacetylase inhibitors. *Proceedings of the National Academy of Sciences of the United States of America, 102*, 673–678.
van der Horst, A., Tertoolen, L. G., de Vries-Smits, L. M., Frye, R. A., Medema, R. H., & Burgering, B. M. (2004). FOXO4 is acetylated upon peroxide stress and deacetylated by the longevity protein hSir2(SIRT1). *The Journal of Biological Chemistry, 279*, 28873–28879.
Wagner, J. M., Hackanson, B., Lubbert, M., & Jung, M. (2010). Histone deacetylase (HDAC) inhibitors in recent clinical trials for cancer therapy. *Clinical Epigenetics, 1*, 117–136.
Walker, J. R., Corpina, R. A., & Goldberg, J. (2001). Structure of the Ku heterodimer bound to DNA and its implications for double-strand break repair. *Nature, 412*, 607–614.
Wang, J., & Chen, J. (2010). SIRT1 regulates autoacetylation and histone acetyltransferase activity of TIP60. *The Journal of Biological Chemistry, 285*, 11458–11464.
Wang, H., Rosidi, B., Perrault, R., Wang, M., Zhang, L., Windhofer, F., et al. (2005). DNA ligase III as a candidate component of backup pathways of nonhomologous end joining. *Cancer Research, 65*, 4020–4030.
Wang, R. H., Sengupta, K., Li, C., Kim, H. S., Cao, L., Xiao, C., et al. (2008). Impaired DNA damage response, genome instability, and tumorigenesis in SIRT1 mutant mice. *Cancer Cell, 14*, 312–323.
Wang, M., Wu, W., Rosidi, B., Zhang, L., Wang, H., & Iliakis, G. (2006). PARP-1 and Ku compete for repair of DNA double strand breaks by distinct NHEJ pathways. *Nucleic Acids Research, 34*, 6170–6182.
Weichert, W., Roske, A., Gekeler, V., Beckers, T., Stephan, C., Jung, K., et al. (2008). Histone deacetylases 1, 2 and 3 are highly expressed in prostate cancer and HDAC2 expression is associated with shorter PSA relapse time after radical prostatectomy. *British Journal of Cancer, 98*, 604–610.
Wyatt, M. D., & Pittman, D. L. (2006). Methylating agents and DNA repair responses: Methylated bases and sources of strand breaks. *Chemical Research in Toxicology, 19*, 1580–1594.
Xie, A., Kwok, A., & Scully, R. (2009). Role of mammalian Mre11 in classical and alternative nonhomologous end joining. *Nature Structural & Molecular Biology, 16*, 814–818.
Xu, W., Ngo, L., Perez, G., Dokmanovic, M., & Marks, P. A. (2006). Intrinsic apoptotic and thioredoxin pathways in human prostate cancer cell response to histone deacetylase inhibitor. *Proceedings of the National Academy of Sciences of the United States of America, 103*, 15540–15545.
Xu, W. S., Parmigiani, R. B., & Marks, P. A. (2007). Histone deacetylase inhibitors: Molecular mechanisms of action. *Oncogene, 26*, 5541–5552.
Yamamori, T., DeRicco, J., Naqvi, A., Hoffman, T. A., Mattagajasingh, I., Kasuno, K., et al. (2010). SIRT1 deacetylates APE1 and regulates cellular base excision repair. *Nucleic Acids Research, 38*, 832–845.
Yang, X. J., & Seto, E. (2008). Lysine acetylation: Codified crosstalk with other posttranslational modifications. *Molecular Cell, 31*, 449–461.
Yuan, H., Wang, Z., Li, L., Zhang, H., Modi, H., Horne, D., et al. (2012). Activation of stress response gene SIRT1 by BCR-ABL promotes leukemogenesis. *Blood, 119*, 1904–1914.
Yuan, Z., Zhang, X., Sengupta, N., Lane, W. S., & Seto, E. (2007). SIRT1 regulates the function of the Nijmegen breakage syndrome protein. *Molecular Cell, 27*, 149–162.
Zhang, Z., Yamashita, H., Toyama, T., Sugiura, H., Ando, Y., Mita, K., et al. (2005). Quantitation of HDAC1 mRNA expression in invasive carcinoma of the breast. *Breast Cancer Research and Treatment, 94*, 11–16.

HDAC Inhibitor Modulation of Proteotoxicity as a Therapeutic Approach in Cancer

David J. McConkey[*,†,1], Matthew White[*,†], Wudan Yan[*,†]
*Department of Urology, U.T. M.D. Anderson Cancer Center, Houston, Texas, USA
†Department of Cancer Biology, U.T. M.D. Anderson Cancer Center, Houston, Texas, USA
[1]Corresponding author: e-mail address: dmcconke@mdanderson.org

Contents

Abstract

The strong clinical activity of the proteasome inhibitor bortezomib (Velcade) in multiple myeloma and other hematological malignancies has focused considerable attention on its mechanisms of action. Although NFκB inhibition was initially the mechanism in focus, accumulating evidence indicates that misfolded protein accumulation leading to proteotoxicity plays an even more important role in cell killing. Proteotoxicity that occurs as a consequence of protein aggregate accumulation has long been associated with the development of neurodegenerative diseases, and a large and growing body of literature has documented how protein aggregates are handled and disposed of via evolutionarily conserved mechanisms involving cross talk between the proteasome and autophagy in normal cells. The type II histone deacetylase HDAC6 plays important roles in these processes and HDAC6 inhibition enhances proteotoxicity. These observations served as the basis for the development of HDAC6-specific chemical inhibitors that are now being evaluated in combination with proteasome inhibitors in preclinical models. Nonetheless, there is also strong evidence that the more classical, chromatin-associated (type I) HDACs are also involved in the regulation of proteotoxicity, although the

Advances in Cancer Research, Volume 116
ISSN 0065-230X
http://dx.doi.org/10.1016/B978-0-12-394387-3.00004-5

biochemical mechanisms underlying their effects are not well defined. Importantly, emerging evidence indicates that subsets of tumor cells contain defects in these protein quality control pathways, which may underlie their vulnerability to proteasome inhibitor-induced death. In addition, our clearer understanding of cytoprotective protein quality control responses is identifying novel candidate targets for therapeutic intervention. In this chapter, we present an overview of protein quality control mechanisms in normal tissues and describe how this information is informing our development of proteasome inhibitors and other agents that impact upon these pathways for cancer therapy.

1. BACKGROUND: THE CELLULAR RESPONSE TO PROTEOTOXICITY

1.1. Proteotoxicity and endoplasmic reticulum stress

The protein aggregation that occurs as a consequence of errors during protein translation and folding and in response to endogenous and exogenous stress can produce significant cytotoxicity. The term "proteotoxicity" was originally coined to describe conditions associated with damage to proteins (i.e., exposure to heat, oxidants, certain amino acid analogs, puromycin, ethanol, heavy metals, arsenicals, tissue explantation, and exposure to viruses, among others) (Hightower, 1991; Morimoto, 1993, 2011; Westerheide & Morimoto, 2005). Proteins become vulnerable to aggregation when hydrophobic domains that are normally buried within their cores become surface exposed (Goldberg, 2003). This results in rapid recruitment of specific molecular chaperones (heat shock proteins (HSPs) and their orthologs) that bind to these exposed hydrophobic regions and promote protein refolding or degradation. Organelle-specific paralogs of the oldest and most highly conserved HSP, HSP70, appear to play central roles in reacting to misfolded protein stress and in coordinating downstream response pathway activation (Ahn, Kim, Yoon, & Vacratsis, 2005; Bertolotti, Zhang, Hendershot, Harding, & Ron, 2000; Daugaard, Rohde, & Jaattela, 2007; Kaser & Langer, 2000; Shi, Mosser, & Morimoto, 1998).

Among the various examples of proteotoxicity, the cellular response to heat shock was the first to be studied extensively. The heat shock response is among the most highly conserved pathways in existence and is present in organisms ranging from bacteria through plants to vertebrates (Hightower, 1991). At the core of the heat shock response are transcription factors (σ^{32} in *Escherichia coli*, heat shock factor-1 (HSF1) in vertebrates) that are activated by unfolded proteins and/or protein aggregates and mediate a cytoprotective response that promotes the rapid induction of protein chaperones (HSPs) that bind to these damaged proteins and either promote their refolding or target them for degradation (Hightower, 1991; Morimoto, 1993, 2011). In unstressed cells, HSF1 is maintained in a poised but inactive state via binding to the protein chaperones (HSPs) HSP90 and HSC70 (Morimoto, 1993) (the constitutively expressed relative of the inducible HSP70). Denatured proteins appear to compete with HSF1 for binding to HSC70 and HSP90, resulting in HSF1 release, multimerization, and activation via posttranslational modification (phosphorylation, sumoylation, and acetylation) (Morimoto, 1993, 2011). The inducible form of HSP70 (formally known as HSP72) is one of HSF1's target genes, and HSP72 can also bind HSF-1 and inhibit its activity, thereby resulting in feedback inhibition of the heat shock response (Westerheide & Morimoto, 2005).

HSC70, HSP70, and HSP90 all appear to localize primarily within cytoplasm and nucleus, but homologues of these proteins are found constitutively within mitochondria and the endoplasmic reticulum (ER), where they also play critical cytoprotective roles in the response to proteotoxic stress (Burbulla et al., 2010 Czarnecka, Campanella, Zummo, & Cappello, 2006; Kopito, 2000; Siegelin et al., 2011; Webster, Naylor, Hartman, Hoj, & Hoogenraad, 1994). Mitochondria are an important source of intracellular reactive oxygen species (ROS), so mitochondrial proteins may be particularly vulnerable to oxidant damage. Recent studies indicate that tumor cell mitochondria contain a mitochondrial form of HSP90 and the HSP90 homologue TNF receptor-associated protein-1 (TRAP-1) (Kang et al., 2007). There, they interact with cyclophilin D, a component of the mitochondrial permeability transition (PT) pore, opening of which can lead to cytochrome c release and caspase activation (Kang et al., 2007). Investigators have developed small molecule inhibitors of HSP90 (gamitrinibs) that can disrupt these interactions, triggering rapid cytochrome c release and death in a variety of different tumor models (Kang & Altieri, 2009; Kang et al., 2007, 2009; Siegelin et al., 2011). They have suggested that HSP90 and TRAP-1 play important roles in regulating a mitochondrial "unfolded protein response" (UPR) that would promote apoptosis and autophagy when levels of misfolded proteins within the organelle become excessive (Siegelin et al., 2011). Importantly, whether cytotoxicity results from a generalized buildup of misfolded proteins or more specifically from cyclophilin D dysregulation leading to permeability transition is not clear. Other chaperones (i.e., mtHSP70/mortalin and HSP60) (Burbulla et al., 2010; Czarnecka et al., 2006; Webster et al., 1994) must also be important but precisely how the control the acute response to mitochondrial proteotoxicity has not been established.

The more familiar UPR is activated by protein folding problems within the ER, a phenomenon that is commonly referred to as "ER stress" (Lee & Hendershot, 2006; Ron & Hubbard, 2008; Ron & Walter, 2007; Szegezdi, Lobgue, Gorman, & Samali, 2006; Tabas & Ron, 2011; Walter & Ron, 2011). Perhaps the most important immediate consequence of ER stress-induced UPR activation is to rapidly and dramatically decrease bulk protein synthesis and to inhibit cell cycle progression to prevent exacerbation of proteotoxicity (Wek & Cavener, 2007). Subsequent effects include stimulating increases in ER protein chaperone levels (to promote refolding and/or removal of existing protein aggregates) and inducing expression of genes involved in transfer of protein aggregates out of the ER to the proteasome and proteasome- and autophagy-mediated protein degradation. Finally, the UPR can promote apoptosis if cytoprotective mechanisms are overwhelmed (Szegezdi et al., 2006). All of these effects are mediated by three parallel signaling pathways that are controlled by the ER-localized HSP70 paralog, glucose-regulated protein of 78 kDa (GRP78, also known as BiP) (Ron & Walter, 2007). As HSC70 does with HSF-1 in the cytosol and nucleus, GRP78 associates with the ER luminal domains of the apical components of the three pathways (PERK, IRE1, and ATF6), maintaining them in "poised" but inactive conformations (Szegezdi et al., 2006). Misfolded proteins compete with GRP78 for binding to these signaling intermediates (Ron & Walter, 2007). Once GRP78 leaves them three, they dimerize, become activated, and stimulate downstream cytoprotective and/or apoptotic responses (Ron & Walter, 2007).

Decreasing bulk protein synthesis and blocking cell cycle progression are the most critical, immediate cytoprotective functions of the UPR, because cycling cells have higher rates of translation than resting cells do, and ongoing translation is the most direct input that can exacerbate proteotoxicity (Fig. 4.1). During ER stress, both are accomplished via release of the ER-resident serine/threonine kinase, PERK, from GRP78 and PERK-mediated phosphorylation of the translation initiation factor eIF2α at S51 (S52 in the murine protein) (Fig. 4.1; Harding, Zhang, Bertolotti, Zeng, & Ron, 2000; Harding, Zhang, & Ron, 1999; Shi et al., 1998). This results in a dramatic decrease in global, cap-dependent translation, reducing rates of bulk protein synthesis by over 70% in comparison to baseline levels. The transcript encoding cyclin D is extremely sensitive to increased eIF2α phosphorylation (Brewer & Diehl, 2000), and the cyclin D protein also has

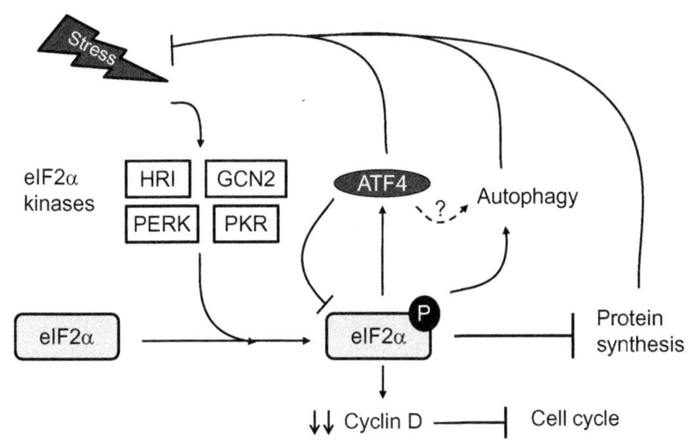

Figure 4.1 Stress-induced phosphorylation of eIF2α. In response to sufficient stress, one of the four known eIF2α kinases becomes activated and phosphorylates the alpha subunit of eIF2 at the S51 residue. This serves to block cap-dependent protein synthesis that underlies the expression of the bulk of proteins within the cell and selectively upregulates ATF4, a transcription factor required for coordinating the integrated stress response (ISR). Together, these two arms—inhibition of protein synthesis and increased translation of ATF4—serve to help the cell alleviate and recover from the stress.

a very short half-life, so as a result, cyclin D protein levels drop very rapidly following UPR activation, and experiments with mouse embryonic fibroblasts (MEFs) expressing a knocked-in, phosphorylation-defective (S51A) form of eIF2α have confirmed that phosphorylation is required for cyclin D downregulation and cell cycle arrest (Hamanaka, Bennett, Cullinan, & Diehl, 2005). Phosphorylation of eIF2α also directly promotes the activation of autophagy (Talloczy et al., 2002; Zhu, Dunner, & McConkey, 2010), although the molecular mechanisms involved have not been precisely defined. In addition, hypoxia also activates PERK (Bi et al., 2005; Blais et al., 2004; Fels & Koumenis, 2006; Koritzinsky et al., 2006; Koumenis et al., 2002; Liu et al., 2006; Wouters et al., 2005), although the biochemical mechanisms involved are still being elucidated.

As most protein synthesis decreases a subset of transcripts is translated at higher levels (Harding, Nova, et al., 2000). Among these, the one that has been studied most extensively is the C/EBP family transcription factor, ATF4 (Fig. 4.1). ATF4 promotes the expression of GRP78 and a variety of other genes that ameliorate proteotoxic and oxidative stress (Harding et al., 2003). However, ATF4 also promotes the expression of GADD153 (CHOP), another C/EBP family transcription factor that has been implicated in ER stress-induced apoptosis (Tabas & Ron, 2011; Zhang et al., 2003).

PERK is a member of a larger family of four homologous serine/threonine kinases that includes the virus- and interferon-regulated kinase PKR (Meurs et al., 1990), the metabolism-sensitive kinase GCN2 (Harding, Novoa, et al., 2000; Sood, Porter, Olsen, Cavener, & Wek, 2000), and the iron-regulated kinase HRI (Chen et al., 1991). Like PERK, all of the other members of the family phosphorylate the S51 position of eIF2α and reduce global protein synthesis (Fig. 4.2; Franco, Hogg, & Martelo, 1981; Gross, Olin, Hessefort, & Bender, 1994; Gross, Rynning, & Knish, 1981; Harding, Novoa, et al., 2000; Kramer, Henderson, Pinphanichakarn, Wallis, & Hardesty, 1977; Mellor, Flowers, Kimball, & Jefferson, 1994). This activity of PKR plays a critical role in preventing the replication and promoting the autophagic destruction of intracellular DNA viruses (Talloczy et al., 2002), and one of the primary functions of GCN2 is to "sense" intracellular amino acid pools; the kinase is activated by direct binding to uncharged tRNAs (Ramirez, Wek, & Hinnebusch, 1991; Wek, Jackson, & Hinnebusch, 1989; Wek, Zhu, & Wek, 1995), so when free amino acid levels decline below a critical threshold, GCN2 promotes translational arrest by phosphorylating eIF2α (Harding, Novoa, et al.,

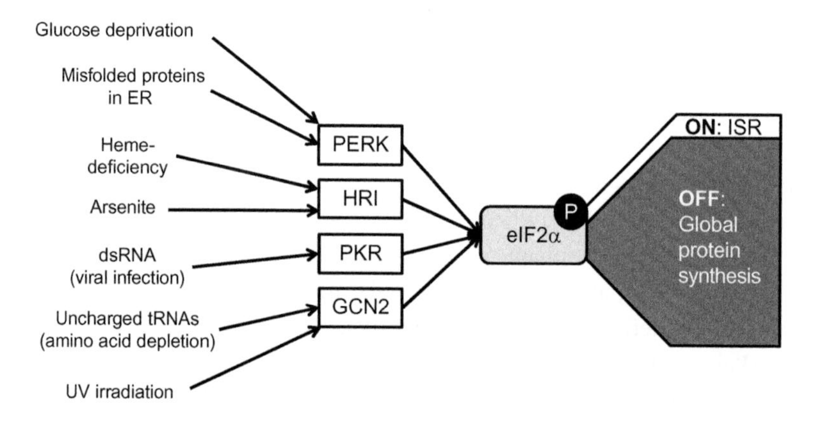

Figure 4.2 Upstream stimuli that induce eIF2α phosphorylation. Phosphorylation of eIF2α represents a major node in the ISR by coordinating translational arrest with cell cycle arrest, autophagy, and ATF4 induction. The diverse upstream signals that converge on the four known eIF2α kinases are shown. It should be emphasized that the upstream signals that activate the pathway may involve more than one type of stress, and therefore more than one eIF2a kinase may contribute to eIF2α phosphorylation. This certainly appears to be the case with proteasome inhibitors, where the acute effects appear to primarily involve GCN2/cytosolic stress but expand to include PERK/ER stress as well.

2000). GCN2 is also activated by UV irradiation (Deng et al., 2002), but whether or not this is tied to uncharged tRNAs is not clear. Finally, heme deficiency and metals such as arsenic activate HRI (Lu, Han, & Chen, 2001; McEwen et al., 2005; Wehner, Schutz, & Sarnow, 2010; Zhan et al., 2002). The fact that diverse stimuli channeled through four different kinases all converge on eIF2α has prompted many investigators to prefer the term "integrated stress response" (ISR) to UPR to reflect the fact that unfolded proteins are not the only stimuli that stimulate these effects.

A second arm of the UPR is controlled by ATF6, a transcription factor that interacts with so-called ER stress response elements to drive the expression of GRP78, other ER chaperones, and CHOP (Hetz, 2012; Szegezdi et al., 2006). In addition, active ATF6 promotes the expression of the precursor form of XBP1 (Szegezdi et al., 2006), which is an important component of the third arm of the UPR that is controlled by IRE1. ATF6 normally resides in the ER, but upon release from GRP78, it migrates to the Golgi apparatus, where it is cleaved by proteases to become active and then translocates to the nucleus.

The third and most ancient arm of the UPR is based on activation of the dual-function protein, IRE1 (Cox, Shamu, & Walter, 1993). IRE1 functions as a serine/threonine kinase whose only confirmed substrate is itself. Trans-autophosphorylation promotes its second activity as an endoribonuclease that splices the mRNA encoding the XBP1 transcription factor, enabling it to become a functional protein (Yoshida, Matsui, Yamamoto, Okada, & Mori, 2001). Interestingly, IRE1 has been shown to interact directly with proapoptotic members of the BCL-2 family of apoptosis regulators (Hetz et al., 2006), and it also physically associates with apoptosis signaling kinase-1 via TRAF2 to promote activation of the proapoptotic Jun N-terminal kinase (JNK) kinase (Nishitoh et al., 2002). Nonetheless, recent studies have demonstrated that XBP1 controls the expression of hundreds of stress-responsive genes involved in cytoprotective responses that circumvent lethal cell injury (Thibault, Ismail, & Ng, 2011). Therefore, like the other arms of the UPR, IRE1–XBP1 signaling may serve a primarily cytoprotective role in cells exposed to ER stress.

1.2. Misfolded proteins: Routes of disposal

The proteasome is a barrel-shaped structure containing 14 dimeric enzymatic subunits capped on either end by two 19S complexes (Adams, 2004; Goldberg, 2003). The expression of many important signal transduction and cell cycle intermediates (IκBα, HIF1, p53, p21, cyclin D, etc) is controlled via proteasome-mediated degradation following recognition by specific E3 ubiquitin ligases, of which there are probably hundreds in mammalian cells (Petroski & Deshaies, 2005). Ubiquitin conjugation allows for binding of substrates to the 19S cap complexes, which bind the attached ubiquitin chains and remove them and then unwind the target proteins via an ATP-dependent mechanism in order to feed them into the 20S catalytic core (Smith, Benaroudj, & Goldberg, 2006; Smith et al., 2007; Smith, Fraga, Reis, Kafri, & Goldberg, 2011). The selectivity that characterizes this type of proteasome-mediated protein degradation is engendered by the posttranslational modifications that promote substrate recognition by the substrate's corresponding E3 ligase (i.e., IKK-dependent phosphorylation on S32 and S36 in the case of IκBα, or proline hydroxylation in the case of VHL). Proteasome inhibitors will block the degradation of these short-lived proteins, so it seems reasonable to assume that inhibiting the proteasome would have broad effects on signal transduction and cell cycle control.

The proteasome is also the first line of defense against the buildup of misfolded proteins (Fig. 4.3; Goldberg, 2003). Ubiquitylation still serves as the mechanism that targets misfolded proteins to the proteasome for

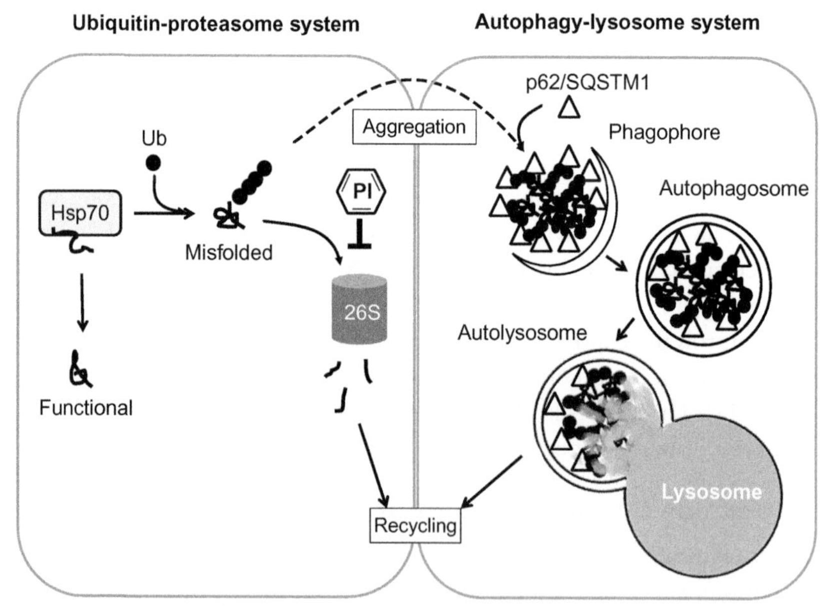

Figure 4.3 Control of misfolded protein degradation by the proteasome and autophagy. In response to chaperone-mediated ubiquitylation, misfolded protein monomers are first sent to the proteasome for degradation. However, when chaperone levels and/or proteasome capacity is overwhelmed, p62 and NBR1 promote aggregate formation and serve as adaptors that mediate binding of ubiquitylated proteins to central components of the autophagy machinery.

degradation, but the recognition mechanism is highly promiscuous and allows for the targeting of any misfolded protein that originates within the cell, even if it is of viral or bacterial origin. The biochemical mechanism involved is elegant in its simplicity and complements the mechanisms that lead to activation of HSF-1 or the UPR. The protein chaperones that are attracted to misfolded proteins carry with them broad-specificity ubiquitin ligases, known as C-terminal HSC70-interacting protein (CHIP) (Connell et al., 2001; McDonough & Patterson, 2003; Murata, Chiba, & Tanaka, 2003) and Parkin (Chin, Olzmann, & Li, 2010; Imai et al., 2002), which promote ubiquitylation of the bound clients. Therefore, rather than being stimulated by changes in protein posttranslational modification, misfolded protein substrate ubiquitylation is facilitated by their interactions with HSPs.

Misfolded proteins that accumulate within the ER are also degraded by the proteasome via a process that is known as ER-associated degradation, or ERAD (Smith, Ploegh, & Weissman, 2011). However, in order to reach the proteasome, misfolded ER proteins must first be retrotranslocated to the cytosol. A protein complex consisting of specific transport proteins (the Derlins) (Eura et al., 2012; Oda et al., 2006) and a transmembrane E3 ligase (HRD1) (Bays, Gardner, Seelig, Joazeiro, & Hampton, 2001) coordinates protein transfer and ubiquitylation. Predictably, there is evidence that GRP78 associates with HRD1 during this process (Hosokawa et al., 2008), strongly suggesting that it works in a very similar manner to HSP70/CHIP-dependent ubiquitylation in the cytosol.

Importantly, because the proteasome's core is spatially restrictive, it can only degrade protein monomers, and only after they have been unwound via an energy-dependent mechanism (Goldberg, 2003). Because of these structural restrictions, protein aggregates can actually block proteasome function (Bennett, Bence, Jayakumar, & Kopito, 2005; Bennett et al., 2007; Hipp et al., 2012), and this proteasome inhibition (rather than more direct cytotoxic effects of protein aggregates) may be the ultimate cause of cell death in neurodegenerative diseases. Proteasome inhibition occurs because these polyubiquitylated aggregates can still interact with ubiquitin receptors in the proteasome's cap complexes, but they are much more resistant to unwinding and cannot be as readily inserted into the 20S core. Therefore, when chaperone levels become overwhelmed and proteins start to form larger aggregates, they are redirected to lysosomes for degradation via autophagy (Fig. 4.3; Kirkin, McEwan, Novak, & Dikic, 2009).

Macroautophagy (hereafter referred to as autophagy) is ideally suited as a complement to the proteasome in misfolded protein degradation because it can accommodate very large substrates, including whole organelles (mitochondria, peroxisomes) and intracellular pathogens (bacteria) (Mizushima, Levine, Cuervo, & Klionsky, 2008). Autophagy involves the encapsulation of targets within double-membrane structures that then merge with lysosomes, resulting in hydrolysis of their contents (Mizushima et al., 2008). This process is most familiar for its role as an evolutionarily conserved survival mechanism during periods of starvation, where indiscriminate digestion of intracellular macromolecules allows cells to "recycle" them to recreate ATP. However, in this context, ubiquitin appears to serve as a specific moiety that allows protein aggregates to be selectively targeted for degradation (Kirkin, McEwan, et al., 2009).

The critical importance of autophagy in homeostatic clearance of protein aggregates has been clearly demonstrated in mice in which autophagy pathway genes (ATG5, ATG7) have been ablated in the brain; these mice develop protein inclusions and neurodegenerative disease that are very similar to what is observed in patients with Alzheimer's or Parkinson's diseases (Hara et al., 2006; Klionsky, 2006; Komatsu et al., 2006). Indeed, inhibition of the proteasome results in the upregulation of ATG5 and ATG7 via a phospho-eIF2α-dependent mechanism (Zhu et al., 2010; Zhu, Chan, Heymach, Wilkinson, & McConkey, 2009). Inhibiting autophagy with chemical inhibitors or ATG5/7 knockdown results in increased formation of ubiquitin-positive inclusion bodies (protein aggregates) and increased cell death (Zhu et al., 2010, 2009).

Strikingly, it appears that ubiquitylation is also critically important for the recognition of protein aggregates (and probably organelles and intracellular pathogens) for degradation via autophagy (Kirkin, McEwan, et al., 2009). When proteasomal capacity becomes overwhelmed, the ubiquitylated protein aggregates that would normally interact with ubiquitin receptors within the proteasome's 19S cap complex become available to interact with two bifunctional proteins (p62/SQSTM1 and NBR1) that function as bridges between the protein aggregates and membrane-associated components of the autophagic degradation machinery (ATG12 and ATG8 in yeast, known as LC3 and GABARAP in mammalian cells) (Kirkin, Lamark, et al., 2009; Kirkin, McEwan, et al., 2009). Both p62 and NBR1 contain C-terminal LC3-interacting regions and ubiquitin-associated domains that enable them to simultaneously interact with LC3/GABARAP and ubiquitylated protein aggregates, respectively (Kirkin, McEwan, et al., 2009). They also contain N-terminal Phox and Bem1p domains that promote their oligomerization, which is also critical for the formation of visible intracellular protein aggregates and for autophagy (Kirkin, McEwan, et al., 2009). During the process of aggregate transfer, p62 itself is degraded via autophagy, so decreased p62 expression is often used as a surrogate for autophagic protein flux (Klionsky et al., 2008). p62 is also interchangeable with ubiquitin as a marker of intracellular protein aggregate accumulation, and their expression is entirely overlapping in cells exposed to proteasome inhibitors or in preclinical models of neurodegenerative disease.

1.3. HDAC6 and aggresome formation

The aggresome appears to be the product of the most extreme form of protein aggregation that is observed in cells whose proteasome function is chronically blocked (Kopito, 2000). Central to the process of aggresome formation is the type II "histone" deacetylase, HDAC6 (Kawaguchi et al., 2003). Two HDAC catalytic domains and a C-terminal BUZ domain that can directly interact with ubiquitin are important characteristics of the structure of HDAC6 (Kirkin, McEwan, et al., 2009; Matthias, Yoshida, & Khochbin, 2008). The protein also interacts directly with tubulin-associated dynein motor proteins, and its interaction with dynein is critical for transfer of protein aggregates to juxtanuclear microtubule organizing centers, where aggresomes are invariably found in cells that contain them (Kopito, 2000). Importantly, HDAC6 does not contain any domains that directly interact with the autophagic machinery (Kirkin, McEwan, et al., 2009), and it is not required for autophagy activation (Lee et al., 2010). However, it does appear to be required for protein aggregate degradation (Iwata, Riley, Johnston, & Kopito, 2005; Pandey et al., 2007). Recent work has implicated HDAC6 in the fusion of autophagosomes to lysosomes (Lee et al., 2010), which is ultimately required for protein aggregate degradation.

Identifying the substrates of HDAC6 that underlie its effects on aggresome formation and autophagy is a high priority in ongoing research efforts. The best-described HDAC6 substrate is tubulin itself (Hubbert et al., 2002). Tubulin acetylation appears to be associated with rigid microtubules and is a diagnostic feature of primary cilia, vestigial surface structures that play important roles in Sonic hedgehog signaling (Loktev et al., 2008; Michaud & Yoder, 2006; Quinlan, Tobin, & Beales, 2008; Simpson, Kerr, & Wicking, 2009). Tubulin deacetylation is critical for microtubule dynamics and cell motility, and HDAC6 is therefore also being implicated in cancer cell migration, invasion, and epithelial-to-mesenchymal transition (EMT) (Lafarga, Aymerich, Tapia, Mayor, & Penela, 2012; Penela et al., 2012; Shan et al., 2008; Valenzuela-Fernandez, Cabrero, Serrador, &

Sanchez-Madrid, 2008). HDAC6 and tubulin acetylation status also play an important roles in surface receptor endocytosis as has recently been demonstrated for the epidermal growth factor receptor (Deribe et al., 2009; Gao, Hubbert, & Yao, 2010). The other HDAC6 substrate that might play an important role in regulating proteotoxicity is HSP90, whose chaperone function is inhibited by HDAC6-mediated acetylation (Bali et al., 2005).

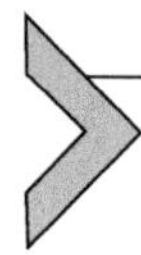

2. PROTEOTOXICITY AS A THERAPEUTIC TARGET IN CANCER

Cancer cells display much higher rates of protein synthesis than do their normal counterparts, probably because increases in proliferation must be associated with increased translation. The PI3 kinase–AKT–mTOR/p70S6 kinase pathway is almost always upregulated in cancer cells, and increased translation is a prominent downstream effect of PI3 kinase/AKT pathway activation. In addition, cancer cells generally express increased ROS levels (particularly if they express mutant Ras; Xu, Trepel, & Neckers, 2011), and ROS are important inducers of protein misfolding and aggregation. Cancers derived from normal cell types that are characterized by well-developed ER/Golgi networks and high secretory capacity (i.e., multiple myeloma (MM) and pancreatic cancer cells) may be particularly vulnerable to disruption of normal protein quality control mechanisms and proteotoxicity. Therefore, like DNA damaging agents, proteotoxic compounds may be selectively toxic to cancer cells, and in particular cancer cells derived from tissues characterized by high secretory capacities (McConkey & Zhu, 2008). Along with their more familiar effects on chromatin, gene expression, and differentiation, HDAC inhibitors exert powerful effects on protein quality control. A wide variety of different anticancer drugs increase ROS levels, and these increases probably contribute to proteotoxicity. However, a discussion of the possible involvement of proteotoxicity in the effects of all of the drugs that cause increased ROS production would expand any discussion well beyond the scope of this review. Therefore, we focus on those drugs that seem to have the most direct and obvious effects on protein quality control, but the concepts can probably be generalized to any stimulus that affects the UPR/ISR, the proteasome, and autophagy.

2.1. Proteasome inhibitors

PS-341 (bortezomib, also known as Velcade) was the first proteasome inhibitor to undergo clinical development for the treatment of cancer (Adams, 2004). Bortezomib is a reversible, peptide boronate inhibitor that must be infused by intravenous injection. The initial scientific rationale for its development was that it blocks degradation of many short-lived proteins that inhibit cancer proliferation, particularly p21, p53, and IκBα (McConkey & Zhu, 2008). The latter was of particular interest because IκBα is the natural inhibitor of the inflammation-associated transcription factor, NFκB, and NFκB promotes cancer cell proliferation, invasion, and survival (McConkey & Zhu, 2008). Furthermore, cancer chemo- and radiotherapy activate NFκB, which limits their cytotoxic effects (Cusack, Liu, & Baldwin, 2000). Therefore, combining bortezomib with conventional therapies was an attractive therapeutic approach. Preclinical studies confirmed that bortezomib inhibited the growth of various solid tumor xenografts alone and in combination with conventional chemotherapy. The effects of bortezomib were also associated with strong inhibition of tumor VEGF production and angiogenesis (Nawrocki et al., 2002; Sunwoo et al., 2001), resulting from downregulation of HIF-1α (Zhu, Chan, Heymach, Wilkinson, & McConkey, 2009). However, NFκB inhibition does not appear to account for the bulk of bortezomib's antitumoral effects (Hideshima et al., 2002, 2009), and bortezomib-based combination therapy with conventional agents has not produced much activity in clinical trials in solid tumors to date, perhaps because bortezomib was used in unselected patients, but perhaps also because the preclinical rationale for developing these combinations was not based on a complete understanding of bortezomib's primary mechanisms of action. Instead, bortezomib produced single agent activity in a subset of patients with MM (Richardson et al., 2003), strongly suggesting that there is something unique about the biology of MM that makes these tumors exquisitely vulnerable to proteasome inhibitor action. Subsequent preclinical studies revealed that bortezomib kills MM cells by inducing proteotoxic (ER) stress (Hideshima et al., 2005; Obeng et al., 2006), providing an attractive explanation for the uniquely strong clinical activity that was observed. Specifically, MM cells are characterized by massive secretion of monoclonal IgM antibody ("M" protein), which serves as the primary serum biomarker that is used to detect the presence of the disease in patients, and laboratory studies confirmed that IgM secretion is directly tied to MM sensitivity to bortezomib-induced cell death (Meister et al., 2007). Based on the success of bortezomib, several other, structurally distinct proteasome inhibitors are now in late preclinical to late clinical development. These compounds include the lactacystin-like molecule NPI-0052 (marizomib) and PR-171 (carfilzomib), which also have effects on the proteasome's three enzymatic activities that are distinct from those produced by bortezomib (Chauhan, Hideshima, & Anderson, 2006; Demo et al., 2007; Moreau et al., 2012; Potts et al., 2011; Ruschak, Slassi, Kay, & Schimmer, 2011).

Subsequent studies were initiated in preclinical models to identify agents that would enhance bortezomib's cytotoxic activity. Several groups discovered that pan-selective HDAC inhibitors (i.e., SAHA/vorinostat) were strongly synergistic with bortezomib in MM cells and other cell types (Dai et al., 2008; Dasmahapatra et al., 2010, 2011; Grant, 2008; Grant & Dent, 2007; Hideshima et al., 2005; Nawrocki et al., 2006; Pei, Dai, & Grant, 2004). Biochemical analyses revealed that SAHA augmented bortezomib-induced ROS production and activation of JNK and that both contributed to cell death. However, given the proteasome's central role in protein quality control, investigators examined whether bortezomib also increased proteotoxic (ER) stress (Nawrocki, Carew, Dunner, et al., 2005; Nawrocki, et al., 2008; Nawrocki, Carew, Pino, et al., 2005; Obeng et al., 2006). Studies concluded that bortezomib-induced increases in downstream targets of the UPR (GRP78 and CHOP), ER dilation, and ER Ca^{2+} mobilization, all of which strongly suggested the involvement of ER stress. In addition, and consistent with previous findings in models of neurodegenerative disease, bortezomib-induced aggresome formation in some cell lines, and aggresomes were disrupted in cells exposed to bortezomib plus HDAC inhibitors (Hideshima et al., 2005; Nawrocki et al., 2006), presumably because of the well-established role of HDAC6 in aggresome formation (Fig. 4.4; Kawaguchi et al., 2003). The effects of pan-selective HDAC inhibitors could be mimicked by the HDAC6-selective compound, tubacin (Schreiber, Anderson PNAS) (Haggarty, Koeller, Wong, Grozinger, & Schreiber, 2003) and direct silencing of HDAC6 (Hideshima et al., 2005; Nawrocki et al., 2006), prompting several companies to initiate programs to develop selective small molecule HDAC6 inhibitors as proteotoxic compounds for cancer therapy. Using tubacin as a platform, Acetylon, Inc. (Cambridge, MA) is the first company to report on the effects of a small molecule HDAC6-selective inhibitor, which produced synergistic cell killing when it is combined with proteasome inhibitors in MM cells *in vitro* and *in vivo* (Santo et al., 2012).

Importantly, HDAC inhibitors are also clinically available that selectively target the class I (chromatin associated) HDACs 1–3. Strikingly, one of these drugs (MS-275, now known as SNDX275) produces synergistic cytotoxicity when it is combined with bortezomib or other proteasome inhibitors

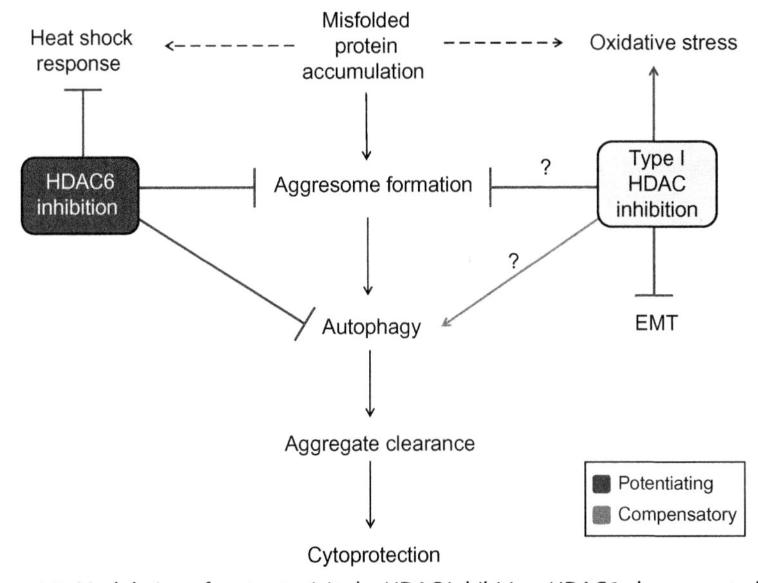

Figure 4.4 Modulation of proteotoxicity by HDAC inhibition. HDAC6 plays a central role in aggresome formation, a cytoprotective response that facilitates protein aggregate degradation via autophagy. In addition, recent work from our laboratory indicates that type I HDACs also play role(s) in aggresome formation, either directly or via indirect effects on chromatin/gene expression. Aside from their effects on misfolded protein clearance, type I HDAC inhibitors also increase ROS production and promote EMT reversal, effects that can more broadly enhance the cytotoxicity of other agents. Conversely, proteasome inhibitors downregulate the expression of type I HDACs and HDAC6 (not shown), so the enhanced cytotoxicity observed in cells exposed to combinations of proteasome, and HDAC inhibitors could be more closely related to the cytotoxic effects of HDAC inhibitors (including TRAIL production) rather than to proteotoxicity in some cell types. (See Page 6 in Color Section at the back of the book.)

(Jona et al., 2011; Miller et al., 2007). We have confirmed that SNDX275 has no effect on tubulin acetylation at concentrations that are higher than are required to increase bortezomib-induced apoptosis (W. Yan, manuscript in preparation), indicating that HDAC6 is not involved in its effects. Nonetheless, SNDX275 still interferes with aggresome formation (W. Yan et al., manuscript in preparation), strongly suggesting that the type I HDACs also play important, unappreciated roles in protein quality control (Fig. 4.4). Whether these effects are direct or indirect (i.e., via very rapid changes in gene expression) is still not clear, although any direct mechanism must account for the fact that the type I HDACs are thought to reside primarily within the nucleus.

Given their broad biological effects on cells, it is not surprising that alternative explanations have been advanced to explain the synergy between proteasome and HDAC inhibitors (Fig. 4.4). Proteasome inhibitors downregulate HDAC1 (Kikuchi et al., 2010; McConkey, 2010) and HDAC6 (W. Yan, manuscript in preparation), so it is also possible that the combinations are synergistic because proteasome inhibitors are modulating the cytotoxic effects of HDAC inhibitors, which include increased expression of TRAIL (Nebbioso et al., 2005), and proteasome inhibitors synergize strongly with TRAIL to produce apoptosis in a variety of different cancer cells (Lashinger et al., 2005; Sayers et al., 2003). Proteasome and HDAC inhibitors both induce ROS (Pei et al., 2004), so the higher levels of ROS produced by the combination could directly promote cytochrome c release and other events independently of effects on protein aggregation (Fig. 4.4). HDAC inhibitors can activate NFκB (via p65 acetylation) (Dai et al., 2008; Dai, Rahmani, Dent, & Grant, 2005), which might limit their cytotoxic activities, and proteasome inhibitors might block these cytoprotective responses. Finally, a developmental program known as EMT that plays important roles in the biology of cancer stem cells and metastatic progression is an important mechanism of global drug resistance (Kalluri & Weinberg, 2009; Roussos et al., 2010). HDAC inhibitors can inhibit or reverse EMT and promote drug sensitivity (Fig. 4.4; Witta et al., 2006).

Human cancer cell lines display marked heterogeneity in their sensitivities to proteasome inhibitor-induced apoptosis (Kamat et al., 2004; Nawrocki et al., 2002). Defining the biological basis of sensitivity and resistance could lead to the development of biomarkers that could be used to prospectively identify the subsets of MM and potentially other patients who would most benefit from proteasome inhibitor therapy and resistance pathways that could become important therapeutic targets in the future. In our hands, the patterns of proteasome inhibitor sensitivity and resistance are indistinguishable in cells exposed to the three major clinically available chemical inhibitors (bortezomib, marizomib, or carfilzomib), suggesting the involvement of shared biology/class effects rather than drug-specific differences. Given the central role of protein synthesis in proteotoxicity, we speculated that variability in basal rates of translation might contribute to this heterogeneity, and using cells engineered to express a conditional Myc gene, we confirmed that oncogene activation was associated with both increased translation and increased sensitivity to bortezomib (Nawrocki et al., 2008). The translation inhibitor cycloheximide is a universal inhibitor of bortezomib-induced cell death (Nawrocki et al., 2006), supporting the idea that basal rates of translation might be important determinants of sensitivity. It seems likely that

Myc is not uniquely capable of doing this, as other oncogenes (Kras, etc) are also known to increase rates of protein synthesis; the effects of Ras could be even more relevant as ROS and proteotoxicity have been directly implicated in the mechanisms underlying its effects on transformation (Xu et al., 2011). We also wondered whether differences in bortezomib-induced eIF2α phosphorylation might also be an important determinant of drug sensitivity. As discussed above, eIF2α phosphorylation seems to be the critical node in the cytoprotective response to proteotoxic stress as it mediates rapid decreases in translation, cell cycle arrest, and autophagy in normal cells. To test this hypothesis, we screened panels of human prostate, bladder, and pancreatic cancer cell lines for their abilities to increase eIF2α phosphorylation and arrest translation in response to exposure to bortezomib or thapsigargin (positive control for ER stress). Consistent with our expectations, bortezomib-resistant cell lines displayed robust bortezomib-induced eIF2α phosphorylation and translational arrest, whereas bortezomib-sensitive cells did not. The differential effects of bortezomib on translational arrest closely paralleled protein aggregate accumulation: cells that failed to increase phosphorylation of eIF2α accumulated aggregates much more rapidly than did the cells that efficiently activated the UPR (M. White, submitted). Importantly, bortezomib-sensitive lines may have higher basal phospho-eIF2α levels than do bortezomib-resistant cells, perhaps because they are experiencing higher baseline levels of stress, although the specific nature of this stress is not clear. (We have compared basal ROS levels and they do not appear to closely correlate with basal eIF2α phosphorylation.) Therefore, eIF2α phosphorylation may not be inducible in the bortezomib-sensitive cells because it has already been driven to maximal levels. The bortezomib-sensitive cells also appeared to be undergoing higher basal rates of autophagy that were not increased further by drug treatment, just as one would predict because of the central involvement of phospho-eIF2α in driving the response (Zhu et al., 2010, 2009). On the other hand, the resistant cells displayed strong increases in autophagy after bortezomib exposure, and in these cells, chemical autophagy inhibitors promoted aggregate accumulation and apoptosis, strongly suggesting that induced autophagy limited aggregate-mediated proteotoxicity in the cells (M. White, submitted; Zhu et al., 2010, 2009). Experiments with the phosphorylation-deficient knock-in MEFs introduced above confirmed that eIF2α phosphorylation was required for bortezomib-induced autophagy (Zhu et al., 2010, 2009). Therefore, eIF2α phosphorylation limits bortezomib-induced, proteotoxic cell death by (1) decreasing protein synthesis and (2) promoting autophagy.

These observations have important implications for the use of HDAC inhibitors as bortezomib-sensitizing agents. Because HDAC6 (and perhaps type I HDACs) appear to be primarily involved in promoting the degradation of misfolded via autophagy (particularly when the proteasome is blocked), combinations of HDAC and proteasome inhibitors may have less benefit in bortezomib-sensitive cells because these cells already possess defects in inducible eIF2α phosphorylation and autophagy. This prediction is consistent with our experience in cell lines, where HDAC inhibitors appear to have the strongest effects in the bortezomib-resistant cells. The model also predicts that combinations of proteasome and autophagy inhibitors will be most valuable in the bortezomib-resistant subset of tumors.

The studies also suggest that inhibition of bortezomib-induced eIF2α phosphorylation could be an even more attractive therapeutic approach than HDAC inhibition. The early studies suggested the ER stress-activated eIF2α kinase, PERK, might be centrally involved (Obeng et al., 2006). However, studies in knockout MEFs suggested that GCN2 might play a more important role than PERK in promoting eIF2α phosphorylation in response to proteasome inhibition (Jiang & Wek, 2005), and the kinase(s) responsible for promoting basal eIF2α phosphorylation in the bortezomib-sensitive cells could also be distinct from the one(s) that responds to proteasome inhibition. Therefore, we systematically examined the effects of RNAi-mediated knockdown of PERK, GCN2, PKR, and HRI on basal eIF2α phosphorylation, bortezomib-induced translational arrest, and aggregate formation in human pancreatic cancer cells. We first performed more careful dose-response and time course experiments than we had undertaken before to identify the minimal concentrations of bortezomib required to kill the drug-sensitive lines (10 nM) and the time at which they first produced visible ubiquitin- and p62-containing protein aggregates (12 h). Strikingly, under these conditions, bortezomib induced marked upregulation of inducible Hsp70 (Hsp72) but had no effect on the levels of Grp78 or GADD153/CHOP, as measured by quantitative RT-PCR (M. White, manuscript in preparation). Furthermore, knockdown of GCN2 (but not PERK or the other kinases) inhibited basal eIF2α phosphorylation in the sensitive cells and inhibited bortezomib-induced translational arrest in the resistant cells. GCN2 knockdown also strongly promoted bortezomib-induced aggregate formation and cytotoxicity. Together, these data strongly suggest that proteasome inhibition primarily induces a cytosolic/nuclear stress response based on GCN2 and HSF-1 activation, and GCN2 inhibitors may be attractive adjuvants to proteasome inhibitor-based therapy.

Our results indicate that cytosolic/nuclear proteotoxicity leading to GCN2 and HSF-1 activation and HSP70 accumulation appears to be more significant than ER stress in the acute effects of bortezomib. But these observations leave unresolved the question of how proteasome inhibitors activate GCN2. As discussed above, the kinase is activated by uncharged tRNAs and "senses" decreases in amino acid availability. Interestingly, the proteasome is required to maintain baseline cellular amino acid pools (Vabulas & Hartl, 2005), so proteasome inhibition may lead to an acute decrease in amino acid availability that is sensed by GCN2. Higher basal rates of (amino acid) metabolism (and higher translation?) may underlie the higher basal GCN2 activation and eIF2α phosphorylation observed in the bortezomib-sensitive cells.

2.2. HSP90 inhibitors

Like proteasome inhibitors, the cytotoxic effects of HSP90 inhibitors may involve substrate selective and more general proteotoxic mechanisms. HSP90 and HSC70 function as co-chaperones that promote the proper folding of many different "client" proteins that play critical roles in cancer cell signal transduction, proliferation, and survival, including steroid (estrogen and androgen) receptors, c-Raf, HER2, CDK4, and AKT (Workman, Burrows, Neckers, & Rosen, 2007). In addition, mutant proteins appear to be particularly dependent on HSP90's chaperone function, so HSP90 inhibitors are attractive candidate therapies for tumors that are driven by mutant oncogenes (i.e., Kras and Braf) (Workman et al., 2007). Indeed, preclinical and clinical studies have demonstrated that HSP90 inhibitors preferentially kill tumor cells that express amplified HER2 or mutant oncogenes (Basso, Solit, Munster, & Rosen, 2002; Mitsiades et al., 2006; Modi et al., 2011; Solit et al., 2002). However, by virtue of its role in promoting CHIP-dependent substrate ubiquitylation, HSP90 also plays a central role in protein quality control that might also be exploitable in vulnerable cancers.

Because they would be expected to impair misfolded protein ubiquitylation, HSP90 inhibitors should impair misfolded protein degradation via either the proteasome or autophagy. However, their functions as inhibitors of proteasome-mediated misfolded protein degradation would not be expected to be as potent as those induced by direct proteasome inhibitors, so benefit might be derived from combining the two classes of drug. Indeed, some studies have demonstrated that the HSP90 inhibitor 17-AAG synergizes with proteasome inhibitors to induce cell death via increased proteotoxicity (Neckers et al., MCT), which is consistent with this idea. Similarly, combining HSP90 inhibitors with HDAC inhibitors might inhibit autophagy-mediated clearance of protein aggregates more efficiently than can be obtained with class of inhibitor alone. Again, preclinical studies have demonstrated that HDAC inhibitors dramatically potentiate the cytotoxicity of 17-AAG. Finally, even greater tumor cell killing might be obtained by combining all three classes of drug, if this can be done without excessive toxicity to normal organs. Precisely, how combinations of HSP90, HDAC, and proteasome inhibitors affect misfolded protein buildup, aggregation, and clearance has not been studied in detail but is an important subject for future investigation.

Another attractive opportunity to enhance the proteotoxic effects of HSP90 inhibitors has been revealed through investigation of the roles of HSP70 family members in regulating HSP90 client protein levels. Because HSP90 and HSC70 function as co-chaperones to promote client protein folding and maintain stability, it might be predicted that HSC70 knockdown might phenocopy the effects of HSP90 inhibitors on these clients. However, RNAi-mediated knockdown of HSC70 failed to cause a detectable decrease in client protein levels, presumably because it resulted in a compensatory, marked increase in HSP72 expression, and HSP72 directly substituted for HSC70 in client protein–HSP90 complexes. Instead, combined knockdown of HSC70 plus HSP70 did produce effects on HSP90 client protein levels that were indistinguishable from those observed with HSP90 inhibitors. Interestingly, HSP90 inhibitors induce strong increases in HSP72 (but not HSC70) mRNA and protein levels, and knockdown of HSP72 (but not HSC70) strongly potentiated the cytotoxic effects of 17-AAG. Therefore, there is a clear disconnect between the effects of 17-AAG and RNAi targeting HSC70, HSP72, or both co-chaperones on client protein accumulation versus cell death. It is tempting to speculate that HSP72 plays a uniquely important role in the proteotoxic effects of 17-AAG and other HSP90 inhibitors and that under these circumstances the proteotoxicity plays a more significant role in cell killing than does client protein degradation. The model would predict that cell death would be triggered by poorly ubiquitylated misfolded proteins that are not efficiently degraded via either the proteasome or autophagy, an hypothesis that will be challenging to test experimentally.

2.3. HSP70 inhibitors

HSP72 is not expressed in normal cells that have not been exposed to proteotoxic stress. However, some tumors (including pancreatic cancers) constitutively express HSP72, so the inducible HSP70 has become a very attractive therapeutic target. Several small molecule inhibitors that block either its upstream activator HSF-1 (i.e., KNK-437, quercetin, triptolide) or HSP72's

ATPase activity (2-phenylethynesulfonamide, or PES) have been developed and tested in preclinical models of human cancer. The results obtained with these inhibitors to date have been promising, but whether or not the tumoricidal effects are due to misfolded protein accumulation or to more direct HSP70-mediated sequestration of proapoptotic proteins (Bax, Apaf-1) has not been determined. Furthermore, HSP70 isoforms appear to have stabilizing effects on lysosomes, so HSP70 inhibitors may disrupt lysosomal function independently of any effects on misfolded proteins. Nonetheless, given its central role in mediating CHIP-dependent ubiquitylation and the strong upregulation of HSP72 observed in cells exposed to proteasome inhibitors and certain other agents, a role for misfolded protein accumulation in the cytotoxic effects of HSP70 inhibitors is very attractive. Whether or not combining HDAC inhibitors with them makes sense is a question that will hopefully be addressed efficiently now that good chemical inhibitors are becoming available.

2.4. Arsenicals

Arsenic trioxide has been used successfully to treat acute promyelocytic leukemia and has more recently demonstrated activity in MM. Like other heavy metals, arsenicals can react directly with protein thiols resulting in direct protein damage and oxidative stress. Consequently, arsenicals produce effects in cells that are consistent with proteotoxicity, including phosphorylation of eIF2α and induction of HSP72, GRP78, ATF4, and CHOP (Agarwal, Roy, Ray, Mazumder, & Bhattacharya, 2009; Mirault, Southgate, & Delwart, 1982; Naranmandura et al., 2012; Tully et al., 2000; Wu, Yen, Lee, & Yih, 2009). Interestingly, even though the effects of arsenicals in some ways resemble the effects of agents that induce "pure" ER stress, eIF2α phosphorylation does not appear to be mediated via PERK, but rather HRI and possibly PKR have been implicated (Zhan et al., 2002). Importantly, whether or not arsenicals act by inducing protein misfolding and/or aggregation has not been well studied, even though this seems highly likely. The clinical activity of arsenicals in patients with MM is also consistent with this idea. Several studies have concluded that HDAC inhibitors promote the cytotoxic effects of arsenicals, although the molecular mechanisms involved have not been described. A role for disruption of autophagy-mediated aggregate clearance appears very attractive.

2.5. Oncolytic viruses

The added burden of viral replication causes a proteotoxicity that is "sensed" by the interferon system and PKR, leading to eIF2α phosphorylation, translational arrest, and autophagy that function to prevent productive replication.

In response, viruses have evolved a number of different mechanisms to subvert this arm of the UPR (as well as IFN production). If these effects can be restricted to cells that contain very specific molecular defects (i.e., cancer cells), then they might be exploitable in cancer therapy. One example of this can be found in ONYX-015, an oncolytic virus that selectively replicates in p53-deficient cells, and viruses which target cells which harbor mutant Kras (Parato, Senger, & Forsyth, Bell, 2005) or defective IFN signaling (Stojdl et al., 2000) have also been developed. Interestingly, HDAC inhibitors can markedly increase the efficacy of oncolytic viral therapy, although these effects may be schedule dependent and have been attributed to increased viral replication rather than to proteotoxicity. Nonetheless, we would suggest that the latter possibility should be explored, and other strategies to enhance proteotoxicity (Mahony et al., 2011) should also be considered.

3. SUMMARY AND FUTURE DIRECTIONS

The biological features of certain subsets of cancers appear to produce high basal levels of proteotoxic stress that makes them particularly vulnerable to agents that disrupt protein quality control mechanisms. The strong clinical activity of bortezomib and other proteasome inhibitors in MM is one clear example of how this vulnerability can be exploited, but there appear to be others, even in solid tumors. HDAC inhibitors produce synergistic cell death when they are combined with proteasome inhibitors, in part because they disrupt aggregate disposal through autophagy, but they affect a variety of other processes, and the relative importance of their proteotoxic effects versus the others (ROS production, EMT reversal) is not yet clear. Furthermore, the clinical value of these combinations has not yet been determined—in preclinical models, the combination of bortezomib plus SAHA produces significant toxicity. Toxicity could prevent the promising preclinical activity from being translated efficiently in patients.

A number of other candidate targets are emerging from preclinical studies. It appears that the important cross talk between the proteasome and autophagy in the disposal of protein aggregates contains many opportunities for therapeutic intervention to improve the antitumoral effects of proteasome inhibitors. The cross talk could be disrupted at the level of protein chaperones (HSP90, HSC70, HSP72), p62/NBR1, and at various points within the central mechanisms involved in autophagy itself. We are also particularly interested in targeting GCN2, which appears to play a critical role in proteasome inhibitor-induced phosphorylation of eIF2α and translational arrest and might be a druggable target. Again, toxicity to normal tissues may present a major barrier to successful exploitation of these potential opportunities, so careful preclinical *in vivo* studies should be performed to rigorously test them.

Finally, more work should be done to define the effects of protein quality control mechanisms in the antitumoral effects of other existing (conventional and investigational) therapies. It appears that the ISR regulates the downstream effects of arsenicals and oncolytic viruses, but precisely how they do needs to be explored further. There are also a large number of agents that target growth factor receptors and the PI3 kinase/AKT/mTOR pathway that are known to affect translation and autophagy that must also have effects on protein quality control that have not been characterized.

REFERENCES

Adams, J. (2004). The development of proteasome inhibitors as anticancer drugs. [Review]. *Cancer Cell*, *5*(5), 417–421.

Agarwal, S., Roy, S., Ray, A., Mazumder, S., & Bhattacharya, S. (2009). Arsenic trioxide and lead acetate induce apoptosis in adult rat hepatic stem cells. *Cell Biology and Toxicology*, *25*, 403–413.

Ahn, S. G., Kim, S. A., Yoon, J. H., & Vacratsis, P. (2005). Heat-shock cognate 70 is required for the activation of heat-shock factor 1 in mammalian cells. *The Biochemical Journal*, *392* (Pt 1), 145–152.

Bali, P., Pranpat, M., Bradner, J., Balasis, M., Fiskus, W., Guo, F., et al. (2005). Inhibition of histone deacetylase 6 acetylates and disrupts the chaperone function of heat shock protein 90: A novel basis for antileukemia activity of histone deacetylase inhibitors. *The Journal of Biological Chemistry*, *280*(29), 26729–26734.

Basso, A. D., Solit, D. B., Munster, P. N., & Rosen, N. (2002). Ansamycin antibiotics inhibit Akt activation and cyclin D expression in breast cancer cells that overexpress HER2. *Oncogene*, *21*, 1159–1166.

Bays, N. W., Gardner, R. G., Seelig, L. P., Joazeiro, C. A., & Hampton, R. Y. (2001). Hrd1p/Der3p is a membrane-anchored ubiquitin ligase required for ER-associated degradation. [Research Support, Non-U.S. Gov't Research Support, U.S. Gov't, P.H.S.]. *Nature Cell Biology*, *3*(1), 24–29.

Bennett, E. J., Bence, N. F., Jayakumar, R., & Kopito, R. R. (2005). Global impairment of the ubiquitin-proteasome system by nuclear or cytoplasmic protein aggregates precedes inclusion body formation. [Research Support, Non-U.S. Gov't Research Support, U.S. Gov't, P.H.S.]. *Molecular Cell*, *17*(3), 351–365.

Bennett, E. J., Shaler, T. A., Woodman, B., Ryu, K. Y., Zaitseva, T. S., Becker, C. H., et al. (2007). Global changes to the ubiquitin system in Huntington's disease. [Research Support, N.I.H., Extramural Research Support, Non-U.S. Gov't]*Nature*, *448*(7154), 704–708.

Bertolotti, A., Zhang, Y., Hendershot, L. M., Harding, H. P., & Ron, D. (2000). Dynamic interaction of BiP and ER stress transducers in the unfolded-protein response. [Research Support, Non-U.S. Gov't Research Support, U.S. Gov't, P.H.S.]. *Nature Cell Biology*, *2*(6), 326–332.

Bi, M., Naczki, C., Koritzinsky, M., Fels, D., Blais, J., Hu, N., et al. (2005). ER stress-regulated translation increases tolerance to extreme hypoxia and promotes tumor growth. *The EMBO Journal*, *24*(19), 3470–3481.

Blais, J. D., Filipenko, V., Bi, M., Harding, H. P., Ron, D., Koumenis, C., et al. (2004). Activating transcription factor 4 is translationally regulated by hypoxic stress. *Molecular and Cellular Biology*, *24*(17), 7469–7482.

Brewer, J. W., & Diehl, J. A. (2000). PERK mediates cell-cycle exit during the mammalian unfolded protein response. *Proceedings of the National Academy of Sciences of the United States of America, 97*(23), 12625–12630.

Burbulla, L. F., Schelling, C., Kato, H., Rapaport, D., Woitalla, D., Schiesling, C., et al. (2010). Dissecting the role of the mitochondrial chaperone mortalin in Parkinson's disease: Functional impact of disease-related variants on mitochondrial homeostasis. *Human Molecular Genetics, 19*(22), 4437–4452.

Chauhan, D., Hideshima, T., & Anderson, K. C. (2006). A novel proteasome inhibitor NPI-0052 as an anticancer therapy. [Research Support, N.I.H., Extramural Research Support, Non-U.S. Gov't Review]. *British Journal of Cancer, 95*(8), 961–965.

Chen, J. J., Throop, M. S., Gehrke, L., Kuo, I., Pal, J. K., Brodsky, M., et al. (1991). Cloning of the cDNA of the heme-regulated eukaryotic initiation factor 2 alpha (eIF-2 alpha) kinase of rabbit reticulocytes: Homology to yeast GCN2 protein kinase and human double-stranded-RNA-dependent eIF-2 alpha kinase. *Proceedings of the National Academy of Sciences of the United States of America, 88*(17), 7729–7733.

Chin, L. S., Olzmann, J. A., & Li, L. (2010). Parkin-mediated ubiquitin signalling in aggresome formation and autophagy. [Research Support, N.I.H., Extramural Review]. *Biochemical Society Transactions, 38*(Pt 1), 144–149.

Connell, P., Ballinger, C. A., Jiang, J., Wu, Y., Thompson, L. J., Hohfeld, J., et al. (2001). The co-chaperone CHIP regulates protein triage decisions mediated by heat-shock proteins. [Research Support, Non-U.S. Gov't Research Support, U.S. Gov't, P.H.S.]. *Nature Cell Biology, 3*(1), 93–96.

Cox, J. S., Shamu, C. E., & Walter, P. (1993). Transcriptional induction of genes encoding endoplasmic reticulum resident proteins requires a transmembrane protein kinase. *Cell, 73*(6), 1197–1206.

Cusack, J. C., Jr., Liu, R., & Baldwin, A. S., Jr. (2000). Inducible chemoresistance to 7-ethyl-10-[4-(1-piperidino)-1-piperidino]-carbonyloxycamptothe cin (CPT-11) in colorectal cancer cells and a xenograft model is overcome by inhibition of nuclear factor-kappaB activation. [Research Support, Non-U.S. Gov't Research Support, U.S. Gov't, P.H.S.]. *Cancer Research, 60*(9), 2323–2330.

Czarnecka, A. M., Campanella, C., Zummo, G., & Cappello, F. (2006). Mitochondrial chaperones in cancer: From molecular biology to clinical diagnostics. *Cancer Biology & Therapy, 5*(7), 714–720.

Dai, Y., Chen, S., Kramer, L. B., Funk, V. L., Dent, P., & Grant, S. (2008). Interactions between bortezomib and romidepsin and belinostat in chronic lymphocytic leukemia cells. [Research Support, N.I.H., Extramural Research Support, Non-U.S. Gov't]. *Clinical cancer research: An official journal of the American Association for Cancer Research, 14*(2), 549–558.

Dai, Y., Rahmani, M., Dent, P., & Grant, S. (2005). Blockade of histone deacetylase inhibitor-induced RelA/p65 acetylation and NF-kappaB activation potentiates apoptosis in leukemia cells through a process mediated by oxidative damage, XIAP downregulation, and c-Jun N-terminal kinase 1 activation. *Molecular and Cellular Biology, 25* (13), 5429–5444.

Dasmahapatra, G., Lembersky, D., Kramer, L., Fisher, R. I., Friedberg, J., Dent, P., et al. (2010). The pan-HDAC inhibitor vorinostat potentiates the activity of the proteasome inhibitor carfilzomib in human DLBCL cells in vitro and in vivo. [In Vitro Research Support, N.I.H., Extramural Research Support, Non-U.S. Gov't].*Blood, 115*(22), 4478–4487.

Dasmahapatra, G., Lembersky, D., Son, M. P., Attkisson, E., Dent, P., Fisher, R. I., et al. (2011). Carfilzomib interacts synergistically with histone deacetylase inhibitors in mantle cell lymphoma cells in vitro and in vivo. [Research Support, N.I.H., Extramural Research Support, Non-U.S. Gov't]. *Molecular Cancer Therapeutics, 10*(9), 1686–1697.

Daugaard, M., Rohde, M., & Jaattela, M. (2007). The heat shock protein 70 family: Highly homologous proteins with overlapping and distinct functions. [Research Support, Non-U.S. Gov't Review]. *FEBS Letters*, *581*(19), 3702–3710.

Demo, S. D., Kirk, C. J., Aujay, M. A., Buchholz, T. J., Dajee, M., Ho, M. N., et al. (2007). Antitumor activity of PR-171, a novel irreversible inhibitor of the proteasome. *Cancer Research*, *67*(13), 6383–6391.

Deng, J., Harding, H. P., Raught, B., Gingras, A. C., Berlanga, J. J., Scheuner, D., et al. (2002). Activation of GCN2 in UV-irradiated cells inhibits translation. *Current Biology*, *12*(15), 1279–1286.

Deribe, Y. L., Wild, P., Chandrashaker, A., Curak, J., Schmidt, M. H., Kalaidzidis, Y., et al. (2009). Regulation of epidermal growth factor receptor trafficking by lysine deacetylase HDAC6. *Science Signaling*, *2*(102), ra84.

Eura, Y., Yanamoto, H., Arai, Y., Okuda, T., Miyata, T., & Kokame, K. (2012). Derlin-1 deficiency is embryonic lethal, Derlin-3 deficiency appears normal, and Herp deficiency is intolerant to glucose load and ischemia in mice. [Research Support, Non-U.S. Gov't] *PLoS One*, 7(3), e34298.

Fels, D. R., & Koumenis, C. (2006). The PERK/eIF2alpha/ATF4 module of the UPR in hypoxia resistance and tumor growth. *Cancer Biology & Therapy*, *5*(7), 723–728.

Franco, R. S., Hogg, J. W., & Martelo, O. J. (1981). Activation and partial characterization of a human reticulocyte heme-dependent eIF-Z alpha kinase. *American Journal of Hematology*, *11*(1), 9–18.

Gao, Y. S., Hubbert, C. C., & Yao, T. P. (2010). The microtubule-associated histone deacetylase 6 (HDAC6) regulates epidermal growth factor receptor (EGFR) endocytic trafficking and degradation. [Research Support, N.I.H., Extramural Research Support, Non-U.S. Gov't Research Support, U.S. Gov't, Non-P.H.S.]. *The Journal of Biological Chemistry*, *285*(15), 11219–11226.

Goldberg, A. L. (2003). Protein degradation and protection against misfolded or damaged proteins. [Review]. *Nature*, *426*(6968), 895–899.

Grant, S. (2008). Is the focus moving toward a combination of targeted drugs?[Review]. *Best Practice & Research. Clinical Haematology*, *21*(4), 629–637.

Grant, S., & Dent, P. (2007). Simultaneous interruption of signal transduction and cell cycle regulatory pathways: Implications for new approaches to the treatment of childhood leukemias. *Current Drug Targets*, *8*(6), 751–759.

Gross, M., Olin, A., Hessefort, S., & Bender, S. (1994). Control of protein synthesis by hemin. Purification of a rabbit reticulocyte hsp 70 and characterization of its regulation of the activation of the hemin-controlled eIF-2(alpha) kinase. *The Journal of Biological Chemistry*, *269*(36), 22738–22748.

Gross, M., Rynning, J., & Knish, W. M. (1981). Evidence that the phosphorylation of eukaryotic initiation factor 2 alpha by the hemin-controlled translational repressor occurs at a single site. *The Journal of Biological Chemistry*, *256*(2), 589–592.

Haggarty, S. J., Koeller, K. M., Wong, J. C., Grozinger, C. M., & Schreiber, S. L. (2003). Domain-selective small-molecule inhibitor of histone deacetylase 6 (HDAC6)-mediated tubulin deacetylation. *Proceedings of the National Academy of Sciences of the United States of America*, *100*(8), 4389–4394.

Hamanaka, R. B., Bennett, B. S., Cullinan, S. B., & Diehl, J. A. (2005). PERK and GCN2 contribute to eIF2alpha phosphorylation and cell cycle arrest after activation of the unfolded protein response pathway. *Molecular Biology of the Cell*, *16*(12), 5493–5501.

Hara, T., Nakamura, K., Matsui, M., Yamamoto, A., Nakahara, Y., Suzuki-Migishima, R., et al. (2006). Suppression of basal autophagy in neural cells causes neurodegenerative disease in mice. [Research Support, Non-U.S. Gov't]. *Nature*, *441*(7095), 885–889.

Harding, H. P., Novoa, I., Zhang, Y., Zeng, H., Wek, R., Schapira, M., et al. (2000). Regulated translation initiation controls stress-induced gene expression in mammalian cells. *Molecular Cell*, *6*(5), 1099–1108.

Harding, H. P., Zhang, Y., Bertolotti, A., Zeng, H., & Ron, D. (2000). Perk is essential for translational regulation and cell survival during the unfolded protein response. *Molecular Cell, 5*(5), 897–904.

Harding, H. P., Zhang, Y., & Ron, D. (1999). Protein translation and folding are coupled by an endoplasmic-reticulum-resident kinase. *Nature, 397*(6716), 271–274.

Harding, H. P., Zhang, Y., Zeng, H., Novoa, I., Lu, P. D., Calfon, M., et al. (2003). An integrated stress response regulates amino acid metabolism and resistance to oxidative stress. *Molecular Cell, 11*(3), 619–633.

Hetz, C. (2012). The unfolded protein response: Controlling cell fate decisions under ER stress and beyond. *Nature Reviews. Molecular Cell Biology, 13*(2), 89–102.

Hetz, C., Bernasconi, P., Fisher, J., Lee, A. H., Bassik, M. C., Antonsson, B., et al. (2006). Proapoptotic BAX and BAK modulate the unfolded protein response by a direct interaction with IRE1alpha. *Science, 312*(5773), 572–576.

Hideshima, T., Bradner, J. E., Wong, J., Chauhan, D., Richardson, P., Schreiber, S. L., et al. (2005). Small-molecule inhibition of proteasome and aggresome function induces synergistic antitumor activity in multiple myeloma. [Comparative Study Research Support, N.I.H., Extramural Research Support, Non-U.S. Gov't Research Support, U.S. Gov't, P.H.S.]. *Proceedings of the National Academy of Sciences of the United States of America, 102*(24), 8567–8572.

Hideshima, T., Chauhan, D., Richardson, P., Mitsiades, C., Mitsiades, N., Hayashi, T., et al. (2002). NF-kappa B as a therapeutic target in multiple myeloma. [Research Support, Non-U.S. Gov't Research Support, U.S. Gov't, P.H.S.]. *The Journal of Biological Chemistry, 277*(19), 16639–16647.

Hideshima, T., Ikeda, H., Chauhan, D., Okawa, Y., Raje, N., Podar, K., et al. (2009). Bortezomib induces canonical nuclear factor-kappaB activation in multiple myeloma cells. *Blood, 114*(5), 1046–1052.

Hightower, L. E. (1991). Heat shock, stress proteins, chaperones, and proteotoxicity. *Cell, 66*(2), 191–197.

Hipp, M. S., Patel, C. N., Bersuker, K., Riley, B. E., Kaiser, S. E., Shaler, T. A., et al. (2012). Indirect inhibition of 26S proteasome activity in a cellular model of Huntington's disease. [Research Support, N.I.H., Extramural Research Support, Non-U.S. Gov't]. *The Journal of Cell Biology, 196*(5), 573–587.

Hosokawa, N., Wada, I., Nagasawa, K., Moriyama, T., Okawa, K., & Nagata, K. (2008). Human XTP3-B forms an endoplasmic reticulum quality control scaffold with the HRD1-SEL1L ubiquitin ligase complex and BiP. *The Journal of Biological Chemistry, 283*(30), 20914–20924.

Hubbert, C., Guardiola, A., Shao, R., Kawaguchi, Y., Ito, A., Nixon, A., et al. (2002). HDAC6 is a microtubule-associated deacetylase. *Nature, 417*(6887), 455–458.

Imai, Y., Soda, M., Hatakeyama, S., Akagi, T., Hashikawa, T., Nakayama, K. I., et al. (2002). CHIP is associated with Parkin, a gene responsible for familial Parkinson's disease, and enhances its ubiquitin ligase activity. *Molecular Cell, 10*(1), 55–67.

Iwata, A., Riley, B. E., Johnston, J. A., & Kopito, R. R. (2005). HDAC6 and microtubules are required for autophagic degradation of aggregated huntingtin. [Research Support, N.I.H., Extramural Research Support, Non-U.S. Gov't] *The Journal of Biological Chemistry, 280*(48), 40282–40292.

Jiang, H. Y., & Wek, R. C. (2005). Phosphorylation of the alpha-subunit of the eukaryotic initiation factor-2 (eIF2alpha) reduces protein synthesis and enhances apoptosis in response to proteasome inhibition. [Research Support, N.I.H., Extramural Research Support, Non-U.S. Gov't Research Support, U.S. Gov't, P.H.S.]. *The Journal of Biological Chemistry, 280*(14), 14189–14202.

Jona, A., Khaskhely, N., Buglio, D., Shafer, J. A., Derenzini, E., Bollard, C. M., et al. (2011). The histone deacetylase inhibitor entinostat (SNDX-275) induces apoptosis in Hodgkin lymphoma cells and synergizes with Bcl-2 family inhibitors. e1001. *Experimental Hematology, 39*(10), 1007–1017e1001.

Kalluri, R., & Weinberg, R. A. (2009). The basics of epithelial-mesenchymal transition. [Research Support, N.I.H., Extramural Research Support, Non-U.S. Gov't Review]. *The Journal of Clinical Investigation, 119*(6), 1420–1428.

Kamat, A. M., Karashima, T., Davis, D. W., Lashinger, L., Bar-Eli, M., Millikan, R., et al. (2004). The proteasome inhibitor bortezomib synergizes with gemcitabine to block the growth of human 253JB-V bladder tumors in vivo. [Research Support, U.S. Gov't, P.H.S.]. *Molecular Cancer Therapeutics, 3*(3), 279–290.

Kang, B. H., & Altieri, D. C. (2009). Compartmentalized cancer drug discovery targeting mitochondrial Hsp90 chaperones. *Oncogene, 28*(42), 3681–3688.

Kang, B. H., Plescia, J., Dohi, T., Rosa, J., Doxsey, S. J., & Altieri, D. C. (2007). Regulation of tumor cell mitochondrial homeostasis by an organelle-specific Hsp90 chaperone network. *Cell, 131*(2), 257–270.

Kang, B. H., Plescia, J., Song, H. Y., Meli, M., Colombo, G., Beebe, K., et al. (2009). Combinatorial drug design targeting multiple cancer signaling networks controlled by mitochondrial Hsp90. *The Journal of Clinical Investigation, 119*(3), 454–464.

Kaser, M., & Langer, T. (2000). Protein degradation in mitochondria. [Research Support, Non-U.S. Gov't Review]. *Seminars in Cell & Developmental Biology, 11*(3), 181–190.

Kawaguchi, Y., Kovacs, J. J., McLaurin, A., Vance, J. M., Ito, A., & Yao, T. P. (2003). The deacetylase HDAC6 regulates aggresome formation and cell viability in response to misfolded protein stress. *Cell, 115*(6), 727–738.

Kikuchi, J., Wada, T., Shimizu, R., Izumi, T., Akutsu, M., Mitsunaga, K., et al. (2010). Histone deacetylases are critical targets of bortezomib-induced cytotoxicity in multiple myeloma. *Blood, 116*(3), 406–417.

Kirkin, V., Lamark, T., Sou, Y. S., Bjorkoy, G., Nunn, J. L., Bruun, J. A., et al. (2009). A role for NBR1 in autophagosomal degradation of ubiquitinated substrates. [Research Support, Non-U.S. Gov't]. *Molecular Cell, 33*(4), 505–516.

Kirkin, V., McEwan, D. G., Novak, I., & Dikic, I. (2009). A role for ubiquitin in selective autophagy. [Research Support, Non-U.S. Gov't Review]. *Molecular Cell, 34*(3), 259–269.

Klionsky, D. J. (2006). Neurodegeneration: Good riddance to bad rubbish. [Comment News]. *Nature, 441*(7095), 819–820.

Klionsky, D. J., Abeliovich, H., Agostinis, P., Agrawal, D. K., Aliev, G., Askew, D. S., et al. (2008). Guidelines for the use and interpretation of assays for monitoring autophagy in higher eukaryotes. [Research Support, N.I.H., Extramural Review]. *Autophagy, 4*(2), 151–175.

Komatsu, M., Waguri, S., Chiba, T., Murata, S., Iwata, J., Tanida, I., et al. (2006). Loss of autophagy in the central nervous system causes neurodegeneration in mice. [Research Support, Non-U.S. Gov't]. *Nature, 441*(7095), 880–884.

Kopito, R. R. (2000). Aggresomes, inclusion bodies and protein aggregation. *Trends in Cell Biology, 10*(12), 524–530.

Koritzinsky, M., Magagnin, M. G., van den Beucken, T., Seigneuric, R., Savelkouls, K., Dostie, J., et al. (2006). Gene expression during acute and prolonged hypoxia is regulated by distinct mechanisms of translational control. *The EMBO Journal, 25*(5), 1114–1125.

Koumenis, C., Naczki, C., Koritzinsky, M., Rastani, S., Diehl, A., Sonenberg, N., et al. (2002). Regulation of protein synthesis by hypoxia via activation of the endoplasmic reticulum kinase PERK and phosphorylation of the translation initiation factor eIF2alpha. *Molecular and Cellular Biology, 22*(21), 7405–7416.

Kramer, G., Henderson, A. B., Pinphanichakarn, P., Wallis, M. H., & Hardesty, B. (1977). Partial reaction of peptide initiation inhibited by phosphorylation of either initiation factor eIF-2 or 40S ribosomal proteins. *Proceedings of the National Academy of Sciences of the United States of America, 74*(4), 1445–1449.

Lafarga, V., Aymerich, I., Tapia, O., Mayor, F., Jr., & Penela, P. (2012). A novel GRK2/HDAC6 interaction modulates cell spreading and motility. [Research Support, Non-U.S. Gov't]. *The EMBO Journal*, *31*(4), 856–869.

Lashinger, L. M., Zhu, K., Williams, S. A., Shrader, M., Dinney, C. P., & McConkey, D. J. (2005). Bortezomib abolishes tumor necrosis factor-related apoptosis-inducing ligand resistance via a p21-dependent mechanism in human bladder and prostate cancer cells. [Research Support, N.I.H., Extramural Research Support, U.S. Gov't, Non-P.H.S. Research Support, U.S. Gov't, P.H.S.]. *Cancer Research*, *65*(11), 4902–4908.

Lee, A. S., & Hendershot, L. M. (2006). ER stress and cancer. *Cancer Biology & Therapy*, *5*(7), 721–722.

Lee, J. Y., Koga, H., Kawaguchi, Y., Tang, W., Wong, E., Gao, Y. S., et al. (2010). HDAC6 controls autophagosome maturation essential for ubiquitin-selective quality-control autophagy. *The EMBO Journal*, *29*(5), 969–980.

Liu, L., Cash, T. P., Jones, R. G., Keith, B., Thompson, C. B., & Simon, M. C. (2006). Hypoxia-induced energy stress regulates mRNA translation and cell growth. *Molecular Cell*, *21*(4), 521–531.

Loktev, A. V., Zhang, Q., Beck, J. S., Searby, C. C., Scheetz, T. E., Bazan, J. F., et al. (2008). A BBSome subunit links ciliogenesis, microtubule stability, and acetylation. [Research Support, N.I.H., Extramural]. *Developmental Cell*, *15*(6), 854–865.

Lu, L., Han, A. P., & Chen, J. J. (2001). Translation initiation control by heme-regulated eukaryotic initiation factor 2alpha kinase in erythroid cells under cytoplasmic stresses. *Molecular and Cellular Biology*, *21*(23), 7971–7980.

Matthias, P., Yoshida, M., & Khochbin, S. (2008). HDAC6 a new cellular stress surveillance factor. [Research Support, Non-U.S. Gov't Review]. *Cell Cycle*, *7*(1), 7–10.

McConkey, D. (2010). Proteasome and HDAC: Who's zooming who? *Blood*, *116*(3), 308–309.

McConkey, D. J., & Zhu, K. (2008). Mechanisms of proteasome inhibitor action and resistance in cancer. [Review]. *Drug resistance updates: Reviews and commentaries in antimicrobial and anticancer chemotherapy*, *11*(4–5), 164–179.

McDonough, H., & Patterson, C. (2003). CHIP: A link between the chaperone and proteasome systems. [Research Support, Non-U.S. Gov't Research Support, U.S. Gov't, P.H.S. Review]. *Cell Stress & Chaperones*, *8*(4), 303–308.

McEwen, E., Kedersha, N., Song, B., Scheuner, D., Gilks, N., Han, A., et al. (2005). Heme-regulated inhibitor kinase-mediated phosphorylation of eukaryotic translation initiation factor 2 inhibits translation, induces stress granule formation, and mediates survival upon arsenite exposure. *The Journal of Biological Chemistry*, *280*(17), 16925–16933.

Meister, S., Schubert, U., Neubert, K., Herrmann, K., Burger, R., Gramatzki, M., et al. (2007). Extensive immunoglobulin production sensitizes myeloma cells for proteasome inhibition. [Research Support, Non-U.S. Gov't]. *Cancer Research*, *67*(4), 1783–1792.

Mellor, H., Flowers, K. M., Kimball, S. R., & Jefferson, L. S. (1994). Cloning and characterization of cDNA encoding rat hemin-sensitive initiation factor-2 alpha (eIF-2 alpha) kinase. Evidence for multitissue expression. *The Journal of Biological Chemistry*, *269*(14), 10201–10204.

Meurs, E., Chong, K., Galabru, J., Thomas, N. S., Kerr, I. M., Williams, B. R., et al. (1990). Molecular cloning and characterization of the human double-stranded RNA-activated protein kinase induced by interferon. *Cell*, *62*(2), 379–390.

Michaud, E. J., & Yoder, B. K. (2006). The primary cilium in cell signaling and cancer. [Research Support, N.I.H., Extramural Research Support, Non-U.S. Gov't Research Support, U.S. Gov't, Non-P.H.S. Review]. *Cancer Research*, *66*(13), 6463–6467.

Miller, C. P., Ban, K., Dujka, M. E., McConkey, D. J., Munsell, M., Palladino, M., et al. (2007). NPI-0052, a novel proteasome inhibitor, induces caspase-8 and ROS-dependent apoptosis alone and in combination with HDAC inhibitors in leukemia cells. [Research Support, N.I.H., Extramural]. *Blood*, *110*(1), 267–277.

Mirault, M. E., Southgate, R., & Delwart, E. (1982). Regulation of heat-shock genes: A DNA sequence upstream of Drosophila hsp70 genes is essential for their induction in monkey cells. *The EMBO Journal*, *1*, 1279–1285.

Mitsiades, C. S., Mitsiades, N. S., McMullan, C. J., Poulaki, V., Kung, A. L., Davies, F. E., et al. (2006). Antimyeloma activity of heat shock protein-90 inhibition. *Blood*, *107*, 1092–1100.

Mizushima, N., Levine, B., Cuervo, A. M., & Klionsky, D. J. (2008). Autophagy fights disease through cellular self-digestion. [Research Support, N.I.H., Extramural Research Support, Non-U.S. Gov't Review]. *Nature*, *451*(7182), 1069–1075.

Modi, S., Stopeck, A., Linden, H., Solit, D., Chandarlapaty, S., Rosen, N., et al. (2011). HSP90 inhibition is effective in breast cancer: A phase II trial of tanespimycin (17-AAG) plus trastuzumab in patients with HER2-positive metastatic breast cancer progressing on trastuzumab. *Clinical Cancer Research*, *17*, 5132–5139.

Moreau, P., Richardson, P. G., Cavo, M., Orlowski, R. Z., San Miguel, J. F., Palumbo, A., et al. (2012). Proteasome inhibitors in multiple myeloma: Ten years later. *Blood*, *120*(5), 947–959.

Morimoto, R. I. (1993). Cells in stress: Transcriptional activation of heat shock genes. *Science*, *259*(5100), 1409–1410.

Morimoto, R. I. (2011). The heat shock response: Systems biology of proteotoxic stress in aging and disease. *Cold Spring Harbor Symposia on Quantitative Biology*, *76*, 91–99.

Murata, S., Chiba, T., & Tanaka, K. (2003). CHIP: A quality-control E3 ligase collaborating with molecular chaperones. [Review]. *The International Journal of Biochemistry & Cell Biology*, *35*(5), 572–578.

Naranmandura, H., Xu, S., Koike, S., Pan, L. Q., Chen, B., Wang, Y. W., et al. (2012). The endoplasmic reticulum is a target organelle for trivalent dimethylarsinic acid (DMAIII)-induced cytotoxicity. *Toxicology and Applied Pharmacology*, *260*, 241–249.

Nawrocki, S. T., Bruns, C. J., Harbison, M. T., Bold, R. J., Gotsch, B. S., Abbruzzese, J. L., et al. (2002). Effects of the proteasome inhibitor PS-341 on apoptosis and angiogenesis in orthotopic human pancreatic tumor xenografts. *Molecular Cancer Therapeutics*, *1*(14), 1243–1253.

Nawrocki, S. T., Carew, J. S., Dunner, K., Jr., Boise, L. H., Chiao, P. J., Huang, P., et al. (2005). Bortezomib inhibits PKR-like endoplasmic reticulum (ER) kinase and induces apoptosis via ER stress in human pancreatic cancer cells. [Research Support, N.I.H., Extramural Research Support, Non-U.S. Gov't]. *Cancer Research*, *65*(24), 11510–11519.

Nawrocki, S. T., Carew, J. S., Maclean, K. H., Courage, J. F., Huang, P., Houghton, J. A., et al. (2008). Myc regulates aggresome formation, the induction of Noxa, and apoptosis in response to the combination of bortezomib and SAHA. [Research Support, N.I.H., Extramural Research Support, Non-U.S. Gov't]. *Blood*, *112*(7), 2917–2926.

Nawrocki, S. T., Carew, J. S., Pino, M. S., Highshaw, R. A., Andtbacka, R. H., Dunner, K., Jr., et al. (2006). Aggresome disruption: A novel strategy to enhance bortezomib-induced apoptosis in pancreatic cancer cells. [Research Support, N.I.H., Extramural Research Support, Non-U.S. Gov't]. *Cancer Research*, *66*(7), 3773–3781.

Nawrocki, S. T., Carew, J. S., Pino, M. S., Highshaw, R. A., Dunner, K., Jr., Huang, P., et al. (2005). Bortezomib sensitizes pancreatic cancer cells to endoplasmic reticulum stress-mediated apoptosis. *Cancer Research*, *65*(24), 11658–11666.

Nebbioso, A., Clarke, N., Voltz, E., Germain, E., Ambrosino, C., Bontempo, P., et al. (2005). Tumor-selective action of HDAC inhibitors involves TRAIL induction in acute myeloid leukemia cells. [Research Support, Non-U.S. Gov't]. *Nature Medicine*, *11*(1), 77–84.

Nishitoh, H., Matsuzawa, A., Tobiume, K., Saegusa, K., Takeda, K., Inoue, K., et al. (2002). ASK1 is essential for endoplasmic reticulum stress-induced neuronal cell death triggered by expanded polyglutamine repeats. *Genes & Development*, *16*(11), 1345–1355.

Obeng, E. A., Carlson, L. M., Gutman, D. M., Harrington, W. J., Jr., Lee, K. P., & Boise, L. H. (2006). Proteasome inhibitors induce a terminal unfolded protein response in multiple myeloma cells. [Comparative Study Research Support, N.I.H., Extramural Research Support, Non-U.S. Gov't]. *Blood, 107*(12), 4907–4916.

Oda, Y., Okada, T., Yoshida, H., Kaufman, R. J., Nagata, K., & Mori, K. (2006). Derlin-2 and Derlin-3 are regulated by the mammalian unfolded protein response and are required for ER-associated degradation. [Research Support, Non-U.S. Gov't]. *The Journal of Cell Biology, 172*(3), 383–393.

Pandey, U. B., Nie, Z., Batlevi, Y., McCray, B. A., Ritson, G. P., Nedelsky, N. B., et al. (2007). HDAC6 rescues neurodegeneration and provides an essential link between autophagy and the UPS. *Nature, 447*(7146), 859–863.

Parato, K. A., Senger, D., Forsyth, P. A., & Bell, J. C. (2005). Recent progress in the battle between oncolytic viruses and tumours. *Nature Reviews. Cancer, 5*, 965–976.

Pei, X. Y., Dai, Y., & Grant, S. (2004). Synergistic induction of oxidative injury and apoptosis in human multiple myeloma cells by the proteasome inhibitor bortezomib and histone deacetylase inhibitors. [Research Support, Non-U.S. Gov't Research Support, U.S. Gov't, Non-P.H.S. Research Support, U.S. Gov't, P.H.S.]. *Clinical cancer research: An official journal of the American Association for Cancer Research, 10* (11), 3839–3852.

Penela, P., Lafarga, V., Tapia, O., Rivas, V., Nogues, L., Lucas, E., et al. (2012). Roles of GRK2 in cell signaling beyond GPCR desensitization: GRK2-HDAC6 interaction modulates cell spreading and motility. *Science Signaling, 5*(224), pt3.

Petroski, M. D., & Deshaies, R. J. (2005). Function and regulation of cullin-RING ubiquitin ligases. [Research Support, Non-U.S. Gov't Review]. *Nature Reviews. Molecular Cell Biology, 6*(1), 9–20.

Potts, B. C., Albitar, M. X., Anderson, K. C., Baritaki, S., Berkers, C., Bonavida, B., et al. (2011). Marizomib, a proteasome inhibitor for all seasons: Preclinical profile and a framework for clinical trials. [Research Support, N.I.H., Extramural Review]. *Current Cancer Drug Targets, 11*(3), 254–284.

Quinlan, R. J., Tobin, J. L., & Beales, P. L. (2008). Modeling ciliopathies: Primary cilia in development and disease. [Review]. *Current Topics in Developmental Biology, 84*, 249–310.

Ramirez, M., Wek, R. C., & Hinnebusch, A. G. (1991). Ribosome association of GCN2 protein kinase, a translational activator of the GCN4 gene of Saccharomyces cerevisiae. *Molecular and Cellular Biology, 11*(6), 3027–3036.

Richardson, P. G., Barlogie, B., Berenson, J., Singhal, S., Jagannath, S., Irwin, D., et al. (2003). A phase 2 study of bortezomib in relapsed, refractory myeloma. [Clinical Trial Clinical Trial, Phase II Multicenter Study Research Support, Non-U.S. Gov't]. *The New England Journal of Medicine, 348*(26), 2609–2617.

Ron, D., & Hubbard, S. R. (2008). How IRE1 reacts to ER stress. *Cell, 132*(1), 24–26.

Ron, D., & Walter, P. (2007). Signal integration in the endoplasmic reticulum unfolded protein response. *Nature Reviews. Molecular Cell Biology, 8*(7), 519–529.

Roussos, E. T., Keckesova, Z., Haley, J. D., Epstein, D. M., Weinberg, R. A., & Condeelis, J. S. (2010). AACR special conference on epithelial-mesenchymal transition and cancer progression and treatment. [Congresses]. *Cancer Research, 70*(19), 7360–7364.

Ruschak, A. M., Slassi, M., Kay, L. E., & Schimmer, A. D. (2011). Novel proteasome inhibitors to overcome bortezomib resistance. [Research Support, Non-U.S. Gov't Review]. *Journal of the National Cancer Institute, 103*(13), 1007–1017.

Santo, L., Hideshima, T., Kung, A. L., Tseng, J. C., Tamang, D., Yang, M., et al. (2012). Preclinical activity, pharmacodynamic, and pharmacokinetic properties of a selective HDAC6 inhibitor, ACY-1215, in combination with bortezomib in multiple myeloma. *Blood, 119*(11), 2579–2589.

Sayers, T. J., Brooks, A. D., Koh, C. Y., Ma, W., Seki, N., Raziuddin, A., et al. (2003). The proteasome inhibitor PS-341 sensitizes neoplastic cells to TRAIL-mediated apoptosis by reducing levels of c-FLIP. [Research Support, U.S. Gov't, P.H.S.]. *Blood*, *102*(1), 303–310.

Shan, B., Yao, T. P., Nguyen, H. T., Zhuo, Y., Levy, D. R., Klingsberg, R. C., et al. (2008). Requirement of HDAC6 for transforming growth factor-beta1-induced epithelial-mesenchymal transition. [Research Support, N.I.H., Extramural Research Support, Non-U.S. Gov't]. *The Journal of Biological Chemistry*, *283*(30), 21065–21073.

Shi, Y., Mosser, D. D., & Morimoto, R. I. (1998). Molecular chaperones as HSF1-specific transcriptional repressors. [Research Support, Non-U.S. Gov't Research Support, U.S. Gov't, P.H.S.]. *Genes & Development*, *12*(5), 654–666.

Shi, Y., Vattem, K. M., Sood, R., An, J., Liang, J., Stramm, L., et al. (1998). Identification and characterization of pancreatic eukaryotic initiation factor 2 alpha-subunit kinase, PEK, involved in translational control. *Molecular and Cellular Biology*, *18*(12), 7499–7509.

Siegelin, M. D., Dohi, T., Raskett, C. M., Orlowski, G. M., Powers, C. M., Gilbert, C. A., et al. (2011). Exploiting the mitochondrial unfolded protein response for cancer therapy in mice and human cells. *The Journal of Clinical Investigation*, *121*(4), 1349–1360.

Simpson, F., Kerr, M. C., & Wicking, C. (2009). Trafficking, development and hedgehog. [Review]. *Mechanisms of Development*, *126*(5–6), 279–288.

Smith, D. M., Benaroudj, N., & Goldberg, A. (2006). Proteasomes and their associated ATPases: A destructive combination. [Review]. *Journal of Structural Biology*, *156*(1), 72–83.

Smith, D. M., Chang, S. C., Park, S., Finley, D., Cheng, Y., & Goldberg, A. L. (2007). Docking of the proteasomal ATPases' carboxyl termini in the 20S proteasome's alpha ring opens the gate for substrate entry. [Research Support, N.I.H., Extramural Research Support, Non-U.S. Gov't]. *Molecular Cell*, *27*(5), 731–744.

Smith, D. M., Fraga, H., Reis, C., Kafri, G., & Goldberg, A. L. (2011). ATP binds to proteasomal ATPases in pairs with distinct functional effects, implying an ordered reaction cycle. [Research Support, N.I.H., Extramural Research Support, Non-U.S. Gov't]. *Cell*, *144*(4), 526–538.

Smith, M. H., Ploegh, H. L., & Weissman, J. S. (2011). Road to ruin: Targeting proteins for degradation in the endoplasmic reticulum. [Research Support, N.I.H., Extramural Research Support, Non-U.S. Gov't Review]. *Science*, *334*(6059), 1086–1090.

Solit, D. B., Zheng, F. F., Drobnjak, M., Munster, P. N., Higgins, B., Verbel, D., et al. (2002). 17-Allylamino-17-demethoxygeldanamycin induces the degradation of androgen receptor and HER-2/neu and inhibits the growth of prostate cancer xenografts. *Clinical Cancer Research*, *8*, 986–993.

Sood, R., Porter, A. C., Olsen, D. A., Cavener, D. R., & Wek, R. C. (2000). A mammalian homologue of GCN2 protein kinase important for translational control by phosphorylation of eukaryotic initiation factor-2alpha. *Genetics*, *154*(2), 787–801.

Stojdl, D. F., Lichty, B., Knowles, S., Marius, R., Atkins, H., Sonenberg, N., et al. (2000). Exploiting tumor-specific defects in the interferon pathway with a previously unknown oncolytic virus. *Nature Medicine*, *6*, 821–825.

Sunwoo, J. B., Chen, Z., Dong, G., Yeh, N., Crowl Bancroft, C., Sausville, E., et al. (2001). Novel proteasome inhibitor PS-341 inhibits activation of nuclear factor-kappa B, cell survival, tumor growth, and angiogenesis in squamous cell carcinoma. [Research Support, U.S. Gov't, P.H.S.]. *Clinical cancer research: An official journal of the American Association for Cancer Research*, 7(5), 1419–1428.

Szegezdi, E., Lobgue, S. E., Gorman, A. M., & Samali, A. (2006). Mediators of endoplasmic stress-induced apoptosis. *EMBO Reports*, 7, 880–885.

Tabas, I., & Ron, D. (2011). Integrating the mechanisms of apoptosis induced by endoplasmic reticulum stress. *Nature Cell Biology*, *13*(3), 184–190.

Talloczy, Z., Jiang, W., Virgin, H. W., 4th, Leib, D. A., Scheuner, D., Kaufman, R. J., et al. (2002). Regulation of starvation- and virus-induced autophagy by the eIF2alpha kinase signaling pathway. *Proceedings of the National Academy of Sciences of the United States of America, 99*(1), 190–195.

Thibault, G., Ismail, N., & Ng, D. T. (2011). The unfolded protein response supports cellular robustness as a broad-spectrum compensatory pathway. [Research Support, Non-U.S. Gov't]. *Proceedings of the National Academy of Sciences of the United States of America, 108* (51), 20597–20602.

Tully, D. B., Collins, B. J., Overstreet, J. D., Smith, C. S., Dinse, G. E., Mumtaz, M. M., et al. (2000). Effects of arsenic, cadmium, chromium, and lead on gene expression regulated by a battery of 13 different promoters in recombinant HepG2 cells. *Toxicology and Applied Pharmacology, 168*, 79–90.

Vabulas, R. M., & Hartl, F. U. (2005). Protein synthesis upon acute nutrient restriction relies on proteasome function. [Research Support, Non-U.S. Gov't]. *Science, 310*(5756), 1960–1963.

Valenzuela-Fernandez, A., Cabrero, J. R., Serrador, J. M., & Sanchez-Madrid, F. (2008). HDAC6: A key regulator of cytoskeleton, cell migration and cell-cell interactions. [Research Support, Non-U.S. Gov't Review]. *Trends in Cell Biology, 18*(6), 291–297.

Walter, P., & Ron, D. (2011). The unfolded protein response: From stress pathway to homeostatic regulation. *Science, 334*(6059), 1081–1086.

Webster, T. J., Naylor, D. J., Hartman, D. J., Hoj, P. B., & Hoogenraad, N. J. (1994). cDNA cloning and efficient mitochondrial import of pre-mtHSP70 from rat liver. *DNA and Cell Biology, 13*(12), 1213–1220.

Wehner, K. A., Schutz, S., & Sarnow, P. (2010). OGFOD1, a novel modulator of eukaryotic translation initiation factor 2alpha phosphorylation and the cellular response to stress. *Molecular and Cellular Biology, 30*(8), 2006–2016.

Wek, R. C., & Cavener, D. R. (2007). Translational control and the unfolded protein response. *Antioxidants & Redox Signaling, 9*(12), 2357–2371.

Wek, R. C., Jackson, B. M., & Hinnebusch, A. G. (1989). Juxtaposition of domains homologous to protein kinases and histidyl-tRNA synthetases in GCN2 protein suggests a mechanism for coupling GCN4 expression to amino acid availability. *Proceedings of the National Academy of Sciences of the United States of America, 86*(12), 4579–4583.

Wek, S. A., Zhu, S., & Wek, R. C. (1995). The histidyl-tRNA synthetase-related sequence in the eIF-2 alpha protein kinase GCN2 interacts with tRNA and is required for activation in response to starvation for different amino acids. *Molecular and Cellular Biology, 15* (8), 4497–4506.

Westerheide, S. D., & Morimoto, R. I. (2005). Heat shock response modulators as therapeutic tools for diseases of protein conformation. *The Journal of Biological Chemistry, 280*(39), 33097–33100.

Witta, S. E., Gemmill, R. M., Hirsch, F. R., Coldren, C. D., Hedman, K., Ravdel, L., et al. (2006). Restoring E-cadherin expression increases sensitivity to epidermal growth factor receptor inhibitors in lung cancer cell lines. [Research Support, N.I.H., Extramural Research Support, Non-U.S. Gov't]. *Cancer Research, 66*(2), 944–950.

Workman, P., Burrows, F., Neckers, L., & Rosen, N. (2007). Drugging the cancer chaperone HSP90: Combinatorial therapeutic exploitation of oncogene addiction and tumor stress. *Annals of the New York Academy of Sciences, 1113*, 202–216.

Wouters, B. G., van den Beucken, T., Magagnin, M. G., Koritzinsky, M., Fels, D., & Koumenis, C. (2005). Control of the hypoxic response through regulation of mRNA translation. *Seminars in Cell & Developmental Biology, 16*(4–5), 487–501.

Wu, Y. C., Yen, W. Y., Lee, T. C., & Yih, L. H. (2009). Heat shock protein inhibitors, 17-DMAG and KNK437, enhance arsenic trioxide-induced mitotic apoptosis. *Toxicology and Applied Pharmacology, 236*, 231–238.

Xu, W., Trepel, J., & Neckers, L. (2011). Ras, ROS and proteotoxic stress: A delicate balance. *Cancer Cell*, *20*(3), 281–282.
Yoshida, H., Matsui, T., Yamamoto, A., Okada, T., & Mori, K. (2001). XBP1 mRNA is induced by ATF6 and spliced by IRE1 in response to ER stress to produce a highly active transcription factor. *Cell*, *107*(7), 881–891.
Zhan, K., Vattem, K. M., Bauer, B. N., Dever, T. E., Chen, J. J., & Wek, R. C. (2002). Phosphorylation of eukaryotic initiation factor 2 by heme-regulated inhibitor kinase-related protein kinases in Schizosaccharomyces pombe is important for fesistance to environmental stresses. *Molecular and Cellular Biology*, *22*(20), 7134–7146.
Zhang, Y., Li, N., Caron, C., Matthias, G., Hess, D., Khochbin, S., et al. (2003). HDAC-6 interacts with and deacetylates tubulin and microtubules in vivo. *The EMBO Journal*, *22*(5), 1168–1179.
Zhu, K., Chan, W., Heymach, J., Wilkinson, M., & McConkey, D. J. (2009). Control of HIF-1alpha expression by eIF2 alpha phosphorylation-mediated translational repression. *Cancer Research*, *69*(5), 1836–1843.
Zhu, K., Dunner, K., Jr., & McConkey, D. J. (2010). Proteasome inhibitors activate autophagy as a cytoprotective response in human prostate cancer cells. *Oncogene*, *29*(3), 451–462.

Intrinsic and Extrinsic Apoptotic Pathway Signaling as Determinants of Histone Deacetylase Inhibitor Antitumor Activity

Geoffrey M. Matthews[*,†,2], **Andrea Newbold**[*,†,2], **Ricky W. Johnstone**[*,†,1]

[*]Cancer Therapeutics Program, Gene Regulation Laboratory, The Peter MacCallum Cancer Centre, St. Andrews Place, East Melbourne, Victoria, Australia
[†]The Sir Peter MacCallum Department of Oncology, University of Melbourne, Parkville, Victoria, Australia
[1]Corresponding author: e-mail address: ricky.johnstone@petermac.org
[2]These authors contributed equally to this work.

Contents

Advances in Cancer Research, Volume 116
ISSN 0065-230X
http://dx.doi.org/10.1016/B978-0-12-394387-3.00005-7

Abstract

Histone deacetylase inhibitors (HDACi) can elicit a range of biological responses that impede the growth and/or survival of tumor cells. Depending on the physiological context, HDACi can induce apoptosis via two well-defined apoptotic pathways; the intrinsic/mitochondrial pathway and the death receptor (DR)/extrinsic pathway. A number of groups have demonstrated that overexpression of prosurvival Bcl-2 family members significantly reduces HDACi-mediated tumor cell death and therapeutic efficacy in preclinical models. In many cases, HDACi activate the intrinsic pathway via upregulation of a number of proapoptotic BH3-only Bcl-2 family genes including *Bim*, *Bid*, and *Bmf*. Additionally, HDACi can engage the extrinsic pathway through upregulation of DR expression, reductions in c-FLIP, and upregulation of ligands such as TRAIL. Overall, it appears that activation of the intrinsic apoptotic pathway is the predominant mechanism of HDACi-induced tumor cell death; however, the DR pathway may also be engaged, either to amplify the apoptotic signal through the intrinsic pathway or to directly induce cell death.

1. INTRODUCTION

Histone deacetylase inhibitors (HDACi) can elicit a range of biological responses including inhibition of cell proliferation, induction of cellular differentiation, modulation of the immune response, suppression of angiogenesis, and the induction of cell death via apoptosis, autophagy, or necrosis (Bolden, Peart, & Johnstone, 2006; Lindemann, Gabrielli, & Johnstone, 2004; Marks et al., 2001; Monneret, 2005). While normal cells are relatively resistant to the apoptotic effects of HDACi, these agents are potent inducers of tumor cell death *in vitro* and *in vivo*

(Dokmanovic & Marks, 2005). The mechanistic basis for the tumor-selective effects of HDACi is poorly understood but does not appear to be due to differential HDAC inhibitory activity as histones and other proteins are equivalently hyperacteylated in both normal and tumor cells following HDACi treatment (Lindemann et al., 2007; Marks et al., 2001). Depending on the physiological context, HDACi-induced apoptosis can occur via two well-defined apoptotic pathways; the intrinsic pathway and the death receptor (DR)/extrinsic pathway (Frew et al., 2008). Herein, we discuss the ability of HDACi to utilize either or both apoptotic pathways as well as their effects on the individual genes and proteins within these pathways, in particular, the Bcl-2 family.

2. THE INTRINSIC APOPTOSIS PATHWAY

The intrinsic apoptotic pathway is initiated by diverse "cell stress" stimuli including UV- and γ-irradiation, chemotherapeutic drugs, hypoxia, hyperthermia, viral infections, free radicals, and the removal of certain growth factors, hormones, and cytokines (Cory & Adams, 2002; Elmore, 2007). This pathway is regulated by pro- and antiapoptotic Bcl-2 superfamily proteins (Cory & Adams, 2002). The integrity of the mitochondrial membrane is maintained by the prosurvival Bcl-2 proteins (Bcl-2, Bcl-X_L, Bcl-w, Mcl-1, and Bcl-2A1 (A1)) (Cory, Huang, & Adams, 2003) that directly or indirectly constrain the activity of proapoptotic Bcl-2 proteins. The proapoptotic Bcl-2 family of proteins consists of two groups: (i) multidomain proteins Bax, Bak, and Bok that contain three Bcl-2 homology (BH) domains (BH1–3); (ii) "BH3-only" family of proteins (Bim, Bad, Bid, Bik, Bmf, Puma, Noxa, and Hrk) that share only the BH3 interaction domain (Adams & Cory, 1998). The balance between expression and activity of the pro- and antiapoptotic Bcl-2 family proteins sets the apoptotic threshold within a given cell and ultimately can determine the fate of that cell. As sensors of cell stress, activated BH3-only proteins initiate the intrinsic apoptosis pathway either by binding to prosurvival proteins, thereby unleashing Bax and/or Bak (Chen et al., 2005; Kim et al., 2006), or by directly binding to and activating the multidomain proapoptotic Bcl-2 proteins (Ren et al., 2010; Vogel et al., 2012; Weber et al., 2007). Once activated, Bax and Bak form oligomers and participate in the formation of pores within intracellular membranes including the nucleus, endoplasmic reticulum, and mitochondrial outer membrane (Brunelle & Letai, 2009).

Mitochondrial outer membrane permeabilization results in the release of proapoptotic proteins including cytochrome *c*, second mitochondria-derived activator of caspase (Smac) (also known as Diablo), and endonuclease G into the cytosol (Johnstone, 2002). SMAC is able to bind and suppress the inhibitors of apoptosis (IAP) family of proteins (Duckett, 2005). Cytochrome *c* associates with procaspase-9 and the adaptor molecule Apaf-1 to form the "apoptosome." The apoptosome is then able to initiate full caspase-9 activation. Caspase-9 subsequently cleaves and activates downstream effector caspases (e.g., caspase-3 and caspase-7), which induces the morphological changes associated with apoptosis (Johnstone, 2002; Peart et al., 2003). In addition to activation of the intrinsic pathway by stress stimuli, DRs (e.g., DR-4/DR-5) can also activate the intrinsic pathway through caspase 8-mediated cleavage and activation of Bid (Strasser, O'Connor, & Dixit, 2000). In this way, Bid serves as a molecular link between the two pathways and can amplify the death-receptor-stimulated response. A more detailed explanation of Bid and the extrinsic pathway including the effects of HDACi on this pathway will be discussed below.

Many chemotherapeutic drugs induce tumor cell death through activation of the intrinsic pathway and genetic lesions within this pathway, including overexpression of prosurvival Bcl-2 family proteins or loss-of-function/expression of proapoptotic members, is often tumor-promoting (Johnstone, Ruefli and Lowe, 2002). Accordingly, inactivation of the intrinsic pathway can have the dual effect of driving tumorigenesis and concomitantly mediating drug resistance. This constitutes a unique form of intrinsic or acquired chemoresistance, separate from previously identified mechanisms such as drug efflux, drug metabolism, and drug inactivation (Johnstone et al., 2002).This also explains why elevated expression of a number of Bcl-2-family proteins can be found in many different types of cancers upon relapse after initial rounds of chemotherapy (Kaufmann et al., 1998; Tu et al., 1998; Wuilleme-Toumi et al., 2005).

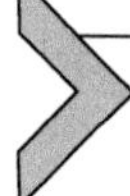

3. HDACi INDUCE APOPTOSIS VIA THE INTRINSIC APOPTOSIS PATHWAY

HDACi-mediated tumor cell death via the mitochondrial pathway has been reported both *in vitro* and *in vivo* by a number of independent laboratories (Bernhard et al., 1999; Lindemann et al., 2007; Peart et al., 2003;

Ruefli et al., 2001; Shao, Gao, Marks, & Jiang, 2004). For example, HDACi can induce the cleavage of Bid and the release of reactive oxygen species (Ruefli et al., 2001), cause disruption of the mitochondrial membrane and the release of cytochrome *c* (Shao et al., 2004), induce full processing of caspase-2 and activation of the caspase cascade (Peart et al., 2003), and/or induce the upregulation of a number of proapoptotic genes such as *Bim*, *Bax*, and *Bak* (Zhang, Gillespie, Borrow, & Herse, 2004).

Blocking the intrinsic pathway via the overexpression of prosurvival Bcl-2 family members inhibits HDACi-induced tumor cell death in multiple cell types (Bernhard et al., 1999; Peart et al., 2003; Vrana et al., 1999). Mitsiades and colleagues demonstrated that overexpression of Bcl-2 in patient-derived B-cell tumor cell lines completely abrogated vorinostat-induced apoptosis and blocked vorinostat-induced cleavage of caspases-8, -9, and -3 (Mitsiades et al., 2003). A second group reported that while SBHA could induce caspase-dependent apoptosis and downregulate the expression of Bcl-X_L and Mcl-1, this effect was lost following the overexpression of Bcl-2 in their system (Zhang, Gillespie, et al., 2004) while overexpression of Bcl-X_L could inhibit sodium butyrate-induced apoptosis via blocking of the mitochondrial pathway in mesothelioma cell lines (Cao, Mohuiddin, Ece, McConkey, & Smythe, 2001). Moreover, we have demonstrated that overexpression of Bcl-2 in an *in vivo* model of Burkitt's lymphoma renders the tumors completely resistant to HDACi treatment (Ellis et al., 2009; Lindemann et al., 2007). Taken together these studies indicate that HDACi can induce cell death via the mitochondrial pathway both *in vitro* and *in vivo* and indicate a functional relationship between HDACi and regulated expression or activity of Bcl-2 family proteins.

It appears that HDACi can activate the intrinsic apoptosis cascade through multiple different mechanisms. Microarray studies assessing the global changes in gene expression following HDACi treatment indicated that the selective up- and downregulated expression of a number genes and proteins in a manner that favors a proapoptotic biological response may be a common mechanism of action of HDACi (Glaser et al., 2003; Mitsiades et al., 2004; Peart et al., 2005). As detailed below, a range of studies using a variety of different cell lines and HDACi have indicated the potential importance of distinct pro- and antiapoptotic Bcl-2 family proteins in mediating the anti-tumor effects of these agents.

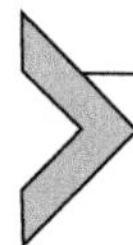

4. THE EFFECTS OF HDACi ON ANTIAPOPTOTIC BCL-2 FAMILY GENES

Overexpression of Bcl-2 or Bcl-X_L antiapoptotic family members can block HDACi-induced apoptosis (Guo et al., 2004; Johnstone, 2002; Lindemann et al., 2007; Wiegmans et al., 2011). Accordingly, HDACi can transcriptionally repress the expression of prosurvival genes encoding BCL-2, BCL-X_L, and MCL-1 (Rosato et al., 2006; Zhang, Adachi, Zhao, Kawamura, & Imai, 2004). Furthermore, gene expression analysis of human acute T-cell leukemia cell lines and chronic lymphocytic leukemia (CLL) cell lines, showed that prior to vorinostat-induced apoptosis, mRNA encoding survivin, BCL-W, and c-FLIP were downregulated (Sanda et al., 2007). Nishioka et al. demonstrated that MS-275 decreased expression of BCL-2 and MCL-1 in acute myeloid leukemia (AML) cells (Nishioka et al., 2008), while we demonstrated the downregulation of BCL-2 in cycling but not arrested T-cell leukemia cell lines following vorinostat treatment (Peart et al., 2005). Other studies indicated that butyric acid treatment of murine splenic T cells could decrease the expression of Bcl-2 and Bcl-X_L (Kurita-Ochiai, Ochiai, & Fukushima, 2000) and treatment of human leukemia cell lines (U937 and Jurkat) with LAQ824 resulted in decreased expression of X-IAP and Mcl-1.

While these studies collectively hint that decreased expression of prosurvival Bcl-2 family genes may be a trigger for HDACi-induced apoptosis, upregulation of prosurvival genes following HDACi treatment may dampen the apoptotic effects of these agents. For example, Inoue and colleagues reported that MCL-1 was upregulated in K562 cells following treatment with LBH-589, leading to a resistance to apoptosis (Inoue, Walewska, Dyer, & Cohen, 2008). Therefore, while HDACi-mediated apoptosis is often associated with decreased expression of antiapoptotic BCL-2 family genes, these agents are often not able to induce apoptosis in tumors overexpressing Bcl-2 prosurvival proteins. Hence, it is not surprising that small molecule inhibitors of Bcl-2 and Bcl-X_L have now been tested in combination with HDACi to provide synergistic tumor cell death. The BH3-only mimetic ABT-737 can sensitize *Eμ-myc* lymphomas overexpressing Bcl-2 or Bcl-X_L to apoptosis mediated by vorinostat or valproic acid (VPA) following a combination treatment of ABT-737 with vorinostat or VPA (Whitecross et al., 2009). Further studies have shown that the effectiveness of this combination was due to concomitant upregulation of the BH3-only gene *BMF* following HDACi treatment and inhibition of Bcl-2 and Bcl-X_L by ABT-737 (Wiegmans et al., 2011). Other groups have also demonstrated that combining small molecule inhibitors of prosurvival Bcl-2 family proteins with HDACi treatment can induce synergistic cell death (Martin et al., 2009; Wei et al., 2010). Wei et al. have recently shown that GX15-070, which inhibits BCL-2, BCL-X_L, and MCL-1, in combination with two different HDACi (MGCD0103 and vorinostat) results in enhanced killing of primary AML cells compared to single agent treatment (Wei et al., 2010). Furthermore, the effects of GX15-070 treatment could be phenocopied through RNAi-mediated knockdown of BCL-2, BCL-X_L, or MCL-1. Interestingly, the combination treatment was more effective following knockdown of all three Bcl-2 family members together rather than a single knockdown of each gene separately (Martin et al., 2009).

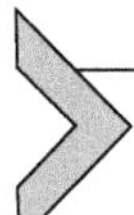

5. EFFECTS OF HDACi ON PROAPOPTOTIC BCL-2 FAMILY MEMBERS

Transcriptional and posttranscriptional induction of the proapoptotic Bcl-2 family proteins Bid, Bim, Bmf, Puma, Noxa, Bad, Bax, and Bak, following HDACi treatment has been reported by numerous groups (Bolden et al., 2006; Lindemann et al., 2007; Pacey et al., 2010; Peart et al., 2005; Puthalakath et al., 2007; Puthalakath & Strasser, 2002; Xargay-Torrent et al., 2011; Zhang & Fang, 2004). Moreover, we and others have shown that knockout/knockdown of key BH3-only genes can reduce the apoptotic and therapeutic effects of HDACi (Lindemann et al., 2007; Wiegmans et al., 2011; Xargay-Torrent et al., 2011). Therefore, the following section will describe the current literature regarding the effects of HDACi on various BH3-only proapoptotic molecules (Table 5.1).

6. EFFECTS OF HDACi ON BID

A major role for Bid in HDACi-mediated apoptosis has been reported (Emanuele et al., 2007; Fantin & Richon, 2007; Lindemann et al., 2007; Xargay-Torrent et al., 2011; Yang, Balch, et al., 2009). Exposure of human leukemia cells to vorinostat in combination with recombinant tumor necrosis factor (TNF)-related apoptosis-inducing ligand (TRAIL) revealed enhanced Bid activation (Rosato et al., 2003), and gene

Table 5.1 Proapoptotic family members involved in HDACi-induced activation of the intrinsic apoptosis pathway

Key molecule(s)	Comment	Functional importance confirmed	References
Bid	HDACi in combination with recombinant TRAIL enhances Bid activation	Yes	Rosato, Almenara, Dai, and Grant (2003)
	HDACi treatment can result in induction of the DR5 death receptor resulting in Bid cleavage, activation of caspases-8 and -10 and sensitivity to TRAIL-induced apoptosis		Nakata et al. (2004)
	The protease responsible for Bid cleavage following HDACi treatment remains unknown		Mitsiades et al. (2004)
Bim	Upregulated by different HDACi in numerous different tumor cell types	Yes	Dai et al. (2011), Inoue, Mai, Dyer, and Cohen (2006), Lindemann et al. (2007), Zhang, Adachi et al. (2004),
	BIM has been linked to HDACi by the direct hyperacetylation of histone H4 within the BIM promoter		Xargay-Torrent et al. (2011)
	Bim can be induced by indirect effects on transcription factors such as E2F-1, FoxO1		Yang, Zhao, et al. (2009), Zhao et al. (2005)
	Bim can also be regulated at a posttranslational level via phosphorylation		Ley, Ewings, Hadfield, and Cook (2005), Nishioka et al. (2008)

Table 5.1 Proapoptotic family members involved in HDACi-induced activation of the intrinsic apoptosis pathway—cont'd

Key molecule(s)	Comment	Functional importance confirmed	References
Bmf	Absence of Bmf results in a severely attenuated apoptotic response to a HDACi	Yes	Labi et al. (2008) and Wiegmans et al. (2011)
	Transcriptional activation of BMF has been found to be due to H3 and/or H4 hyperacetylation in the promoter region of the *BMF* gene		Wiegmans et al. (2011), Xargay-Torrent et al. (2011)
	HDACi may induce Bmf expression via the transcription factor Smad4		Ramjaun et al. (2006)
Bad	HDACi can induce or repress expression of Bad depending on the cell type	No	Baumann et al. (2012), Maiso et al. (2009)
	The activation of Bad is primarily caused by posttranslational events		Cory et al. (2003), Zacharias et al. (2011)
	Enhanced transcription of Bad has been demonstrated in CLL cells with vorinostat		Pérez-Perarnau et al. (2011)
	Conversely, treatment of OPM-2 cells with CR2408 induced a strong downregulation of Bad protein expression		Baumann et al. (2012)
Noxa/ Puma	Upregulated following treatment with a number of structurally diverse HDACi	Yes (Noxa)	Shankar et al. (2005), and Singh, Shankar, and Srivastava (2005)
	HDACi induce histone H3 and/or H4		Kim et al. (2010), Ramsey, He, Forster,

Continued

Table 5.1 Proapoptotic family members involved in HDACi-induced activation of the intrinsic apoptosis pathway—cont'd

Key molecule(s)	Comment	Functional importance confirmed	References
.	acetylation in the promoter regions of both PUMA and NOXA		Ory, and Ellisen (2011), and Xargay-Torrent et al. (2011)
	HDACi-mediated acetylation of p53 can lead to upregulation of Puma/Noxa and subsequent induction of apoptosis	No (Puma)	Wiegmans et al. (2011) and Yamaguchi et al. (2009)
Bax/Ba	HDACi mediate their effects via indirect rather than direct mechanisms	No	Chirakkal et al. (2006), Kerr et al. (2012), Waby et al. (2010), and Yamaguchi et al. (2009)
	Transcriptional upregulation via acetylation of SP1/SP3 in the BAK promoter region has been reported		Chirakkal et al. (2006)
	The acetylation of Ku70 promotes Bax translocation to the mitochondrial and induction of apoptosis		Amsel, Rathaus, Kronman, and Cohen (2008)
	p53 acetylation also disrupts the Ku70-Bax complex enabling its translocation from the nucleus to the cytoplasm		Yamaguchi et al. (2009)

expression analysis following LAQ824 treatment of B16 melanoma cells also showed an upregulation of Bid (Vo et al., 2006). A number of HDACi have been shown to mediate Bid cleavage and activation prior to mitochondrial membrane damage (Mitsiades et al., 2003; Peart et al., 2003; Ruefli et al., 2001) indicating that this is a trigger for induction of the intrinsic apoptosis pathway following HDACi treatment. Furthermore, Mitsiades et al. showed that vorinostat-induced apoptosis in multiple myeloma

(MM) cells promoted cleavage of Bid, and that cleavage of Bid occurred in the absence of caspase activity (Mitsiades et al., 2004). The protease responsible for Bid cleavage following HDACi treatment remains unknown; however, there is circumstantial evidence to suggest that an HDACi-responsive gene is directly or indirectly involved in this activating event. For example, treatment of tumor cells with inhibitors of gene transcription (actinomycin D) or translation (cycloheximide) prevented vorinostat-mediated Bid cleavage (Ruefli et al., 2001). In addition, HDACi treatment can result in induction of the DR5 DR resulting in Bid cleavage, activation of caspases-8 and -10, and sensitivity to TRAIL-induced apoptosis (Nakata et al., 2004).

The importance of Bid in mediating HDACi-induced apoptosis has been established by studies demonstrating that genetic knockout of Bid in the *Eμ-myc* model of Burkitts lymphoma inhibited the apoptotic and therapeutic effects of vorinostat (Lindemann et al., 2007). These tumors retained their sensitivity to etoposide clearly demonstrating the selective importance of Bid for vorinostat-mediated apoptosis (Lindemann et al., 2007).

7. EFFECTS OF HDACi ON BIM

Bim is upregulated by different HDACi in numerous different tumor cell types including (i) *Eμ-myc* lymphomas with vorinostat (Lindemann et al., 2007); (ii) AML and acute lymphoblastic leukemia cells with LBH-589 (Dai et al., 2011); (iii) primary CLL cells, Jurkat, and K562 cells with Trichostatin A (TSA) (Inoue et al., 2006); (iv) suberic bishydroxamic acid (SBHA)- and oxamflatin-treatment of MM cells (Chen, Dai, Pei, & Grant, 2009); and (iv) gastrointestinal adenocarcinomas treated with *m*-carboxycinnamic acid bis-hydroxamide (CBHA), romidepsin, and MS-275 (Zhang, Adachi et al., 2004). In all examples, upregulated expression of BIM mRNA correlated with increased HDACi-induced apoptosis. In addition, enhanced levels of three Bim isoforms were observed following treatment of human hepatoma cell lines with vorinostat (Emanuele et al., 2007).

7.1. Bim induction by direct promoter hyperacetylation

HDACi-induced expression of Bim may occur through direct hyper-acetylation at the BIM promoter (Zhang, Yuan, et al., 2005) or via the recruitment of a number of different transcription complexes to its promoter region(s) (Zhao et al., 2005). For example, increased Bim expression in mantle cell lymphoma (MCL) following treatment with vorinostat was concomitant with histone H4 acetylation within the BIM promoter region and this correlated with death of these tumor cells (Xargay-Torrent et al., 2011). In contrast, no change to BIM mRNA expression was detected in vorinostat-treated cells that were insensitive to this agent (Xargay-Torrent et al., 2011). Chromatin immunoprecipitation (ChIP) assays for hyperacetylated histone H4 demonstrated a two- to sixfold increase in histone acetylation at the BIM promoter following treatment of sensitive MCL cell lines with vorinostat, while very little hyperacetylation of the BIM promoter regions were observed in insensitive cell lines (Xargay-Torrent et al., 2011). Taken together, these results indicate that the vorinostat-mediated transcriptional activation of BIM may occur through histone H4 hyperacetylation within the gene promoter region and that this was tightly associated with sensitivity of MCL cells to HDACi-induced cell death.

7.2. Bim induction by indirect effects on transcription factors

HDACi may also indirectly influence Bim expression. For example, TSA and vorinostat can promote the recruitment of E2F-1 to the BIM promoter leading to the upregulation of Bim expression (Zhao et al., 2005). While acetylation of E2F-1 by HDACi was not reported, it was posited that hyperacetylation of additional proteins associated directly with E2F-1 or increased recruitment of histone acetyltransferases may underpin *Bim* gene induction through E2F-1 (Zhao et al., 2005). In another study, romidepsin-induced expression of Bim was dependent on acetylation of forkhead box class O1 (FoxO1) transcription factor (Yang, Zhao, et al., 2009). FoxO1 has been shown to regulate genes involved in cell cycle arrest, apoptosis, and DNA repair, among others. Importantly, FoxO1 has been recently shown to be critical in regulating Bim-induced cell death (Yang, Zhao, et al., 2009). Yang et al. demonstrated that when FoxO1 was mutated at specific lysine residues, treatment with romidepsin was unable to induce FoxO1 expression and mutant FoxO1 was unable to transactivate a reporter gene containing BIM promoter elements (Yang, Zhao, et al., 2009). These results showed for the first time that HDACi could induce apoptosis through FoxO1-mediated induction of Bim.

7.3. Induction/activation of Bim posttranslationally

Bim can be regulated at a posttranslational level via phosphorylation (Ley et al., 2005). Of particular relevance to HDACi is the tyrosine kinase, Erk. Although the precise mechanism is unknown, Erk is thought to play a role in the regulation of Bim via the Erk1/2 MAP kinase pathway

(Ley et al., 2005). Nishioka et al. reported that the levels of phospho-Erk in AML patient samples treated with MS-275 were reduced (Nishioka et al., 2008). Therefore, MS-275 may inhibit the turnover of Bim via down-regulation of Erk and possibly provide yet another HDACi-mediated mechanism for Bim induction.

7.4. Functional importance of Bim in HDACi-induced cell death

Expression profiling assays demonstrated that human leukemia and MM cells treated with SBHA could upregulate Bim, Puma, and Noxa associated with Bax/Bak activation and apoptosis induction (Chen et al., 2009). The activation of Bax/Bak was abrogated by shRNA-mediated knockdown of BIM and overexpression of Bcl-2, Bcl-X_L, or Mcl-1. Importantly, the same effects were not seen with knockdown of PUMA or NOXA indicating that BIM, but not PUMA or NOXA were important for the apoptotic effects of SBHA (Chen et al., 2009). In addition, others have also reported the suppression of HDACi-mediated apoptosis by knockdown of BIM using siRNA and concluded that Bim is one of the key intrinsic pathway proteins responsible for HDACi-mediated apoptosis (Schwulst et al., 2008; Zhao et al., 2005). In support of this data, we have also demonstrated the importance of Bim in mediating the apoptotic effects of vorinostat *in vivo*. In mice bearing *Eμ-myc* lymphomas, vorinostat induced the expression of both Bim mRNA and Bim_L protein. Knockout of *Bim* in these lymphomas resulted in a decrease in vorinostat-mediated apoptosis *in vitro* and *in vivo* and suppression of the therapeutic effects of vorinostat (Lindemann et al., 2007). Conversely, vorinostat-induced expression of *Bim* in *Eμ-myc/Bcl-2* lymphomas was not required to "prime" these cells for apoptosis mediated by ABT-737 (Wiegmans et al., 2011). These studies highlight the specific roles that BH3-only proteins play in HDACi-mediated apoptosis as single agents, compared to their effects when HDACi are used in combination with agents such as ABT-737.

8. EFFECTS OF HDACi ON BMF

Preferential upregulation of Bmf by HDACi has been reported in a broad range of cancer cells including; *Eμ-myc* lymphoma (Wiegmans et al., 2011), melanoma, colorectal adenocarcinoma, and esophageal squamous cell carcinoma cell lines (Zhang, Adachi, Kawamura, & Imai, 2005; Zhang et al., 2006). In addition, Inoue et al. reported that LBH-589- and vorinostat-induced apoptosis in CLL cells was preceded by the upregulation of Bmf protein (Inoue, Riley, Gant, Dyer, & Cohen, 2007). Moreover, it has been shown that vorinostat may actually "prime" tumors overexpressing Bcl-2 for rapid ABT-737-mediated apoptosis by inducing expression of *Bmf* in *Eμ-myc* lymphomas (Wiegmans et al., 2011).

8.1. Bmf induction by direct promoter hyperacetylation

HDACi may induce expression of Bmf through direct or indirect effects on the promoter (Ralli et al., 2012; Wiegmans et al., 2011; Zhang et al., 2006; Zhang, Adachi et al., 2005). Treatment of a human squamous carcinoma cell line with romidepsin or CBHA resulted in enhanced BMF transcription concomitant with enhanced histone acetylation within the promoter region (Zhang, Adachi et al., 2005). In addition, we reported an upregulation of the *BMF* gene following treatment with vorinostat in *Eμ-myc* cell lines overexpressing Bcl-2 (Wiegmans et al., 2011). ChIP assays performed on the *Eμ-myc/Bcl-2* cells demonstrated robust histone H3 hyperacetylation at the BMF promoter. This data suggested that upregulation of BMF following vorinostat treatment occurs via direct histone hyperacetylation at the promoter (Wiegmans et al., 2011). Similarly, treatment of MCL lines with vorinostat resulted in enhanced BMF mRNA expression that correlated with hyperacetylation of histone H4 in the *BMF* gene promoter region (Xargay-Torrent et al., 2011).

8.2. Functional importance of Bmf in HDACi-induced cell death

There are numerous reports demonstrating that expression of Bmf is vital for the induction of HDACi-mediated apoptosis (Labi et al., 2008; Wiegmans et al., 2011; Zhang et al., 2006). In one study, knockdown of Bmf in human squamous carcinoma cells resulted in decreased tumor cell apoptosis mediated by romidepsin and CBHA (Zhang et al., 2006). In a second study, knockdown of *Bmf* in *Eμ-myc/Bcl-2* lymphomas resulted in a severely attenuated apoptotic response to a vorinostat/ABT-737 combination (Wiegmans et al., 2011). Finally, another study analyzing the development of *Bmf*$^{-/-}$ mice reported that *Bmf*-deficient thymocytes which were normally sensitive to most apoptotic stimuli tested were abnormally resistant to vorinostat (Labi et al., 2008). Indicative of the possible cell-type dependent role of different BH3-only proteins in mediating HDACi-induced apoptosis, knockdown of BMF in CLL cells using siRNA did not prevent the HDACi-induced loss of mitochondrial membrane potential, caspase processing, or phosphatidylserine externalization following treatment with either LBH-589 or vorinostat (Inoue et al., 2007).

9. EFFECTS OF HDACi ON BAD

HDACi can either induce or repress expression of Bad depending on the cell type (Baumann et al., 2012; Khandelwal, Gediya, & Njar, 2008; Maiso et al., 2009; Pérez-Perarnau et al., 2011; Sawa et al., 2001; Singh et al., 2005; Strait et al., 2005). HDACi can induce Bad transcription in a number of T-cell leukemia cell lines (Peart et al., 2003); however, it is not known whether this occurs through direct promoter hyperacetylation or by indirect effects on currently unidentified transcription factors. The activity of Bad is primarily regulated by posttranslational events (Cory et al., 2003). Survival signals stimulating the activation of Akt/protein kinase B have been reported to result in the phosphorylation of Bad, thus allowing the binding of Bad to 14-3-3. Once Bad is complexed with 14-3-3 in the cytosol, it is then sequestered by Bcl-X_L/Bcl-2 and unable to induce apoptosis (Fu, Subramanian, & Masters, 2000). It is has recently been reported that VPA and vorinostat can increase the kinase activity of Akt thus leading to a decrease in Bad protein expression and an increased antiapoptotic response. This would imply that in this particular scenario, HDACi can have indirect effects on posttranslational changes in Bad expression (Zacharias et al., 2011).

Maiso et al. treated AML cells with the pan-HDACi, LBH-589, and observed significant losses to the viability of both cell lines and primary AML cells from patients (Maiso et al., 2009). Importantly, apoptosis induction in this study was associated with modest increases in Bad (and Bak) protein expression. Unfortunately, it was not determined whether these changes to Bad expression were related to transcriptional or posttranscriptional modification by HDACi treatment. Two earlier studies reported TSA, vorinostat, MS-275, and/or CBHA-induced expression of Bad in ovarian and breast cancer cells (Singh et al., 2005; Strait et al., 2005). The enhanced expression of BAD mRNA has also been demonstrated following treatment of CLL cells with vorinostat (Pérez-Perarnau et al., 2011). Recently, the apoptosis-inducing properties of the novel pan-HDACi CR2408 was compared to vorinostat in MM cell lines and in primary human MM cells. Initial assessment proved that CR2408 was significantly more potent than vorinostat in inhibiting all HDACs (HDACs 1–11) and led to the induction of apoptosis in a concentration-dependent manner.

Interestingly, treatment of OPM-2 cells with CR2408 induced a strong downregulation of Bad protein levels by Western blot (while Bim was upregulated) (Baumann et al., 2012). Therefore, it appears that the regulation of Bad by HDACi does not always correlate with the apoptotic effects of these agents. To date, no studies have been reported assessing the functional importance of Bad following HDACi-mediated induction of apoptosis through gene knockout/knockdown studies.

10. EFFECTS OF HDACi ON NOXA/PUMA

A number of studies have reported the upregulation of Noxa and/or Puma in cell lines following treatment with a number of structurally diverse HDACi (Inoue et al., 2007; Shankar et al., 2005; Singh et al., 2005; Xargay-Torrent et al., 2011).

10.1. Noxa/Puma induction by direct promoter hyperacetylation

HDACi can induce gene expression of NOXA and PUMA concomitant with histone hyperacetylation within their promoter regions (Ramsey et al., 2011; Xargay-Torrent et al., 2011). A significant correlation has been found between the induction of NOXA mRNA levels and an increase in H4 acetylation within the NOXA promoter region following treatment with vorinostat (Xargay-Torrent et al., 2011). Similarly, another group found that while TSA treatment did not affect the expression of PUMA in head and neck squamous cell carcinoma cell lines, TSA did increase the expression of NOXA. The authors also reported that TSA upregulated NOXA expression possibly through hyperacetylation of histone H3 at the promoter (Kim et al., 2010). Fritsche and colleagues also reported an increase in NOXA mRNA in conjunction with an increase in NOXA promoter activity following treatment with VPA (Fritsche et al., 2009). The authors used ChIP to demonstrate increased levels of acetylated histone H3 and binding of RNA polymerase II to the NOXA promoter (Fritsche et al., 2009).

10.2. Noxa/Puma induction by indirect effects on transcription factors

PUMA and NOXA are direct transcriptional targets of p53 (Nakano & Vousden, 2001; Oda et al., 2000). While HDACi can induce tumor cell apoptosis independently of p53 (Lindemann et al., 2007), these agents can induce hyperacetylation and subsequent activation of p53. Activated p53 can induce apoptosis through the induction of PUMA and NOXA (Yamaguchi et al., 2009). It is therefore possible that HDACi can induce NOXA/PUMA expression indirectly through the acetylation and activation of p53. In agreement with this, it has been shown that lymphomas with knockout or mutation of p53 showed significantly reduced expression of both NOXA and PUMA following vorinostat treatment (Wiegmans et al., 2011). Induction of PUMA mRNA by TSA also correlated with increased histone H4 acetylation at a p63 binding site within the PUMA promoter (Ramsey et al., 2011). Gene knockdown studies demonstrated that HDACi-induced PUMA upregulation was p63 dependent (Ramsey et al., 2011).

10.3. Functional importance of Noxa/Puma in HDACi-induced cell death

Knockdown of NOXA protected MCL cells from vorinostat-induced apoptosis (Xargay-Torrent et al., 2011) and similarly Inoue et al. reported that TSA- or LBH-589-mediated apoptosis in primary CLL cells was suppressed using siRNA specific for NOXA (Inoue et al., 2007). We also demonstrated the importance of NOXA in mediating the apoptotic effects of vorinostat combined with ABT-737 through knockout of *Noxa* in *Eμ-myc/Bcl-2* cells that suppressed the apoptotic response to the combination treatment (Wiegmans et al., 2011). To date, the effects of HDACi-induced apoptosis on Puma knockout/knockdown cells or *Puma* knockout mice have not been described.

11. EFFECTS OF HDACi ON BAX AND/OR BAK

In addition to changes in expression of BH3-only proapoptotic proteins by HDACi, HDACi-mediated regulation of Bax and Bak have also been reported (Chirakkal et al., 2006; Maiso et al., 2009; Premkumar, Jane, Agostino, Didomenico, & Pollack, 2011; Rosato et al., 2008; Waby et al., 2010). Specifically, Bax and/or Bak can be activated/upregulated in response to a variety of HDACi in a cell-type-specific manner (Kerr et al., 2012; Srivastava, Kurzrock, & Shankar, 2010; Thomas, Thurn, Bicaku, Marchion, & Munster, 2011; Waby et al., 2010). However, it appears that HDACi mediate their effects on Bax and Bak via indirect, rather than direct, mechanisms as outlined below.

11.1. Indirect effects of HDACi on Bax and Bak

An increase in the levels of both BAX and BAK mRNA have been observed in colon cancer cell lines following treatment with butyrate (Chirakkal et al., 2006). The coincubation of HCT116 cells with cycloheximide and butyrate prevented the upregulation of BAK and significantly reduced the level of HDACi-mediated apoptosis. The region of the BAK promoter associated with butyrate-induced transcription was related to an SP1/SP3 binding site and butyrate treatment appears to upregulate BAK expression by enhanced binding of SP3 to this promoter sequence. This has recently been confirmed by ChIP assays using both butyrate and TSA (Waby et al., 2010). Using ChIP, Waby et al. showed an association between HDACs 1 and 2 with SP1 and SP3 (Waby et al., 2010). Taken together, this suggests a specific role for class I HDACs in the deacetylation of SP1 and SP3 and the induction of HDACi-mediated apoptosis.

11.2. Indirect posttranscriptional activation of Bax/Bak

Various HDACi, including vorinostat, are reported to mediate a conformational change in Bax associated with its translocation into the mitochondrial membrane (Rahmani et al., 2003) and recent investigations have suggested a role for Ku70 acetylation in this process (Amsel et al., 2008; Kerr et al., 2012; Yamaguchi et al., 2009). The Ku70–Bax complex can be disrupted by acetylation of Ku70, promoting Bax translocation to the mitochondria and the induction of apoptosis (Amsel et al., 2008; Rosato et al., 2008). Moreover, p53 acetylation disrupts the Ku70–Bax complex leading to a Bax conformational change followed by its rapid translocation from the nucleus to the cytoplasm. Yamaguchi et al. demonstrated that the expression of various p53 mutants in p53 null cells were able to enhance the onset of vorinostat- or LAQ824-mediated apoptosis (Yamaguchi et al., 2009). In contrast, knockdown of mutant p53 significantly inhibited the apoptotic effect (Yamaguchi et al., 2009). The authors also assessed apoptosis induction in cells with wild type and mutant p53 in association with loss of Ku70 expression. It was noted that when Ku70 was knocked down, HDACi were able to induce a significant degree of apoptosis in p53-null cells, but not cells expressing mutant p53 (Yamaguchi et al., 2009).

11.3. Functional importance of Bax/Bak in HDACi-induced cell death

Treatment of $Bax^{-/-}$ HCT116 cells with butyrate resulted in less apoptosis than observed in parental HCT116 cells; however, following longer-term treatment, the level of apoptosis induced by butyrate was similar between both cell types (Chirakkal et al., 2006). This infers that Bax may regulate the rate of HDACi-induced apoptosis without affecting the ultimate fate of the cells following treatment. Similar results have been demonstrated using BAX siRNA (Meng et al., 2012). In a recent study, the functional importance of Bax and/or Bak in apoptosis mediated by the combination of the HDACi SBHA with ABT-737 was measured in $Bax^{-/-}$, $Bak^{-/-}$, or $Bax^{-/-}$ $Bak^{-/-}$ double knockout mouse embryonic fibroblasts (MEFs) (Chen et al., 2009). Single knockout MEFs were only partially resistant to cell death induced by SBHA/ABT-737 combination treatment (single agent results not given), while $Bax^{-/-}$ $Bak^{-/-}$ double knockout MEFs were completely resistant to this treatment (Chen et al., 2009). It appears therefore that both Bax and Bak play critical roles in mediating the onset of apoptosis following HDACi combination therapy through their conformational change and translocation to the mitochondria or their upregulation, or both, in an HDACi- and cell-type specific manner. Unfortunately, no studies report the effects of single agent HDACi treatment in $Bax^{-/-}$ $Bak^{-/-}$ double knockout cells.

12. THE EXTRINSIC APOPTOSIS PATHWAY

HDACi have been shown to enhance signaling through the extrinsic apoptosis pathway via diverse mechanisms including upregulation of DRs, reductions in the level of cytoplasmic FLICE-like inhibitory protein (c-FLIP), and upregulation of TRAIL (also known as Apo2L and TNFS10), among others (Johnstone, Frew, & Smyth, 2008). Importantly, the effects of HDACi on the extrinsic apoptosis pathway can enhance the sensitivity of many tumor cell types to activators of this pathway, such as TRAIL.

The extrinsic apoptosis pathway is activated by the binding of FasL, TRAIL/Apo2L, and TNF-α, or agonistic monoclonal antibodies (mAb) to TRAIL-R1 (DR-4) or TRAIL-R2 (DR-5) on human cells, or DR-5 on murine cells (Fandy, Shankar, Rossy, Sausvilley, & Srivastava, 2005; Johnstone et al., 2008; Kaminskyy, Surova, Vaculova, & Zhivotovsky, 2011; Wang, 2008; Yu & Shi, 2008). TRAIL also interacts with decoy receptors TRAIL-R3 (DcR1), TRAIL-R4 (DcR2), and osteoprotegerin although these receptors are unable to transmit apoptotic signals (Gonzalvez & Ashkenazi, 2010; Johnstone et al., 2008). Oligomerization of DRs at the cell membrane leads to signal internalization and recruitment of the adapter protein FAS-associated death domain (FADD) together with procaspase-8 or procaspase-10, forming the death-inducing signaling complex (DISC). In so-called type I cells, DISC formation results in the initiation of apoptosis through the cleavage of caspase 8/10, inducing the downstream "canonical" activation of effector caspase,

caspase-3 and ultimately, the biochemical and morphological hallmarks of apoptosis. In "type II" cells, the intrinsic pathway becomes activated by caspase-8-mediated cleavage of Bid into t-Bid (Johnstone et al., 2008; Kaminskyy et al., 2011). Therefore, dependent on the cell type, the "noncanonical" activation of caspase-3 may act as the primary mechanism of TRAIL-induced cell death or may allow coactivation of both the extrinsic and intrinsic apoptosis pathways.

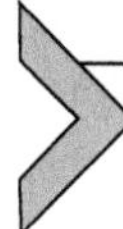

13. EFFECTS OF HDACi ON THE EXTRINSIC APOPTOSIS PATHWAY

Many investigators have observed a significant enhancement in DR-mediated apoptotic signaling in tumor cells treated with HDACi (Frew, Johnstone, & Bolden, 2009; Guo et al., 2004; Johnstone et al., 2008; Kauh et al., 2010; Vanoosten, Moore, Ludwig, & Griffith, 2005). Importantly, HDACi-mediated upregulation/activation of the extrinsic pathway is thought to result in the enhanced activation of the intrinsic apoptotic pathway, through Bid cleavage, leading to an amplification of the apoptotic response.

14. EFFECTS OF HDACi ON DR AND THEIR LIGANDS

HDACi can augment extrinsic apoptotic signaling by increasing cell surface DR and/or ligand expression (Fandy & Srivastava, 2006; Gomez-Benito, Martinez-Lorenzo, Anel, Marzo, & Naval, 2007; Nebbioso et al., 2005; Shankar et al., 2009; Vanoosten et al., 2005). For example, the enhanced expression of TRAIL following HDACi treatment has been shown to occur in a tumor-selective manner (Nebbioso et al., 2005). Treatment of the human leukemic monocyte lymphoma cell line, U937, with VPA, MS-275, or vorinostat resulted in time- and concentration-dependent inhibition of cell growth and induction of apoptosis concomitant with strong induction of TNFS10, the gene encoding TRAIL, leading to increased TRAIL protein expression.

14.1. Direct transcriptional effects of HDACi on DR5/TRAIL expression

Gene reporter assays using the TNFS10 promoter identified MS-275-responsive elements within the transcription start site (Nebbioso et al., 2005). ChIP assays demonstrated binding of HDACs 1 and 2 to the TNFS10 promoter, while HDACs 3, 4, and 5 were much less abundant. Interestingly, treatment with MS-275 resulted in histone H3 acetylation at the TNFS10 promoter and reduced occupancy of HDAC1 (Nebbioso et al., 2005). Further, ChIP assays demonstrated binding of the acetyltransferase CBP at the TNFS10 promoter after treatment with MS-275 followed by recruitment of RNA polymerase II. Promoter analyses identified a role for acetylated SP1 and SP3 binding to the TNFS10 promoter to mediate transcriptional activation of the gene following HDACi treatment. Importantly, primary AML blasts sensitive to MS-275-induced apoptosis also showed induction of TNFS10 mRNA and TRAIL protein. These results suggest that one mechanism by which HDACi may act, at least in AML, is by acetylation of SP1 and SP3 at the TNFS10 promoter driving expression of the proapoptotic protein TRAIL. In addition, ChIP assays demonstrated that enhanced mRNA expression of DR-5 by HDACi was associated with increased levels of hyperacetylated histones H3 and H4 at the DR5 promoter (Guo et al., 2004). This highlights the ability of HDACi to enhance apoptotic signaling through the extrinsic pathway by augmenting transcription of DR and their ligands through histone hyperacetylation within their gene promoters.

14.2. Posttranscriptional induction/activation of DR/ligands

Localization of DR within membrane lipid rafts can enhance apoptotic signaling (Cottin, Doan, & Riches, 2002; Hueber, Bernard, Herincs, Couzinet, & He, 2002; Vanoosten et al., 2005), and treatment of TRAIL-resistant prostate tumor cells with romidepsin resulted in increased levels of TRAIL-R1 within membrane lipid rafts and significantly increased sensitivity to TRAIL-induced apoptosis (Vanoosten et al., 2005). Indeed, this response to HDACi has also been demonstrated in MM, where TRAIL-resistant cells were sensitized to TRAIL by VPA coincident with relocalization of DRs to lipid rafts (Gomez-Benito et al., 2007). The exact mechanism leading to the relocalization of DRs to lipid rafts by HDACi remains unknown; however, posttranslational events including N-glycosylation and palmitoylation are reported to play major roles in DR4 and DR5 clustering and DISC assembly (Gonzalvez & Ashkenazi, 2010).

14.3. Functional importance of DRs/ligands in HDACi-induced cell death

The role of DRs and their ligands in MS-275-mediated apoptosis was investigated by Nebbioso et al. in AML cells using RNAi against TRAIL (Nebbioso et al., 2005). The authors showed that the therapeutic effects of MS-275 were reduced in cells with knockdown of TRAIL. In agreement with these studies, pretreatment of cancer cells with DR5-Fc significantly prevented vorinostat-induced cell death indicating a role for autocrine or paracrine TRAIL–TRAIL-R signaling in HDACi-induced cell death (Yeh, Deng, Sha, Hsiao, & Kuo, 2009). In contrast to these studies, we have observed no requirement for a functional DR pathway for HDACi-induced death of *Eμ-myc* lymphoma cells (Ellis et al., 2009; Lindemann et al., 2007). Using the *Eμ-myc* model, we demonstrated that knockout of TRAIL or overexpression of the viral serpin CrmA that inhibits caspase-8 activation downstream of DR engagement had little or no effect on the apoptotic and therapeutic activities of vorinostat or LBH-589 (Ellis et al., 2009; Lindemann et al., 2007). Therefore, the involvement of the extrinsic pathway in HDACi-induced apoptosis may be cell-type dependent.

14.4. Effects of HDACi on DISC components: FADD, caspase 8, and c-FLIP

HDACi are reported to induce recruitment of FADD and caspase-8 into the DISC (Guo et al., 2004; Inoue, Harper, Walewska, Dyer, & Cohen, 2009; Inoue et al., 2004). Treatment of TRAIL-resistant CLL cells with TSA and romidepsin, at concentrations unable to induce apoptosis led to a significant sensitization of primary CLL cells to TRAIL-induced apoptosis by facilitating the formation of an active DISC causing rapid activation of caspase-8 and apoptosis. The enhanced TRAIL-mediated apoptosis in CLL cells following HDACi treatment was associated with an upregulation of TRAIL-R2 expression (Inoue et al., 2004). However, the sensitization of CLL cells to TRAIL also occurred in the absence of TRAIL-R2 upregulation as cotreatment with cycloheximide to prevent the HDACi-mediated upregulation of TRAIL-R2 had no effect on TRAIL sensitivity. This suggests that the increase in TRAIL-R2 by HDACi was not required to enhance the degree of apoptosis in CLL and U937 cells and that other mechanisms, including recruitment of FADD and caspase-8 activation, may be more important.

HDACi have also been tested as epigenetic modulators of caspase-8-expression, particularly in certain tumor cell types where evasion of apoptosis is caused by the epigenetic silencing of this caspase (Fulda, 2008; Hacker et al., 2009; Johnstone et al., 2008; Kaminskyy et al., 2011; Lemaire et al., 2012; Subramanian, Opipari, Bian, Castle, & Kwok, 2005). In neuroblastoma and medulloblastoma, it has been reported that the inactivation of caspase-8 is a hallmark of advanced disease (Hacker et al., 2009; Johnstone et al., 2008; Subramanian et al., 2005). In addition, the silencing of caspase-8 in small cell lung cancer (SCLC), although initially thought to be methylation related, was restored by treatment with HDACi (Kaminskyy et al., 2011).

The effects of HDACi on the stabilization and degradation of FLIP has been described in MM, mesothelioma, thyroid cancer cells, among others (Carlisi et al., 2009; Frew et al., 2008; Kauh et al., 2010; Kerr et al., 2012; Martin et al., 2011; Mitsiades et al., 2005). We treated a murine breast cancer cell line, 4T1.2, with vorinostat and observed significantly reduced levels of c-FLIP at the protein level and marginal reductions at the mRNA level (Frew et al., 2008). Additionally, Carlisi and colleagues observed marked reductions in protein levels of c-FLIP in HepG2 cells following treatment with vorinostat that was not attenuated following caspase inhibition (Carlisi et al., 2009).

14.5. Direct transcriptional effects of HDACi on caspase-8, FADD, and FLIP expression

VPA and CI-994 can induce expression of caspase-8 at both the mRNA and protein levels in SCLC cells that correlated with significant histone H3 hyperacetylation (Kaminskyy et al., 2011). Recent studies demonstrated that silencing of HDAC 2 in pancreatic tumor cells led to the enhanced cleavage of caspase-8 and Bid in TRAIL-treated cells (Schuler et al., 2010). Similar results were seen when these cells were cotreated with VPA (class I selective) and TRAIL suggesting that HDAC 2 is a major determinant of TRAIL sensitivity. In SCLC, however, Kaminskyy et al. demonstrated that silencing of HDAC 2 did not significantly increase caspase-8 levels suggesting that multiple HDACs may be involved (Kaminskyy et al., 2011).

14.6. Posttranscriptional effects of HDACi on caspase-8, FADD, and FLIP

It was reported that increased recruitment of FADD to the DISC was associated with neither enhanced TRAIL binding nor oligomerization or internalization of TRAIL-R1 following HDACi treatment but occurred through increased phosphorylation of FADD (Inoue et al., 2009). The destabilization of FLIP following HDACi treatment is reported to occur via the direct acetylation of Ku70 that results in release of FLIP from Ku70, polyubiquitination, and subsequent proteosome-mediated degradation (Kerr et al., 2012). This is in agreement with recent results documenting vorinostat- and LBH-589-induced downregulation of FLIP that was associated with increased levels of ubiquitinated FLIP, loss of FLIP stability, and that FLIP downregulation is inhibited by proteosome inhibitors (Frew et al.,

2008; Hurwitz et al., 2012; Kauh et al., 2010; Safa & Pollok, 2011; Yerbes & Lopez-Rivas, 2012).

14.7. Functional importance of caspase-8/FADD/c-FLIP in HDACi-induced cell death

Knockdown of both FLIP and caspase-8 inhibited vorinostat-mediated apoptosis providing a confusing view of the functional importance of DR signaling by HDACi (Hurwitz et al., 2012). Moreover, expression of dominant negative-FADD had no significant effect on the level of apoptosis induced by a range of HDACi (Singh et al., 2005). In mouse models of lymphoma and breast carcinoma, we have demonstrated that overexpression of c-FLIP or CrmA had no effect on vorinostat-, LBH-589-, or LAQ824-induced apoptosis (Ellis et al., 2008; Frew et al., 2008; Lindemann et al., 2007). These studies suggest that although c-FLIP may often be reduced following HDACi treatment, the downregulation is not critical for these agents to kill their tumor targets.

15. CONCLUDING REMARKS

Herein, we have described that dependent on the physiological context, HDACi-induced apoptosis can occur via two functionally separate yet molecularly linked apoptotic pathways; the intrinsic pathway and the extrinsic/DR pathway (Frew et al., 2008). While both pathways may be activated, either singularly or together, functional analysis appears to favor the intrinsic pathway as the most utilized apoptotic mechanism for HDACi. Recent studies have now demonstrated that HDACi have the ability to effect global changes to prosurvival and proapoptotic Bcl-2 family gene expression indicative of a proapoptotic biological response (Nishioka et al., 2008; Sanda et al., 2007). Why HDACi appear to selectively induce proapoptotic Bcl-2 family genes while repressing expression of prosurvival genes remains unclear at present, and whether this occurs mainly through direct changes in the acetylation status at the promoter regions of these genes, or through indirect effects mediated by transcription factors and chromatin-modifying enzymes that respond to HDACi treatment remains to be ascertained.

ACKNOWLEDGMENTS

R. W. J. is a Principal Research Fellow of the National Health and Medical Research Council of Australia (NHMRC) and supported by NHMRC Program and Project Grants, Cancer Council Victoria, The Leukemia Foundation of Australia, Victorian

Breast Cancer Research Consortium, and Victorian Cancer Agency. A. N. is supported by a fellowship from the Cancer Council Victoria.

REFERENCES

Adams, J. M., & Cory, S. (1998). The Bcl-2 protein family: Arbiters of cell survival. *Science, 281*, 1322–1326.

Amsel, A. D., Rathaus, M., Kronman, N., & Cohen, H. Y. (2008). Regulation of the proapoptotic factor Bax by Ku70-dependent deubiquitylation. *Proceedings of the National Academy of Science of the United States of America, 105*, 5117–5122.

Baumann, P., Junghanns, C., Mandl-Weber, S., Strobl, S., Oduncu, F., & Schmidmaier, R. (2012). The pan-histone deacetylase inhibitor CR2408 disrupts cell cycle progression, diminishes proliferation and causes apoptosis in multiple myeloma cells. *British Journal of Haematology, 156*, 633–642.

Bernhard, D., Ausserlechner, M. J., Tonko, M., Loffler, M., Hartmann, B. L., Csordas, A., & Kofler, R. (1999). Apoptosis induced by the histone deacetylase inhibitor sodium butyrate in human leukemic lymphoblasts. *The FASEB Journal, 13*, 1991–2001.

Bolden, J. E., Peart, M. J., & Johnstone, R. W. (2006). Anticancer activities of histone deacetylase inhibitors. *Nature Reviews. Drug Discovery, 5*, 769–784.

Brunelle, J. K., & Letai, A. (2009). Control of mitochondrial apoptosis by the Bcl-2 family. *Journal of Cell Science, 122*, 437–441.

Cao, X. X., Mohuiddin, I., Ece, F., McConkey, D. J., & Smythe, W. R. (2001). Histone deacetylase inhibitor downregulation of bcl-xl gene expression leads to apoptotic cell death in mesothelioma. *American Journal of Respiratory Cell and Molecular Biology, 25*, 562–568.

Carlisi, D., Lauricella, M., D'Anneo, A., Emanuele, S., Angileri, L., Di Fazio, P., Santulli, A., Vento, R., & Tesoriere, G. (2009). The histone deacetylase inhibitor suberoylanilide hydroxamic acid sensitises human hepatocellular carcinoma cells to TRAIL-induced apoptosis by TRAIL-DISC activation. *European Journal of Cancer, 45*, 2425–2438.

Chen, S., Dai, Y., Pei, X.-Y., & Grant, S. (2009). Bim upregulation by histone deacetylase inhibitors mediates interactions with the Bcl-2 antagonist ABT-737: Evidence for distinct roles for Bcl-2, Bcl-xL, and Mcl-1. *Molecular and Cellular Biology, 29*, 6149–6169.

Chen, L., Willis, S. N., Wei, A., Smith, B. J., Fletcher, J. I., Hinds, M. G., Colman, P. M., Day, C. L., Adams, J. M., & Huang, D. C. S. (2005). Differential targeting of prosurvival Bcl-2 proteins by their BH3-only ligands allows complementary apoptotic function. *Molecular Cell, 17*, 393–403.

Chirakkal, H., Leech, S. H., Brookes, K. E., Prais, A. L., Waby, J. S., & Corfe, B. M. (2006). Upregulation of BAK by butyrate in the colon is associated with increased Sp3 binding. *Oncogene, 25*, 7192–7200.

Cory, S., & Adams, J. M. (2002). The Bcl2 family: Regulators of the cellular life-or-death switch. *Nature Reviews. Cancer, 2*, 647–656.

Cory, S., Huang, D. C. S., & Adams, J. M. (2003). The Bcl-2 family: Roles in cell survival and oncogenesis. *Oncogene, 22*, 8590–8607.

Cottin, V., Doan, J. E., & Riches, D. W. (2002). Restricted localization of the TNF receptor CD120a to lipid rafts: A novel role for the death domain. *Journal of Immunology, 168*, 4095–4102.

Dai, Y., Chen, S., Wang, L., Pei, X.-Y., Kramer, L. B., Dent, P., & Grant, S. (2011). Bortezomib interacts synergistically with belinostat in human acute myeloid leukaemia and acute lymphoblastic leukaemia cells in association with perturbations in NF-κB and Bim. *British Journal of Haematology, 153*, 222–235.

Dokmanovic, M., & Marks, P. A. (2005). Prospects: Histone deacetylase inhibitors. *Journal of Cellular Biochemistry, 96*, 293–304.

Duckett, C. S. (2005). IAP proteins: Sticking it to Smac. *Biochemistry Journal, 385*, e1–e2.

Ellis, L., Bots, M., Lindemann, R. K., Bolden, J. E., Newbold, A., Cluse, L. A., Scott, C. L., Strasser, A., Atadja, P., Lowe, S. W., & Johnstone, R. W. (2009). The histone deacetylase inhibitors LAQ824 and LBH589 do not require death receptor signaling or a functional apoptosome to mediate tumor cell death or therapeutic efficacy. *Blood, 114*, 380–393.

Ellis, L., Pan, Y., Smyth, G. K., George, D. J., McCormack, C., Williams-Truax, R., Mita, M., Beck, J., Burris, H., Ryan, G., Atadja, P., Butterfoss, D., Dugan, M., Culver, K., Johnstone, R. W., & Prince, H. M. (2008). Histone deacetylase inhibitor panobinostat induces clinical responses with associated alterations in gene expression profiles in cutaneous T-cell lymphoma. *Clinical Cancer Research, 14*, 4500–4510.

Elmore, S. (2007). Apoptosis: A review of programmed cell death. *Toxicologic Pathology, 35*, 495–516.

Emanuele, S., Lauricella, M., Carlisi, D., Vassallo, B., D'Anneo, A., Di Fazio, P., Vento, R., & Tesoriere, G. (2007). SAHA induces apoptosis in hepatoma cells and synergistically interacts with the proteasome inhibitor Bortezomib. *Apoptosis, 12*, 1327–1338.

Fandy, T. E., Shankar, S., Rossy, D. D., Sausvilley, E., & Srivastava, R. K. (2005). Interactive effects of HDAC inhibitors and TRAIL on apoptosis are associated with changes in mitochondrial functions and expressions of cell cycle regulatory genes in multiple myeloma. *Neoplasia, 7*, 646–657.

Fandy, T. E., & Srivastava, R. K. (2006). Trichostatin A sensitizes TRAIL-resistant myeloma cells by downregulation of the antiapoptotic Bcl-2 proteins. *Cancer Chemotherapy and Pharmacology, 58*, 471–477.

Fantin, V. R., & Richon, V. M. (2007). Mechanisms of resistance to histone deacetylase inhibitors and their therapeutic implications. *Clinical Cancer Research, 13*, 7237–7242.

Frew, A. J., Johnstone, R. W., & Bolden, J. E. (2009). Enhancing the apoptotic and therapeutic effects of HDAC inhibitors. *Cancer Letters, 280*, 125–133.

Frew, A. J., Lindemann, R. K., Martin, B. P., Clarke, C. J. P., Sharkey, J., Anthony, D. A., Banks, K.-M., Haynes, N. M., Gangatirkar, P., Stanley, K., Bolden, J. E., Takeda, K., Yagita, H., Secrist, J. P., Smyth, M. J., & Johnstone, R. W. (2008). Combination therapy of established cancer using a histone deacetylase inhibitor and a TRAIL receptor agonist. *Proceedings of the National Academy of Sciences of the United States of America, 105*, 11317–11322.

Fritsche, P., Gottlicher, M., Saur, D., Schmid, R. M., Schneider, G., Schnieke, A., Schuler, S., & Seidler, B. (2009). HDAC2 mediates therapeutic resistance of pancreatic cancer cells via the BH3-only protein NOXA. *Gut, 58*, 1399–1409.

Fu, H., Subramanian, R. R., & Masters, S. C. (2000). 14-3-3 Proteins: Structure, function, and regulation. *Annual Reviews in Pharmacology and Toxicology., 40*, 617–647.

Fulda, S. (2008). Modulation of TRAIL-induced apoptosis by HDAC inhibitors. *Current Cancer Drug Targets, 8*, 132–140.

Glaser, K. B., Staver, M. J., Waring, J. F., Stender, J., Ulrich, R. G., & Davidsen, S. K. (2003). Gene expression profiling of multiple histone deacetylase (HDAC) inhibitors: Defining a common gene set produced by HDAC inhibition in T24 and MDA carcinoma cell lines. *Molecular Cancer Therapeutics, 2*, 151–163.

Gomez-Benito, M., Martinez-Lorenzo, M. J., Anel, A., Marzo, I., & Naval, J. (2007). Membrane expression of DR4, DR5 and caspase-8 levels, but not Mcl-1, determine sensitivity of human myeloma cells to Apo2L/TRAIL. *Experimental Cell Research, 313*, 2378–2388.

Gonzalvez, F., & Ashkenazi, A. (2010). New insights into apoptosis signaling by Apo2L/TRAIL. *Oncogene, 29*, 4752–4765.

Guo, F., Sigua, C., Tao, J., Bali, P., George, P., Li, Y., Wittmann, S., Moscinski, L., Atadja, P., & Bhalla, K. (2004). Cotreatment with histone deacetylase inhibitor LAQ824 enhances Apo-2L/tumor necrosis factor-related apoptosis inducing ligand-induced death inducing signaling complex activity and apoptosis of human acute leukemia cells. *Cancer Research, 64*, 2580–2589.

Hacker, S., Dittrich, A., Mohr, A., Schweitzer, T., Rutkowski, S., Krauss, J., Debatin, K. M., & Fulda, S. (2009). Histone deacetylase inhibitors cooperate with IFN-gamma to restore caspase-8 expression and overcome TRAIL resistance in cancers with silencing of caspase-8. *Oncogene, 28*, 3097–3110.

Hueber, A. O., Bernard, A. M., Herincs, Z., Couzinet, A., & He, H. T. (2002). An essential role for membrane rafts in the initiation of Fas/CD95-triggered cell death in mouse thymocytes. *EMBO Reports, 3*, 190–196.

Hurwitz, J. L., Stasik, I., Kerr, E. M., Holohan, C., Redmond, K. M., McLaughlin, K. M., Busacca, S., Barbone, D., Broaddus, V. C., Gray, S. G., O'Byrne, K. J., Johnston, P. G., Fennell, D. A., & Longley, D. B. (2012). Vorinostat/SAHA-induced apoptosis in malignant mesothelioma is FLIP/caspase 8-dependent and HR23B-independent. *European Journal of Cancer, 48*, 1096–1107.

Inoue, S., Harper, N., Walewska, R., Dyer, M. J., & Cohen, G. M. (2009). Enhanced Fas-associated death domain recruitment by histone deacetylase inhibitors is critical for the sensitization of chronic lymphocytic leukemia cells to TRAIL-induced apoptosis. *Molecular Cancer Therapy, 8*, 3088–3097.

Inoue, S., MacFarlane, M., Harper, N., Wheat, L. M., Dyer, M. J., & Cohen, G. M. (2004). Histone deacetylase inhibitors potentiate TNF-related apoptosis-inducing ligand (TRAIL)-induced apoptosis in lymphoid malignancies. *Cell Death and Differentiation, 11*(Suppl. 2), S193–S206.

Inoue, S., Mai, A., Dyer, M. J. S., & Cohen, G. M. (2006). Inhibition of histone deacetylase Class I but not Class II is critical for the sensitization of leukemic cells to tumor necrosis factor—Related apoptosis-inducing ligand-induced apoptosis. *Cancer Research, 66*, 6785–6792.

Inoue, S., Riley, J., Gant, T. W., Dyer, M. J. S., & Cohen, G. M. (2007). Apoptosis induced by histone deacetylase inhibitors in leukemic cells is mediated by Bim and Noxa. *Leukemia, 21*, 1773–1782.

Inoue, S., Walewska, R., Dyer, M. J. S., & Cohen, G. M. (2008). Downregulation of Mcl-1 potentiates HDACi-mediated apoptosis in leukemic cells. *Leukemia, 22*, 819–825.

Johnstone, R. W. (2002). Histone-deacetylase inhibitors: Novel drugs for the treatment of cancer. *Nature Reviews. Drug Discovery, 1*, 287–299.

Johnstone, R. W., Frew, A. J., & Smyth, M. J. (2008). The TRAIL apoptotic pathway in cancer onset, progression and therapy. *Nature Reviews. Cancer, 8*, 782–798.

Johnstone, R. W., Ruefli, A. A., & Lowe, S. W. (2002). Apoptosis: A link between cancer genetics and chemotherapy. *Cell, 108*, 153–164.

Kaminskyy, V. O., Surova, O. V., Vaculova, A., & Zhivotovsky, B. (2011). Combined inhibition of DNA methyltransferase and histone deacetylase restores caspase-8 expression and sensitizes SCLC cells to TRAIL. *Carcinogenesis, 32*, 1450–1458.

Kaufmann, S. H., Karp, J. E., Svingen, P. A., Krajewski, S., Burke, P. J., Gore, S. D., & Reed, J. C. (1998). Elevated expression of the apoptotic regulator Mcl-1 at the time of leukemic relapse. *Blood, 91*, 991–1000.

Kauh, J., Fan, S., Xia, M., Yue, P., Yang, L., Khuri, F. R., & Sun, S. Y. (2010). c-FLIP degradation mediates sensitization of pancreatic cancer cells to TRAIL-induced apoptosis by the histone deacetylase inhibitor LBH589. *PLoS One, 5*, e10376.

Kerr, E., Holohan, C., McLaughlin, K. M., Majkut, J., Dolan, S., Redmond, K., Riley, J., McLaughlin, K., Stasik, I., Crudden, M., Van Schaeybroeck, S., Fenning, C., O'Connor, R., Kiely, P., Sgobba, M., Haigh, D., Johnston, P. G., & Longley, D. B. (2012). Identification of an acetylation-dependant Ku70/FLIP complex that regulates FLIP expression and HDAC inhibitor-induced apoptosis. *Cell Death and Differertiation, 19*, 1317–1327.

Khandelwal, A., Gediya, L. K., & Njar, V. C. O. (2008). MS-275 synergistically enhances the growth inhibitory effects of RAMBA VN/66-1 in hormone-insensitive PC-3 prostate cancer cells and tumours. *British Journal of Cancer, 98*, 1234–1243.

Kim, J., Guan, J., Chang, I., Chen, X., Han, D., & Wang, C.-Y. (2010). PS-341 and histone deacetylase inhibitor synergistically induce apoptosis in head and neck squamous cell carcinoma cells. *Molecular Cancer Therapeutics, 9*, 1977–1984.

Kim, H., Rafiuddin-Shah, M., Tu, H.-C., Jeffers, J. R., Zambetti, G. P., Hsieh, J. J. D., & Cheng, E. H. Y. (2006). Hierarchical regulation of mitochondrion-dependent apoptosis by BCL-2 subfamilies. *Nature Cell Biology, 8*, 1348–1358.

Kurita-Ochiai, T., Ochiai, K., & Fukushima, K. (2000). Butyric-acid-induced apoptosis in murine thymocytes and splenic T- and B-cells occurs in the absence of p53. *Journal of Dental Research, 79*, 1948–1954.

Labi, V., Erlacher, M., Kiessling, S., Manzl, C., Frenzel, A., O'Reilly, L., Strasser, A., & Villunger, A. (2008). Loss of the BH3-only protein Bmf impairs B cell homeostasis and accelerates γ irradiation-induced thymic lymphoma development. *The Journal of Experimental Medicine, 205*, 641–655.

Lemaire, M., Fristedt, C., Agarwal, P., Menu, E., Valckenborgh, E. V., De Bruyne, E., Osterborg, A., Atadja, P., Larsson, O., Axelson, M., Van Camp, B., Jernberg-Wiklund, H., & Vanderkerken, K. (2012). The HDAC inhibitor LBH589 enhances the antimyeloma effects of the IGF-1RTK inhibitor Picropodophyllin. *Clinical Cancer Research, 18*, 2230–2239.

Ley, R., Ewings, K. E., Hadfield, K., & Cook, S. J. (2005). Regulatory phosphorylation of Bim: Sorting out the ERK from the JNK. *Cell Death and Differentiation, 12*, 1008–1014.

Lindemann, R. K., Gabrielli, B., & Johnstone, R. W. (2004). Histone-deacetylase inhibitors for the treatment of cancer. *Cell Cycle, 3*, 777–786.

Lindemann, R. K., Newbold, A., Whitecross, K. F., Cluse, L. A., Frew, A. J., Ellis, L., Williams, S., Wiegmans, A. P., Dear, A. E., Scott, C. L., Pellegrini, M., Wei, A., Richon, V. M., Marks, P. A., Lowe, S. W., Smyth, M. J., & Johnstone, R. W. (2007). Analysis of the apoptotic and therapeutic activities of histone deacetylase inhibitors by using a mouse model of B cell lymphoma. *Proceedings of the National Academy of Sciences of the United States of America, 104*, 8071–8076.

Maiso, P., Colado, E., Ocio, E. M., Garayoa, M., Martin, J., Atadja, P., Pandiella, A., & San-Miguel, J. F. (2009). The synergy of panobinostat plus doxorubicin in acute myeloid leukemia suggests a role for HDAC inhibitors in the control of DNA repair. *Leukemia, 23*, 2265–2274.

Marks, P. A., Rifkind, R. A., Richon, V. M., Breslow, R., Miller, T., & Kelly, W. K. (2001). Histone deacetylases and cancer: Causes and therapies. *Nature Reviews. Cancer, 1*, 194–202.

Martin, B. P., Frew, A. J., Bots, M., Fox, S., Long, F., Takeda, K., Yagita, H., Atadja, P., Smyth, M. J., & Johnstone, R. W. (2011). Antitumor activities and on-target toxicities mediated by a TRAIL receptor agonist following cotreatment with panobinostat. *International Journal of Cancer, 128*, 2735–2747.

Martin, A. P., Park, M. A., Mitchell, C., Walker, T., Rahmani, M., Thorburn, A., Häussinger, D., Reinehr, R., Grant, S., & Dent, P. (2009). BCL-2 Family inhibitors enhance histone deacetylase inhibitor and Sorafenib lethality via autophagy and overcome blockade of the extrinsic pathway to facilitate killing. *Molecular Pharmacology, 76*, 327–341.

Meng, J., Zhang, H. H., Zhou, C. X., Li, C., Zhang, F., & Mei, Q. B. (2012). The histone deacetylase inhibitor trichostatin A induces cell cycle arrest and apoptosis in colorectal cancer cells via p53-dependent and -independent pathways. *Oncology Reports, 28*, 384–388.

Mitsiades, C. S., Mitsiades, N. S., McMullan, C. J., Poulaki, V., Shringarpure, R., Hideshima, T., Akiyama, M., Chauhan, D., Munshi, N., Gu, X., Bailey, C., Joseph, M., Libermann, T. A., Richon, V. M., Marks, P. A., & Anderson, K. C. (2004). Transcriptional signature of histone deacetylase inhibition in multiple myeloma: Biological and clinical implications. *Proceedings of the National Academy of Sciences of the United States of America, 101*, 540–545.

Mitsiades, N., Mitsiades, C. S., Richardson, P. G., McMullan, C., Poulaki, V., Fanourakis, G., Schlossman, R., Chauhan, D., Munshi, N. C., Hideshima, T., Richon, V. M., Marks, P. A., & Anderson, K. C. (2003). Molecular sequelae of histone deacetylase inhibition in human malignant B cells. *Blood, 101*, 4055–4062.

Mitsiades, C. S., Poulaki, V., McMullan, C., Negri, J., Fanourakis, G., Goudopoulou, A., Richon, V. M., Marks, P. A., & Mitsiades, N. (2005). Novel histone deacetylase inhibitors in the treatment of thyroid cancer. *Clinical Cancer Research, 11*, 3958–3965.

Monneret, C. (2005). Histone deacetylase inhibitors. *European Journal of Medicinal Chemistry, 40*, 1–13.

Nakano, K., & Vousden, K. H. (2001). PUMA, a novel proapoptotic gene, is induced by p53. *Molecular Cell, 7*, 683–694.

Nakata, S., Yoshida, T., Horinaka, M., Shiraishi, T., Wakada, M., & Sakai, T. (2004). Histone deacetylase inhibitors upregulate death receptor 5//TRAIL-R2 and sensitize apoptosis induced by TRAIL//APO2-L in human malignant tumor cells. *Oncogene, 23*, 6261–6271.

Nebbioso, A., Clarke, N., Voltz, E., Germain, E., Ambrosino, C., Bontempo, P., Alvarez, R., Schiavone, E. M., Ferrara, F., Bresciani, F., Weisz, A., de Lera, A. R., Gronemeyer, H., & Altucci, L. (2005). Tumor-selective action of HDAC inhibitors involves TRAIL induction in acute myeloid leukemia cells. *Nature Medicine, 11*, 77–84.

Nishioka, C., Ikezoe, T., Yang, J., Takeuchi, S., Phillip Koeffler, H., & Yokoyama, A. (2008). MS-275, a novel histone deacetylase inhibitor with selectivity against HDAC1, induces degradation of FLT3 via inhibition of chaperone function of heat shock protein 90 in AML cells. *Leukemia Research, 32*, 1382–1392.

Oda, E., Ohki, R., Murasawa, H., Nemoto, J., Shibue, T., Yamashita, T., Tokino, T., Taniguchi, T., & Tanaka, N. (2000). Noxa, a BH3-only member of the Bcl-2 family and candidate mediator of p53-induced apoptosis. *Science, 288*, 1053–1058.

Pacey, S., Gore, M., Chao, D., Banerji, U., Larkin, J., Sarker, S., Owen, K., Asad, Y., Raynaud, F., Walton, M., Judson, I., Workman, P., & Eisen, T. (2010). A Phase II trial of 17-allylamino, 17-demethoxygeldanamycin (17-AAG, tanespimycin) in patients with metastatic melanoma. *Investigational New Drugs, 30*, 341–349.

Peart, M. J., Smyth, G. K., van Laar, R. K., Bowtell, D. D., Richon, V. M., Marks, P. A., Holloway, A. J., & Johnstone, R. W. (2005). Identification and functional significance of genes regulated by structurally different histone deacetylase inhibitors. *Proceedings of the National Academy of Sciences of the United States of America, 102*, 3697–3702.

Peart, M. J., Tainton, K. M., Ruefli, A. A., Dear, A. E., Sedelies, K. A., O'Reilly, L. A., Waterhouse, N. J., Trapani, J. A., & Johnstone, R. W. (2003). Novel mechanisms of apoptosis induced by histone deacetylase inhibitors. *Cancer Research, 63*, 4460–4471.

Pérez-Perarnau, A., Coll-Mulet, L., Rubio-Patiño, C., Iglesias-Serret, D., Cosialls, A. M., González-Gironès, D. M., de Frias, M., Fernández de Sevilla, A., de la Banda, E., Pons, G., & Gil, J. (2011). Analysis of apoptosis regulatory genes altered by histone deacetylase inhibitors in chronic lymphocytic leukemia cells. *Epigenetics, 6*, 1228–1235.

Premkumar, D. R., Jane, E. P., Agostino, N. R., Didomenico, J. D., & Pollack, I. F. (2011). Bortezomib-induced sensitization of malignant human glioma cells to vorinostat-induced apoptosis depends on reactive oxygen species production, mitochondrial dysfunction, Noxa upregulation, Mcl-1 cleavage, and DNA damage. *Molecular Carcinogenesis*, http://dx.doi.org/10.1002/mc.21835.

Puthalakath, H., O'Reilly, L. A., Gunn, P., Lee, L., Kelly, P. N., Huntington, N. D., Hughes, P. D., Michalak, E. M., McKimm-Breschkin, J., Motoyama, N., Gotoh, T., Akira, S., Bouillet, P., & Strasser, A. (2007). ER stress triggers apoptosis by activating BH3-only protein Bim. *Cell*, *129*, 1337–1349.

Puthalakath, H., & Strasser, A. (2002). Keeping killers on a tight leash: Transcriptional and post-translational control of the pro-apoptotic activity of BH3-only proteins. *Cell Death and Differentiation*, *9*, 505–512.

Rahmani, M., Yu, C., Dai, Y., Reese, E., Ahmed, W., Dent, P., & Grant, S. (2003). Coadministration of the heat shock protein 90 antagonist 17-allylamino-17-demethoxygeldanamycin with suberoylanilide hydroxamic acid or sodium butyrate synergistically induces apoptosis in human leukemia cells. *Cancer Research*, *63*, 8420–8427.

Ralli, R., Banks, K. M., Wiegmans, A. P., Carney, D., Seymour, J. F., Johnstone, R. W., & Alsop, A. E. (2012). Histone deacetylase inhibitors are unable to synergize with ABT-737 in killing primary chronic lymphocytic leukaemia cells in vitro. *Leukemia*, *26*, 1433–1435.

Ramjaun, A. R., Tomlinson, S., Eddaoudi, A., & Downward, J. (2006). Upregulation of two BH3-only proteins, Bmf and Bim, during TGF[beta]-induced apoptosis. *Oncogene*, *26*, 970–981.

Ramsey, M. R., He, L., Forster, N., Ory, B., & Ellisen, L. W. (2011). Physical association of HDAC1 and HDAC2 with p63 mediates transcriptional repression and tumor maintenance in squamous cell carcinoma. *Cancer Research*, *71*, 4373–4379.

Ren, D., Tu, H.-C., Kim, H., Wang, G., Bean, G., Takeuchi, O., Jeffers, J., Zambetti, G., Hsieh, J.-D., & Cheng, E. H. (2010). BID, BIM, and PUMA are essential for activation of the BAX- and BAK-dependent cell death program. *Science*, *330*, 1390–1393.

Rosato, R. R., Almenara, J. A., Dai, Y., & Grant, S. (2003). Simultaneous activation of the intrinsic and extrinsic pathways by histone deacetylase (HDAC) inhibitors and tumor necrosis factor-related apoptosis-inducing ligand (TRAIL) synergistically induces mitochondrial damage and apoptosis in human leukemia cells. *Molecular Cancer Therapeutics*, *2*, 1273–1284.

Rosato, R. R., Almenara, J. A., Maggio, S. C., Coe, S., Atadja, P., Dent, P., & Grant, S. (2008). Role of histone deacetylase inhibitor-induced reactive oxygen species and DNA damage in LAQ-824/fludarabine antileukemic interactions. *Molecular Cancer Therapeutics*, 7, 3285–3297.

Rosato, R. R., Maggio, S. C., Almenara, J. A., Payne, S. G., Atadja, P., Spiegel, S., Dent, P., & Grant, S. (2006). The histone deacetylase inhibitor LAQ824 induces human leukemia cell death through a process involving XIAP down-regulation, oxidative injury, and the acid sphingomyelinase-dependent generation of ceramide. *Molecular Pharmacology*, *69*, 216–225.

Ruefli, A. A., Ausserlechner, M. J., Bernhard, D., Sutton, V. R., Tainton, K. M., Kofler, R., Smyth, M. J., & Johnstone, R. W. (2001). The histone deacetylase inhibitor and chemotherapeutic agent suberoylanilide hydroxamic acid (SAHA) induces a cell-death pathway characterized by cleavage of Bid and production of reactive oxygen species. *Proceedings of the National Academy of Sciences of the United States of America*, *98*, 10833–10838.

Safa, A. R., & Pollok, K. E. (2011). Targeting the anti-apoptotic protein c-FLIP for cancer therapy. *Cancers (Basel)*, *3*, 1639–1671.

Sanda, T., Okamoto, T., Uchida, Y., Nakagawa, H., Iida, S., Kayukawa, S., Suzuki, T., Oshizawa, T., Miyata, N., & Ueda, R. (2007). Proteome analyses of the growth inhibitory effects of NCH-51, a novel histone deacetylase inhibitor, on lymphoid malignant cells. *Leukemia*, *21*, 2344–2353.

Sawa, H., Murakami, H., Ohshima, Y., Sugino, T., Nakajyo, T., Kisanuki, T., Tamura, Y., Satone, A., Ide, W., Hashimoto, I., & Kamada, H. (2001). Histone deacetylase inhibitors such as sodium butyrate and trichostatin A induce apoptosis through an increase of the bcl-2-related protein Bad. *Brain Tumor Pathology*, *18*, 109–114.

Schuler, S., Fritsche, P., Diersch, S., Arlt, A., Schmid, R. M., Saur, D., & Schneider, G. (2010). HDAC2 attenuates TRAIL-induced apoptosis of pancreatic cancer cells. *Molecular Cancer*, *9*, 80.

Schwulst, S. J., Muenzer, J. T., Peck-Palmer, O. M., Chang, K. C., Davis, C. G., McDonough, J. S., Osborne, D. F., Walton, A. H., Unsinger, J., McDunn, J. E., & Hotchkiss, R. S. (2008). BIM siRNA decreases lymphocyte apoptosis and improves survival in sepsis. *Shock*, *30*, 127–134.

Shankar, S., Davis, R., Singh, K. P., Kurzrock, R., Ross, D. D., & Srivastava, R. K. (2009). Suberoylanilide hydroxamic acid (Zolinza/vorinostat) sensitizes TRAIL-resistant breast cancer cells orthotopically implanted in BALB/c nude mice. *Molecular Cancer Therapeutics*, *8*, 1596–1605.

Shankar, S., Singh, T. R., Fandy, T. E., Luetrakul, T., Ross, D. D., & Srivastava, R. K. (2005). Interactive effects of histone deacetylase inhibitors and TRAIL on apoptosis in human leukemia cells: Involvement of both death receptor and mitochondrial pathways. *International Journal of Molecular Medicine*, *16*, 1125–1138.

Shao, Y., Gao, Z., Marks, P. A., & Jiang, X. (2004). Apoptotic and autophagic cell death induced by histone deacetylase inhibitors. *Proceedings of the National Academy of Sciences of the United States of America*, *101*, 18030–18035.

Singh, T. R., Shankar, S., & Srivastava, R. K. (2005). HDAC inhibitors enhance the apoptosis-inducing potential of TRAIL in breast carcinoma. *Oncogene*, *24*, 4609–4623.

Srivastava, R. K., Kurzrock, R., & Shankar, S. (2010). MS-275 sensitizes TRAIL-resistant breast cancer cells, inhibits angiogenesis and metastasis, and reverses epithelial-mesenchymal transition in vivo. *Molecular Cancer Therapeutics*, *9*, 3254–3266.

Strait, K. A., Warnick, C. T., Ford, C. D., Dabbas, B., Hammond, E. H., & Ilstrup, S. J. (2005). Histone deacetylase inhibitors induce G2-checkpoint arrest and apoptosis in cisplatinum-resistant ovarian cancer cells associated with overexpression of the Bcl-2–related protein Bad. *Molecular Cancer Therapeutics*, *4*, 603–611.

Strasser, A., O'Connor, L., & Dixit, V. (2000). Apoptosis signaling. *Annual Review of Biochemistry*, *69*, 217–245.

Subramanian, C., Opipari, A. W., Bian, X., Castle, V. P., & Kwok, R. P. S. (2005). Ku70 acetylation mediates neuroblastoma cell death induced by histone deacetylase inhibitors. *Proceedings of the National Academy of Sciences of the United States of America*, *102*, 4842–4847.

Thomas, S., Thurn, K. T., Bicaku, E., Marchion, D. C., & Munster, P. N. (2011). Addition of a histone deacetylase inhibitor redirects tamoxifen-treated breast cancer cells into apoptosis, which is opposed by the induction of autophagy. *Breast Cancer Research and Treatment*, *130*, 437–447.

Tu, Y., Renner, S., Xu, F.-H., Fleishman, A., Taylor, J., Weisz, J., Vescio, R., Rettig, M., Berenson, J., Krajewski, S., Reed, J. C., & Lichtenstein, A. (1998). BCL-X expression in multiple myeloma: Possible indicator of chemoresistance. *Cancer Research*, *58*, 256–262.

Vanoosten, R. L., Moore, J. M., Ludwig, A. T., & Griffith, T. S. (2005). Depsipeptide (FR901228) enhances the cytotoxic activity of TRAIL by redistributing TRAIL receptor to membrane lipid rafts. *Molecular Therapeutics*, *11*, 542–552.

Vo, D. D., Donahue, T. R., de la Rocha, P., Begley, J. L., Yang, M.-Y., Kharazi, P., Prins, R. M., Economou, J. S., Garban, H. J., & Ribas, A. (2006). A histone deacetylase inhibitor (HDACI) sensitizes B16 melanoma to adoptive transfer (AT) immunotherapy. *Journal of Immunotherapy*, *29*, 633. http://dx.doi.org/10.1097/01.cji.0000211343.73588.59.

Vogel, S., Raulf, N., Bregenhorn, S., Biniossek, M. L., Maurer, U., Czabotar, P., & Borner, C. (2012). Cytosolic Bax: Does it require binding proteins to keep its pro-apoptotic activity in check? *Journal of Biological Chemistry*, *287*(12), 9112–9127.

Vrana, J. A., Decker, R. H., Johnson, C. R., Wang, Z., Jarvis, W. D., Richon, V. M., Ehinger, M., Fisher, P. B., & Grant, S. (1999). Induction of apoptosis in U937 human leukemia cells by suberoylanilide hydroxamic acid (SAHA) proceeds through pathways that are regulated by Bcl-2/Bcl-XL, c-Jun, and p21CIP1, but independent of p53. *Oncogene, 18*, 7016–7025.

Waby, J. S., Chirakkal, H., Yu, C., Griffiths, G. J., Benson, R. S., Bingle, C. D., & Corfe, B. M. (2010). Sp1 acetylation is associated with loss of DNA binding at promoters associated with cell cycle arrest and cell death in a colon cell line. *Molecular Cancer, 9*, 275.

Wang, S. (2008). The promise of cancer therapeutics targeting the TNF-related apoptosis-inducing ligand and TRAIL receptor pathway. *Oncogene, 27*, 6207–6215.

Weber, A., Paschen, S. A., Heger, K., Wilfling, F., Frankenberg, T., Bauerschmitt, H., Seiffert, B. M., Kirschnek, S., Wagner, H., & Häcker, G. (2007). BimS-induced apoptosis requires mitochondrial localization but not interaction with anti-apoptotic Bcl-2 proteins. *The Journal of Cell Biology, 177*, 625–636.

Wei, Y., Kadia, T., Tong, W., Zhang, M., Jia, Y., Yang, H., Hu, Y., Tambaro, F. P., Viallet, J., O'Brien, S., & Garcia-Manero, G. (2010). The combination of a histone deacetylase inhibitor with the Bcl-2 homology domain-3 mimetic GX15-070 has synergistic antileukemia activity by activating both apoptosis and autophagy. *Clinical Cancer Research, 16*, 3923–3932.

Whitecross, K. F., Alsop, A. E., Cluse, L. A., Wiegmans, A., Banks, K.-M., Coomans, C., Peart, M. J., Newbold, A., Lindemann, R. K., & Johnstone, R. W. (2009). Defining the target specificity of ABT-737 and synergistic antitumor activities in combination with histone deacetylase inhibitors. *Blood, 113*, 1982–1991.

Wiegmans, A. P., Alsop, A., Bots, M., Cluse, L. A., Williams, S. P., Banks, K.-M., Ralli, R., Scott, C. L., Frenzel, A., Villunger, A., & Johnstone, R. W. (2011). Deciphering the molecular events necessary for synergistic tumor cell apoptosis mediated by the histone deacetylase inhibitor vorinostat and the BH3 mimetic ABT-737. *Cancer Research, 71*, 3603–3615.

Wuilleme-Toumi, S., Robillard, N., Gomez, P., Moreau, P., Le Gouill, S., Avet-Loiseau, H., Harousseau, J. L., Amiot, M., & Bataille, R. (2005). Mcl-1 is overexpressed in multiple myeloma and associated with relapse and shorter survival. *Leukemia, 19*, 1248–1252.

Xargay-Torrent, S., López-Guerra, M., Saborit-Villarroya, I., Rosich, L., Campo, E., Roué, G., & Colomer, D. (2011). Vorinostat-induced apoptosis in mantle cell lymphoma is mediated by acetylation of proapoptotic BH3-only gene promoters. *Clinical Cancer Research, 17*, 3956–3968.

Yamaguchi, H., Woods, N. T., Piluso, L. G., Lee, H.-H., Chen, J., Bhalla, K. N., Monteiro, A., Liu, X., Hung, M.-C., & Wang, H.-G. (2009). p53 Acetylation is crucial for its transcription-independent proapoptotic functions. *Journal of Biological Chemistry, 284*, 11171–11183.

Yang, Y., Balch, C., Kulp, S., Mand, M., Nephew, K., & Chen, C. (2009). A rationally designed histone deacetylase inhibitor with distinct antitumor activity against ovarian cancer. *Neoplasia, 11*, 552–563.

Yang, Y., Zhao, Y., Liao, W., Yang, J., Wu, L., Zheng, Z., Yu, Y., Zhou, W., Li, L., Feng, J., Wang, H., & Zhu, W.-G. (2009). Acetylation of FoxO1 activates Bim expression to induce apoptosis in response to histone deacetylase inhibitor depsipeptide treatment. *Neoplasia, 11*, 313–324.

Yeh, C. C., Deng, Y. T., Sha, D. Y., Hsiao, M., & Kuo, M. Y. (2009). Suberoylanilide hydroxamic acid sensitizes human oral cancer cells to TRAIL-induced apoptosis through increase DR5 expression. *Molecular Cancer Therapeutics, 8*, 2718–2725.

Yerbes, R., & Lopez-Rivas, A. (2012). Itch/AIP4-independent proteasomal degradation of cFLIP induced by the histone deacetylase inhibitor SAHA sensitizes breast tumour cells to TRAIL. *Investigational New Drugs*, *30*, 541–547.

Yu, J. W., & Shi, Y. (2008). FLIP and the death effector domain family. *Oncogene*, *27*, 6216–6227.

Zacharias, N., Sailhamer, E., Li, Y., Liu, B., Butt, M., Shuja, F., Velmahos, G., de Moya, M., & Alam, H. (2011). Histone deacetylase inhibitors prevent apoptosis following lethal hemorrhagic shock in rodent kidney cells. *Resuscitation*, *82*, 105–109.

Zhang, Y., Adachi, M., Kawamura, R., & Imai, K. (2005). Bmf is a possible mediator in histone deacetylase inhibitors FK228 and CBHA-induced apoptosis. *Cell Death and Differentiation*, *13*, 129–140.

Zhang, Y., Adachi, M., Kawamura, R., Zou, H., Imai, K., Hareyama, M., & Shinomura, Y. (2006). Bmf contributes to histone deacetylase inhibitor-mediated enhancing effects on apoptosis after ionizing radiation. *Apoptosis*, *11*, 1349–1357.

Zhang, Y., Adachi, M., Zhao, X., Kawamura, R., & Imai, K. (2004). Histone deacetylase inhibitors FK228, N-(2-aminophenyl)-4-[N-(pyridin-3-yl-methoxycarbonyl)aminomethyl]benzamide and m-carboxycinnamic acid bis-hydroxamide augment radiation-induced cell death in gastrointestinal adenocarcinoma cells. *International Journal of Cancer*, *110*, 301–308.

Zhang, L., & Fang, B. (2004). Mechanisms of resistance to TRAIL-induced apoptosis in cancer. *Cancer Gene Therapy*, *12*, 228–237.

Zhang, X. D., Gillespie, S. K., Borrow, J. M., & Hersey, P. (2004). The histone deacetylase inhibitor suberic bishydroxamate regulates the expression of multiple apoptotic mediators and induces mitochondria-dependent apoptosis of melanoma cells. *Molecular Cancer Therapeutics*, *3*, 425–435.

Zhang, H. M., Yuan, J., Cheung, P., Chau, D., Wong, B. W., McManus, B. M., & Yang, D. (2005). Gamma interferon-inducible protein 10 induces HeLa cell apoptosis through a p53-dependent pathway initiated by suppression of human papillomavirus type 18 E6 and E7 expression. *Molecular and Cell Biology*, *25*, 6247–6258.

Zhao, Y., Tan, J., Zhuang, L., Jiang, X., Liu, E. T., & Yu, Q. (2005). Inhibitors of histone deacetylases target the Rb-E2F1 pathway for apoptosis induction through activation of proapoptotic protein Bim. *Proceedings of the National Academy of Sciences of the United States of America*, *102*, 16090–16095.

Histone Deacetylase Inhibitors and Rational Combination Therapies

Steven Grant[1], Yun Dai
Division of Hematology/Oncology, Virginia Commonwealth University Health Sciences Center, P.O. Box 980035, Richmond, Virginia, USA
[1]Corresponding author: e-mail address: stgrant@vcu.edu

Contents

Advances in Cancer Research, Volume 116
ISSN 0065-230X
http://dx.doi.org/10.1016/B978-0-12-394387-3.00006-9

Abstract

Histone deacetylase inhibitors (HDACIs) are epigenetically acting agents that modify chromatin structure and by extension, gene expression. However, they may influence the behavior and survival of transformed cells by diverse mechanisms, including promoting expression of death- or differentiation-inducing genes while downregulating the expression of prosurvival genes; acting directly to increase oxidative injury and DNA damage; acetylating and disrupting the function of multiple proteins, including DNA repair and chaperone proteins; and interfering with the function of corepressor complexes. Notably, HDACIs have been shown in preclinical studies to target transformed cells selectively, and these agents have been approved in the treatment of certain hematologic malignancies, for example, cutaneous T-cell lymphoma and peripheral T-cell lymphoma. However, attempts to extend the spectrum of HDACI activity to other malignancies, for example, solid tumors, have been challenging. This has led to the perception that HDACIs may have limited activity as single agents. Because of the pleiotropic actions of HDACIs, combinations with other antineoplastic drugs, particularly other targeted agents, represent a particularly promising avenue of investigation. It is likely that emerging insights into mechanism(s) of HDACI activity will allow optimization of this approach, and hopefully, will expand HDACI approvals to additional malignancies in the future.

ABBREVIATIONS

AML acute myelogenous leukemia
ATM ataxia telangiectasia mutated
Bcl-6 B-cell lymphoma 6 protein
CLL chronic lymphocytic leukemia
CML chronic myelogenous leukemia
ERK1/2 extracellular-regulated kinase 1/2
FADD Fas-associated death domain
HATs histone acetyltransferases
HDACIs histone deacetylase inhibitors
Hsp90 heat shock protein 90
IKK IkappaB kinase
JAK2 Janus kinase 2
JNK c-Jun N-terminal kinase
Mn-TBAP Manganese[III]-tetrakis 4-benzoic acid porphyrin
NAD nicotinamide adenine dinucleotide
NEMO NF-kappa B essential modulator
NF-κB nuclear factor kappa light chain-enhancer of B cells
NHL non-Hodgkin's lymphoma
ROS reactive oxygen species
SIRT silent mating type information regulation 2 homolog
STAT5 signal transducer and transcriptional activator 5
TRAIL TNF-related apoptosis-inducing ligand
TSA trichostatin A
XIAP X-linked inhibitor of apoptosis protein

1. INTRODUCTION

Epigenetic therapies refer to strategies that are aimed at altering the expression of genes implicated in transformation. One of the major targets of this strategy is histones, around which DNA is wrapped to form nucleosomes. Histones are subject to numerous posttranslational modifications, including acetylation, methylation, phosphorylation, and sumoylation, among others (Jenuwein & Allis, 2001). In particular, histone acetylation/deacetylation has been recognized as a critical regulator of chromatin structure and gene expression and has been a major target of epigenetic therapies. These events are reciprocally controlled by histone deacetylases and histone acetyltransferases. In general, acetylation of histones, by neutralizing negatively charged carboxyl groups of DNA, which bind tightly to positively charged lysine residues of histone tails, leads to a more open chromatin structure, which is conducive to the expression of genes involved in cell death and/or differentiation (Marks, Richon, Kelly, Chiao, & Miller, 2004). Thus, interfering with the balance between histone acetylation/deacetylation may shift the balance away from the transformed phenotype.

2. HISTONE DEACETYLASES

HDACs have been shown to be frequently dysregulated in cancer (Marks & Jiang, 2005). Broadly speaking, HDACs have been subdivided into a family of 18 genes that represent 4 classes. Class I (analogous to the yeast reduced potassium dependency 3/Rpd3) and II (histone deacetylase 1/Hda1) HDACs are sensitive to the inhibitor trichostatin A (TSA) and do not require the nicotinamide adenine dinucleotide $(NAD)^+$ cofactor, whereas Class III (analogous to silent information regulator 2/Sir2) HDACs are insensitive to TSA and require NAD^+ (Glozak & Seto, 2007). A recently described HDAC, HDAC11, belongs to a fourth class (Gao, Cueto, Asselbergs, & Atadja, 2002). A subclass of Class II HDACs (Class IIb) is distinguished by its ability to act as a tubulin acetylase (Haggarty, Koeller, Wong, Grozinger, & Schreiber, 2003). In addition to their ability to acetylate histones, HDACs can deacetylate a broad range of proteins and interfering with this capacity may play an important role in the lethality of histone deacetylase inhibitors (HDACIs; Glozak, Sengupta, Zhang, & Seto, 2005).

3. HISTONE DEACETYLASE INHIBITORS

HDACIs include a broad range of compounds which can be classified into the following chemical structures: hydroxamic acids (e.g., vorinostat, panobinostat, belinostat); benzamides (e.g., SNDX-275; entinostat; MGCD0103); short-chain fatty acids (e.g., sodium butyrate, valproic acid, AN-9); cyclic peptides (romidepsin, apicidin); and thiolates, carboxamides, and nonhydroxamic acids (Batty, Malouf, & Issa, 2009; Drummond et al., 2005). These agents vary considerably in their potency and specificity toward individual HDACs. For example, short-chain fatty acids tend to be active at millimolar concentrations, whereas certain hydroxamic acids (e.g., belinostat) and benzamides (e.g., romidepsin) are active at low nanomolar concentrations. In addition, hydroxamic acid HDACIs such as vorinostat and panobinostat act as pan-HDACIs, whereas others (e.g., MGCD0103) primarily target Class I HDACs (Fournel et al., 2008). Of note, specific Class IIb HDACIs (e.g., tubacin) have now been described (Haggarty et al., 2003). Furthermore, clinically relevant Class IIb HDACIs have recently been developed and have now entered the clinical arena. A theoretical advantage of such agents is that they may be relatively free of host toxicities that stem from inhibition of Class I HDACs. One such agent, ACY-1215, has shown promising preclinical interactions with other targeted agents (e.g., bortezomib) in multiple myeloma (Santo et al., 2012), and trials involving this compound are currently underway.

4. DETERMINANTS OF HDAC INHIBITOR LETHALITY

4.1. Downregulation of antiapoptotic proteins

In addition to their capacity to promote gene expression, that is, by inducing a more relaxed chromatin structure and/or by acetylating promoter regions, HDACIs also downregulate the expression of diverse genes (Mitsiades et al., 2004). HDACI-mediated downregulation of antiapoptotic members of the Bcl-2 (B-cell lymphoma 2 protein) family has been implicated in the lethality of these agents toward diverse transformed cell types. For example in human melanoma cell lines, HDACIs have been shown to downregulate Bcl-xL, Mcl-1, and X-linked inhibitor of apoptosis protein (XIAP; Gillespie, Borrow, Zhang, & Hersey, 2006), and similar findings have been observed in human leukemia cells (Rosato et al., 2007, 2006). HDACI-mediated downregulation of XIAP has also been implicated in the lethal effects of HDACIs in Hodgkin's lymphoma cells (Jona et al., 2011). In the case of Mcl-1, such actions may stem from activation of miRNAs (Saito et al., 2009). Silencing of miRNAs such as miRNA29b by HDACIs may play a particularly important role in downregulating Mcl-1 expression in chronic lymphocytic leukemia (CLL) cells, which may in turn increase their sensitivity to other cytotoxic agents (Sampath et al., 2012). Also in CLL cells, induction of apoptosis by depsipeptide has been related, in part, to downregulation of FLIP (Flice-inhibitory protein; Aron et al., 2003). In human glioma cells, HDACIs exposure resulted in diminished expression of survivin and XIAP through a cdc2-dependent mechanism (Kim et al., 2005). Thus, the ability of HDACIs to reduce levels of antiapoptotic proteins, coupled with upregulation of proapoptotic proteins (see below) may shift the balance away from cell survival and toward cell death.

4.2. Upregulation of proapoptotic proteins (Bim)

HDACIs have are known to upregulate the expression of multiple proapoptotic members of the Bcl-2 family, particularly Bim. For example, HDACIs have been shown to promote recruitment of E2F1 to the Bim promoter, leading to increased expression of this protein (Tan et al., 2006; Zhao et al., 2005). Moreover, this phenomenon played a significant functional role in the increased sensitivity of transformed cells exhibiting high E2F1 activity to HDACIs. Upregulation of Bim also played a critical role in sensitization of human leukemia cells to BH3 mimetics (Chen, Dai, Pei, Grant, 2009). In acute lymphoblastic leukemia cells, HDACIs reversed insensitivity to glucocorticoids through a Bim-dependent mechanism (Bachmann et al., 2010) as well as chemoresistance in Burkitt's lymphoma cells (Richter-Larrea et al., 2010). Upregulation of Bim has also been implicated in antileukemic synergism between HDACIs and proteasome inhibitors in CLL and acute myelogenous leukemia (AML) cells (Dai, Chen, Kramer, et al., 2008; Dai et al., 2011) and dual aurora kinase-Bcr/Abl inhibitors in chronic myelogenous leukemia (CML) cells (Dai, Chen, Venditti, et al., 2008). In addition to Bim, other proapoptotic proteins such as Noxa or Bmf may also play a role in HDACI lethality toward transformed cells (Inoue, Riley, Gant, Dyer, & Cohen, 2007); (Wiegmans et al., 2011). In this context, upregulation of

Noxa by HDACIs has been attributed to p53 acetylation (Terui et al., 2003). In mantle cell lymphoma cells, HDACI-mediated lethality was associated with acetylation of promoters for multiple proapoptotic proteins, including Bim, BMF, and Noxa (Xargay-Torrent et al., 2011).

4.3. Activation of the death receptor pathway

Among their diverse actions, HDACIs have been shown to upregulate components of the extrinsic, death receptor pathway, including DR4, DR5, TRAIL (TNF-related apoptosis-inducing ligand), and Fas. For example, in human leukemia cells, HDACIs selectively induced death receptors without exerting comparable effects toward their normal counterparts (Insinga et al., 2005; Nebbioso et al., 2005). Because both HDACIs (Ungerstedt et al., 2005) and TRAIL (Ashkenazi, 2002) selectively target transformed versus normal cells, the notion of combining these agents is an attractive one. Indeed, synergistic interactions between TRAIL and HDACIs in human leukemia cells have been observed and related to inhibition of Class I versus Class II HDACs (Inoue et al., 2004). In human glioblastoma cells, HDACI-mediated sensitization to TRAIL was related to c-Myc-induced downregulation of c-FLIP (Bangert et al., 2012). Degradation of c-FLIP has also been implicated in sensitization of pancreatic cancer cells to TRAIL by HDACIs (Kauh et al., 2010). Numerous examples of cooperation between HDACIs and death receptor agonists have been described in both malignant hematopoietic and epithelial tumor systems (Martin et al., 2011) (and see below).

4.4. Induction of Bid

Bid is a BH3-only member of the Bcl-2 family, that is, primarily activated through the extrinsic apoptotic pathway, for example, via the FADD (Fas-associated death domain)-dependent induction of caspase-8, which cleaves Bid into its active form, tBid (Debatin & Krammer, 2004). Activated tBid in turn triggers mitochondrial injury, culminating in the release of cytochrome C into the cytosol, which cooperates with Apaf-1 to activate caspase-9 followed by induction of executioner caspases (Ott, Norberg, Zhivotovsky, & Orrenius, 2009). The requirement for tBid in the optimal induction of apoptosis has been employed to distinguish between two forms of apoptosis, type I and type II. In Type I cells, apoptosis is largely independent of the mitochondria and proceeds through activation of caspase-8 and FADD, associated with marked DISC formation (Scaffidi et al., 1998) and mitochondrial injury via induction of Bid. In contrast, apoptosis in Type II cells is associated with only modest DISC formation, and activation of caspase-8 occurs as a consequence, rather than a cause, of mitochondrial damage. In this context, Ruefli et al., reported that in human T-cell leukemia cells, the HDACI vorinostat-induced apoptosis through activation of Bid through a caspase-8-independent mechanism that involved oxidative injury and which was blocked by Bcl-2 (Ruefli et al., 2001). Bid cleavage was also implicated, in conjunction with activation of the mitochondrial pathway, in the lethal effects of vorinostat toward head and neck cancer cells (Gillenwater, Zhong, & Lotan, 2007). In *in vivo* studies, vorinostat-mediated lethality toward an Eμ-Myc lymphoma model was shown to be independent of the death receptor pathway, but dependent upon Bid activation and Bim (Lindemann et al., 2007). Collectively, these findings suggest that Bid activation represents a potentially important determinant of HDACI lethality in transformed cells.

4.5. Induction of $p21^{CIP1}$

$p21^{CIP1}$ is an endogenous cyclin-dependent kinase inhibitor that is induced by diverse stimuli, including DNA-damaging agents, ionizing radiation, and differentiation-inducing agents, among others (Ocker & Schneider-Stock, 2007). Although it is classically induced via activation of p53, p53-independent pathways of $p21^{CIP1}$ induction also exist (Besson, Dowdy, & Roberts, 2008). $p21^{CIP1}$ contributes to cell cycle arrest by blocking phosphorylation of the retinoblastoma protein by CDKs, thereby promoting E2F-1 binding which prevents cell cycle progression (Chen, Tsai, Leone, 2009). In addition to its cell cycle-related effects, $p21^{CIP1}$ has also been reported to oppose apoptosis through various mechanisms, including those related to redox events (O'reilly, 2005). $p21^{CIP1}$ was one of the first genes found to be induced by HDACIs, an event associated with marked acetylation of the $p21^{CIP1}$ promoter (Richon, Sandhoff, Rifkind, & Marks, 2000). Thus, $p21^{CIP1}$ belongs to a group of HDACI-inducible genes whose net effect is to diminish HDACI lethality. In support of this concept, various interventions that block $p21^{CIP1}$ induction have been shown to potentiate HDACI lethality (see below) (Almenara, Rosato, & Grant, 2002). On the other hand, in some settings, HDACI-mediated $p21^{CIP1}$ induction may have proapoptotic functions. For example, in colon tumor cells, $p21^{CIP1}$ induction by HDACIs was associated with downregulation of the antiapoptotic protein survivin, which played a key role in potentiating TRAIL lethality (Nawrocki et al., 2007). Although the HDACI vorinostat was relatively ineffective in inducing $p21^{CIP1}$ in T-lymphoblastic leukemia cells, coadministration of the hypomethylating agent decitabine potently derepressed $p21^{CIP1}$ expression, resulting in striking antiproliferative effects (Davies et al., 2011). In this context, transactivation of the p53/p21 pathway by HDACIs such as depsipeptide in transformed cells has been directly related to induction of oxidative injury andDNA damage (Wang et al., 2012) (see below). Finally, enhanced induction of $p21^{CIP1}$ has been invoked as a mechanism for synergism with HDACIs and depletion of the polycomb protein EZH2 in melanoma cells (Fan et al., 2011)

4.6. Induction of oxidative injury

Among their numerous actions, it has long been recognized that HDACIs induce the generation of reactive oxygen species (ROS) in transformed cells (Ruefli et al., 2001). However, the mechanism of ROS generation, as well as their source (e.g., mitochondrial vs. extramitochondrial) has not been determined with certainty. In human lymphoblastic leukemia cells, HDACI-mediated ROS generation was shown to be dependent upon processing of Bid but independent of caspase-3 activation (Ruefli et al., 2001). In this context, the selective toxicity that HDACIs exert toward transformed cells has been attributed to differential induction of antioxidant proteins, including Manganese superoxide dismutase 2 ($MnSOD_2$), and particularly thioredoxin (Ungerstedt et al., 2005). Furthermore, the extent of induction of ROS and thioredoxin was found to be a reliable predictor of HDACI lethality in human prostate cancer cells (Xu, Ngo, Perez, Dokmanovic, & Marks, 2006). Analogously, in human leukemia cells, induction of ROS correlated more closely with HDACI lethality than other classic activities, for example, acetylation of histones H3 and H4 (Rosato, Almenara, & Grant, 2003). Significantly, administration of antioxidants (e.g., TBAP, a cell permeable superoxide dismutase mimetic) substantially protected cells from both HDACI-induced DNA damage and lethality (Dai, Rahmani, Dent, & Grant, 2005). Antioxidants have also been shown to diminish vorinostat-mediated DNA damage, manifested by γH2A.X formation, in human myeloid leukemia cells, accompanied by a reduction in lethality (Petruccelli et al., 2011). Conversely, interference with antioxidant defenses, for example, by phenylethylisothiocyanate, which depletes cellular glutathione, significantly increased the antileukemic effects of vorinostat and overcame resistance in human leukemia cells (Hu et al., 2010). Furthermore,

DNA damage induction associated with HDACI lethality in human leukemia cells (Gaymes et al., 2006) has been attributed to ROS generation (Rosato et al., 2008, 2010). In light of these findings, it is not surprising that multiple attempts to enhance the activity of HDACIs with other targeted agents, including multiple proteasome inhibitors (e.g., bortezomib, carfilzomib) have been shown to involve potentiation of oxidative injury and DNA damage (Dasmahapatra et al., 2010; Miller, Singh, Rivera-Del, Manton, & Chandra, 2011).

4.7. Interference with DNA repair

The major cellular targets of ROS are cellular lipids and DNA. In view of the ability of HDACIs to induce ROS, it is not surprising that DNA damage induction has been strongly implicated the lethality of these agents toward transformed cells. Furthermore, the ability of HDACIs to modulate DNA repair processes may amplify the lethal consequences of ROS generation. In this context, HDACIs have been shown to trigger early double-stranded DNA breaks in cancer cells, reflected by formation of the atypical histone γH2A. X, which participates in both DNA damage sensing and repair, as well as activation of the DNA damage check point protein, ataxia telangiectasia mutated (ATM). Furthermore, these events may occur preferentially in transformed versus normal cells, contributing to HDACI selectivity (Gaymes et al., 2006). The ability of antioxidants to block both HDACI-mediated DNA damage as well as cell death argues for a hierarchy wherein HDACIs trigger ROS resulting in lethal DNA damage (Dai et al., 2005; Rosato et al., 2008). Furthermore, HDACIs may disrupt the DNA repair process by acetylating DNA repair proteins and/or downregulating them, resulting in amplification of lethality. Specifically, HDACIs may induce acetylation of the DNA repair protein Ku70 through a process involving interference with suppression of Bax-induced cell death (Cohen et al., 2004; Subramanian, Opipari, Bian, Castle, & Kwok, 2005). At the same time, they may downregulate the expression of repair proteins including RAD50, RAD51, and MRE11 (Lee, Choy, Ngo, Foster, & Marks, 2010; Rosato et al., 2008). Interestingly, disruption of DNA repair by HDACIs may be more pronounced in transformed versus normal cells, contributing to selectivity (Lee et al., 2010; Rosato et al., 2008). The possibility that HDACs may participate actively in the DNA repair process provides another mechanism by which HDACIs may act. In support of this notion, loss of HDAC3 increases spontaneous DNA damage and interferes with the S-phase checkpoint in MEF cells (Bhaskara et al., 2008). Finally, evidence has recently been presented indicating that HDACs may participate in both homologous and nonhomologous end-joining DNA repair, processes that may be disrupted by HDACIs (Kachhap et al., 2010; Miller et al., 2010). Finally, very recent studies link HDACI actions related to perturbations in DNA repair to modulation of the function of Mec1, the yeast ortholog of ATR, through an autophagy-related process (Robert et al., 2011). Specifically, it has been proposed that HDACIs disrupt the function and stability of critical proteins involved in the resection of DNA ends during repair processes (Shubassi, Robert, Vanoli, Minucci, & Foiani, 2012). Collectively, these findings suggest that HDACI may simultaneously trigger DNA damage while interrupting repair responses, contributing to lethality. They also provide a theoretical basis by which HDACIs may promote the activity of both conventional genotoxic as well as targeted anticancer agents.

4.8. Interference with chaperone function

The HDACIs, particularly those that disrupt the function of Class IIb HDACs, which include tubulin acetylases, can lead to acetylation of multiple proteins, including chaperone proteins such as heat shock protein 90 (Hsp90), which play a major role in the cell's ability to dispose of unwanted or misfolded proteins (Bali et al., 2005). Disruption of Hsp90 can lead in turn to downregulation of diverse Hsp90 client proteins implicated in transformed cell survival and proliferation, including p53, ErbB1, ErbB2, Raf-1, and Akt, among numerous others (Workman, Burrows, Neckers, & Rosen, 2007; Yu et al., 2002). Furthermore, because mutant oncoproteins such as Bcr/Abl and FLT3-ITD are particularly dependent upon Hsp90 for their maintenance, disruption of Hsp90 function and downregulation of these proteins has been invoked to explain potentiation of tyrosine kinase inhibitor lethality by HDACIs (Bali et al., 2004; Fiskus, Pranpat, Bali, et al., 2006) as well as interactions between HDACIs and the dual EGFR/HER2 inhibitor lapatinib (LaBonte et al., 2011). In addition, degradation of FLT3 has been implicated in the lethal effects of the HDACI MS-275 toward FLT3 mutant AML cells (Nishioka, Ikezoe, Yang, Takeuchi, et al., 2008). This is a noteworthy finding, as MS-275 has generally been viewed primarily as an inhibitor of Class I HDACs, and it has been proposed that pan- or Class IIb HDACIs, which target HDAC6, are the most likely to disrupt chaperone function (Bali et al., 2005).

4.9. Interference with the function of corepressors/cofactors

In addition to their ability to modulate gene expression by acetylating gene promoters, HDACs also interact with corepressors to regulate transcription (Hug & Lazar, 2004). In this context, Bcl-6 (B-cell lymphoma 6 protein) is a corepressor implicated in lymphomagenesis (Polo et al., 2004). Recently, disruption of Hsp90 function has been invoked to account for down-regulation of the corepressor Bcl-6 and resulting cell death in diffuse large B-cell lymphoma (DLBCL) cells (Cerchietti et al., 2010). In addition, Hdac3 interactions with the cofactors NCOR1 and SMRT have been shown to play critical roles in maintaining genomic integrity and in ameliorating DNA damage (Bhaskara et al., 2008, 2010). Finally, vorinostat has been shown to interfere with binding of the Sin3 complex to the $p21^{CIP1}$ promoter in 293T cells. This action was attributed to dissociation of the ING2 subunit from the Sin3 complex by HDACIs (Smith, Martin-Brown, Florens, Washburn, & Workman, 2010).

4.10. Proteotoxic and ER stress

Misfolded proteins are disposed of by the cell by dynein motors, a component of a subcellular organelle termed the aggresome. HDACIs, particularly those that block HDAC6, disrupt the function of aggresomes through a yet to be defined mechanism, resulting in accumulation of misfolded proteins (Kawaguchi et al., 2003). This added burden contributes to proteotoxic stress, which can in turn lead to endoplasmic reticulum stress and cell death. Indeed, several studies have suggested that the lethal effects of HDACIs may stem, at least in part, through promotion of ER stress and/or disruption of aggresome function (Nawrocki et al., 2006, 2007). An increase in proapoptotic components of the ER stress response has recently (e.g., CHOP) been linked to the lethal effects of HDACIs toward mantle cell lymphoma cells and breast cancer cells (Rao et al., 2010; Rottenberg et al., 2008). On the other hand, induction of Grp78, a critical sensor of the ER stress response, by HDACIs has been invoked to explain resistance to these agents in colon tumor and other transformed cell lines (Baumeister, Dong, Fu, & Lee, 2009).

4.11. Interference with checkpoint regulation

Cell cycle checkpoints protect transformed cells from external (e.g., genotoxic stresses) and internal (replicative) stress, thereby protecting the integrity of the genome. These checkpoints are frequently dysregulated in cancer (Kastan & Bartek, 2004). HDACIs have been shown to disrupt the G2-phase checkpoint, which causes cells to enter an aberrant mitosis (Warrener et al., 2003). In addition, HDACIs can cause mitotic slippage, thereby interfering with the mitotic spindle checkpoint, that is, by preventing accumulation of chromosomal passenger proteins necessary for maintain mitotic arrest (Stevens, Beamish, Warrener, & Gabrielli, 2008). HDACIs also induce premature sister chromatid separation, which may contribute to lethality (Magnaghi-Jaulin, Eot-Houllier, Fulcrand, & Jaulin, 2007). These events may reflect a requirement for HDAC3 complexes in proper functioning of the mitotic spindle and kinetochore microtubule (Ishii, Kurasawa, Wong, & Yu-Lee, 2008).

4.12. Activation of the stress-related JNK pathway

JNK (c-Jun N-terminal kinase) represents, along with p38 and ERK1/2 (extracellular-regulated kinase 1/2), a member of the three-module MAPK pathway, which plays a critical role in cell survival and proliferation. JNK is activated in response to diverse genotoxic and other stresses, and in general, exerts proapoptotic effects. In some systems, the balance between the outputs of the prodeath JNK/p38 pathways and the antiapoptotic ERK1/2 pathway determines cell fate (Xia, Dickens, Raingeaud, Davis, & Greenberg, 1995). Exposure of transformed cells to HDACIs induces JNK activation (Vrana et al., 1999). Several studies have demonstrated that JNK activation plays a functional role in the lethality of HDACIs, particularly when they are combined with other agents (e.g., IKK (IkappaB kinase) inhibitors or proteasome inhibitors) (Dai et al., 2010a; Dasmahapatra et al., 2010). Recent evidence suggests that in such a setting, JNK activation reflects induction of DNA damage and ROS generation, based on evidence that antioxidants can prevent both JNK activation and lethality (Dai et al., 2005).

4.13. Activation of NF-κB

As noted previously, HDACIs acetylate diverse proteins, including transcription factors such as YY1 and nuclear factor kappa light chain-enhancer of B cells (NF-κB; Glozak et al., 2005). It has long been recognized that acetylation of RelA (p65), which, with p50, forms the predominant NF-κB heterodimer, leads to prolonged NF-κB activation (Chen & Greene, 2004). Acetylation of RelA exerts multiple effects, including increasing RelA nuclear localization, diminishing binding to the RelA antagonist IκBα, and increasing DNA binding and transactivation (Chen, Mu, & Greene, 2002). Conversely, interventions that block RelA acetylation significantly enhance HDACI lethality (Dai et al., 2005). Until recently, the mechanism by which HDACIs initially activated NF-κB remained unknown. However, in human leukemia cells, this has been shown to stem from oxidative injury and induction of NF-κB by the atypical, inside out NF-kappa B essential modulator (NEMO)/SUMOylation DNA damage pathway (Rosato et al., 2010). In several respects, this process mirrors the close relationship that exists between ROS generation, NF-κB induction, and JNK activation observed with other DNA-damaging stimuli (Bubici, Papa, Pham, Zazzeroni, & Franzoso, 2004).

4.14. STAT5 inhibition

Signal transducer and transcriptional activator 5 (STAT5) is a transcription factor implicated in the survival of diverse malignant cell types, including CML. HDACs have been found to be required for the recruitment of and transcriptional transactivation by STAT5 (Rascle, Johnston, & Amati, 2003). In this context, HDACIs have been shown to promote CML cell death in association with downregulation of STAT5 (FiskusPranpat, Balasis, et al., 2006). Furthermore, resistance of cutaneous T-cell lymphoma cells to the HDACI vorinostat has been associated with persistent activation of STAT5 (Fantin et al., 2008). Potentiation of Bcr/Abl kinase inhibitor lethality by HDACIs has been related to inhibition of STAT5 in Bcr/Abl^{+} cells (Nguyen et al., 2011). In human cutaneous T-cell lymphoma cells, induction of cell death by the pan-HDACI panobinostat was related to inactivation of both STAT5 and STAT3 (Shao et al., 2010). Together, these findings raise the possibility that interventions capable of inactivating STAT5 might be employed to enhance the antitumor effects of HDACIs.

4.15. Proteasome inhibition

HDACIs have also been reported to interfere with proteasome function. For example, in multiple myeloma cells, treatment with the pan-HDACI vorinostat resulted in multiple perturbations, including inhibition of proteasome activity and downregulation of proteasome subunits (Mitsiades et al., 2004). In a genome-wide loss-of-function screen, HDACI-mediated apoptosis was associated with alterations in the expression of HR23B, which shuttles ubiquitinated proteins to the proteasome, and which is expressed at high levels in cutaneous T-cell lymphoma

(Fotheringham et al., 2009). The ability of HDACIs to interfere with proteasome function may contribute to interactions with proteasome inhibitors (see below).

4.16. Antiangiogenic effects

In addition to the actions outlined above, HDACIs have been reported to exert antiangiogenic activity in multiple transformed cell types. This capacity has been related, at least in part, to downregulation of VEGF and/or other angiogenic molecules (Deroanne et al., 2002; Gillenwater et al., 2007). Notably, synergistic interactions between HDACIs and VEGF antagonists have been reported in several neoplastic cell types and have been attributed to enhance antiangiogenic activity (Qian et al., 2004).

4.17. Induction of autophagy

Autophagy is an evolutionarily conserved pathway through which cells degrade proteins and cellular organelles to provide the cell with energy under conditions of metabolic stress (Levine & Klionsky, 2004). Like ER stress, autophagy can initially play a cytoprotective role, but when a certain threshold is reached, autophagy can contribute to cell death (White & DiPaola, 2009). In human leukemia cells, HDACIs have been shown to induce autophagy; moreover, inhibition of autophagy, for example, by agents such as chloroquine, promoted HDACI lethality, indicating that in this setting, autophagy primarily plays an antiapoptotic role (Carew et al., 2007). The mechanism by which HDACIs induce autophagy represents an area of continuing investigation. Recent studies suggest that acetylation of Atg3, an autophagy-related ubiquitin carrier protein, is involved in the regulation of Atg8 lipidation, and by extension, the induction of autophagy (Yi et al., 2012). In addition, it has recently been shown that HDACIs induce upregulation of Atg3 while inhibiting mTOR, a well-known stimulus for autophagy (Gammoh et al., 2012). Collectively, these findings raise the possibility that interrupting cytoprotective autophagy pathways in cells exposed to HDACIs may represent a promising strategy for enhancing the antitumor potential of this class of compounds.

5. HDACI STRATEGIES IN CANCER THERAPY

5.1. HDACIs as single agents

To date, the major role for HDACIs has been in hematologic malignancies, and more specifically, in cutataneous T-cell lymphomas. For example, the pan-HDACI vorinostat displayed substantial activity when administered alone in this disease, and was subsequently approved for this indication (Grant, Easley, & Kirkpatrick, 2007). In addition, romidepsin (depsipeptide), which has classically been considered an inhibitor of class I HDACs, exhibited significant activity in this disease (Whittaker et al., 2010) and has also been approved in this setting. In acute leukemia, HDACIs have shown some albeit modest single agent activity in AML (Garcia-Manero, Assouline, et al., 2008; Garcia-Manero, Yang, et al., 2008). Recently, romidepsin has shown significant single agent activity in patients with refractory peripheral T-cell leukemia (Piekarz et al., 2011), leading to accelerated FDA approval for this agent in this disease. In a phase II trial, single agent vorinostat exhibited some, albeit limited, activity in patients with mantle cell lymphoma or marginal zone lymphoma (Kirschbaum et al., 2011). To date, HDACIs have shown relatively modest single agent activity in solid tumor malignancies, although preliminary evidence of activity when combined with other agents has been described (Munster et al., 2011). The basis for the limited activity of HDACIs in solid tumors, or for the emergence of resistance to these agents in malignant T-cell disorders is not currently known, but may be related to activation of the members of the STAT family of transcription factors (Fantin et al., 2008).

5.2. Combination therapies

Given the relatively restricted spectrum of activity of HDACIs when administered as single agents, attention has turned to their use as part of combination regimens. Because the mechanism(s) of HDACI lethality are so diverse, mechanisms underlying their interactions with other agents are not unexpectedly highly complex. In general, two strategies have been taken in the design of such regimens. First, HDACIs have been combined with standard cytotoxic agents in the hope of lowering the apoptotic threshold. Second, considerable attention has focused on attempts to combine HDACIs with other targeted agents. Because such agents, like HDACIs, are often pleiotropic in their actions, multiple mechanisms may underlie interactions with this class of drugs.

5.2.1 Established cytotoxic agents

5.2.1.1 DNA-damaging agents

One attractive concept is that HDACIs, by promoting a more relaxed chromatin structure, may facilitate the action of cytotoxic agents that interact with DNA. For example, it has been shown that pretreatment of tumor cells with an HDACI may render chromatin more open and hence susceptible to the action of agents such as topoisomerase inhibitors (Marchion et al., 2004) or numerous other agents that trigger DNA damage (Thurn, Thomas, Moore, & Munster, 2011). Furthermore, HDACIs may, by acetylating and disrupting the function of proteins involved in DNA repair, including Ku70 (Subramanian et al., 2005) and CtIP (Robert et al., 2011), potentiate the lethal actions of genotoxic agents by amplifying DNA damage (Shubassi et al., 2012). Whether these actions will lead to an improvement in the therapeutic index remains to be determined.

5.2.1.2 Nucleoside analogs

In human leukemia cells, several studies have shown that HDACIs can sensitize human leukemia cells to nucleoside analogs such as ara-C and fludarabine (Rosato et al., 2008). Synergistic interactions between these agents have been related to multiple actions, including downregulation of Bcl-xL and XIAP, potentiation of oxidative injury, and increased DNA damage. In pediatric acute leukemia cells, knockdown or inhibition of HDACs 1 and 6 were found to play significant functional roles in potentiation of ara-C lethality (Xu et al., 2011).

5.2.2 Interactions with targeted agents

The past several years has seen a dramatic increase in interest in combination regimens employing HDACIs with various other targeted agents. The rationale for this strategy is that given the pleiotropic actions of HDACIs, as well as their putative selectivity for transformed cells (Ungerstedt et al., 2005), rational combinations with other novel agents can be envisioned which may be associated with an improved therapeutic index. Moreover, interruption of multiple pathways by such agents and HDACIs may circumvent the problem of loss of addiction of the tumor cell to a single pathway. While this concept has not yet been confirmed, there has been sufficiently promising initial information available to justify continuing investigation of this area.

5.2.2.1 Interactions with hypomethylating agents

By far, the most evidence exists for combining HDACIs with hypomethylating agents as a therapeutic strategy in cancer. Hypomethylating agents, which generally inhibit DNA methyltransferases, reverse gene silencing by DNA methylation of CpG islands, a characteristic of numerous cancers. The concept underlying this approach is that a combined strategy in which HDACIs induce relaxation of chromatin while hypomethylating agents promote gene expression may be particularly effective in reversing the silencing of genes that inhibit transformed cell growth and/or survival, for example, endogenous cyclin-dependent kinase inhibitors such as $p16^{INK}$ (Jones & Baylin, 2007). In support of this notion, numerous studies have shown enhanced antiproliferative effects for regimens combining HDACIs with hypomethylating agents such as 5-azacytidine or 5-deoxyazacytidine (Cameron, Bachman, Myohanen, Herman, & Baylin, 1999). For example, in human lung cancer cells, coadministration of 5-azacytidine and the HDACI entinostat resulted in a marked inhibition of apoptosis and inhibition of proliferation in association with derepression of p21 and p16 (Belinsky et al., 2011). In addition to its therapeutic potential, this strategy was also effective in preventing the development of lung cancer in mice (Belinsky et al., 2003). Of note, in certain hematologic malignancies such as AML and myelodysplastic syndromes (MDSs), combination regimens incorporating HDACIs such as vorinostat or MS-275 with 5-azacytidine have shown promising activity (Gore et al., 2006; Griffiths & Gore, 2008). Such an approach may also have activity in non-Hodgkin's lymphoma (NHL) and solid tumor malignancies (Stathis et al., 2011). Synergistic antilymphoma effects were also observed in DLBCL models exposed to decitabine and panobinostat, associated with specific perturbations in gene expression profiles (Kalac et al., 2011). One question that has not been completely resolved, however, is whether these agents act solely by reinducing expression of genes responsible for cell death or differentiation or whether at least part of their activity stems from cytotoxic interactions. For example, the lethal consequences of DNA damage induction by hypomethylating agents such as decitabine may be amplified by HDACI-mediated downregulation of DNA repair genes (Lee et al., 2010).

5.2.2.2 Interactions with proteasome inhibitors

As noted previously, considerable evidence exists indicating that HDAC and proteasome inhibitors interact to promote cell death in transformed cells (Hideshima et al., 2005; Miller et al., 2007; Nawrocki et al., 2006; Pei, Dai, & Grant, 2004; Yu, Rahmani, Conrad, et al., 2003). Such interactions may occur at multiple levels, which are not mutually exclusive (Hideshima, Richardson, & Anderson, 2011). For example, HDACIs induce activation of NF-κB through a DNA damage pathway (Rosato et al., 2010), leading in turn to induction of antiapoptotic genes such as those encoding Bcl-xL and XIAP. Interruption of this process, either by IKK proteasome inhibitors, results in a dramatic increase in lethality (Dai et al., 2005; Daley, 2003). Alternatively, class II HDACIs (tubulin acetylases) disrupt dynein function and by extension, aggresome actions, which, in combination with interference with protein degradation, results in proteotoxic stress (Bali et al., 2005). The latter process may lead to an endoplasmic reticulum stress-related form of cell death (Nawrocki et al., 2005, 2006). In addition, it is possible that proteasome inhibition by HDACIs may cooperate with that induced by proteasome inhibitors to promote cell death (Fotheringham et al., 2009). Of note, results of recent studies indicate that the novel irreversible proteasome inhibitor carfilzomib interacts synergistically in DLBCL cells, including those resistant to bortezomib (Dasmahapatra et al., 2010). As described above, proteasome inhibitors have been shown to downregulate expression of DNMT1 by an NF-κB- and SP1-dependent mechanism (Liu et al., 2008), raising the possibility that combined HDAC inhibition and hypomethylation may contribute to synergistic interactions between these agents. Finally, in as much as both proteasome inhibitors and HDACIs can promote cell death through a ROS- and DNA damage-dependent mechanism (Ling, Liebes, Zou, & Perez-Soler, 2003; Rosato et al., 2008, 2010; Yu, Rahmani, Dent, & Grant, 2004), the possibility that synergistic interactions may stem from potentiation of these events arises. Indeed, recent studies suggest that in NHL cells, synergism between proteasome and HDAC inhibitors reflects enhanced oxidative injury and DNA damage (Bhalla et al., 2009; Dasmahapatra et al., 2011). Regardless of the mechanism(s) accounting for enhanced lethality, recent clinical studies suggest that a strategy combining bortezomib and vorinostat may have considerable promise in certain hematologic malignancies, that is, multiple myeloma (Badros et al., 2009). Promising results with the combination of romidepsin and bortezomib have also been observed in this disease (Harrison et al., 2011).

5.2.2.3 Interactions with tyrosine kinase inhibitors

A strong rationale exists for combining HDAC with tyrosine kinase inhibitors in cancer chemotherapy. Developing inhibitors of mutant oncogenic kinases has been one of the central goals of cancer drug development over the past decade. A prototypical example of this approach is the development of imatinib mesylate, a potent inhibitor of the Bcr/Abl kinase, a mutant oncoprotein responsible for the development of CML. Imatinib mesylate, as well as its second generation analogs dasatinib and niltonib, have revolutionized the treatment of CML (Druker, 2008). Because mutant oncoproteins are particularly dependent upon chaperone proteins such as Hsp90 for their stabilization (An, Schulte, & Neckers, 2000), and because certain HDACIs acetylate and disrupt Hsp90 function (Glozak et al., 2005), the concept of employing HDACIs in conjunction with inhibitors of oncogenic kinases has arisen. Indeed, multiple studies have demonstrated that coadministration of Bcr/Abl kinase inhibitors with HDACIs leads to enhanced Bcr/Abl inhibition and cell death in CML cells, a phenomenon in part related to downregulation of the Bcr/Abl oncoprotein itself (Fiskus, Pranpat, Balasis, et al., 2006; Fiskus, Pranpat, Bali, et al., 2006; Yu, Rahmani, Almenara, et al., 2003). Similar interactions have been reported in the case of pharmacologic inhibitors of the FLT3 oncoprotein, which is mutated in up to 30% of patients with AML (Bali et al., 2004). More recently, enhanced killing has been observed with combined exposure of cells exhibiting Janus kinase 2 (JAK2) mutations, a characteristic of myeloproliferative neoplasms, to JAK2 inhibitors in combination with HDACIs (Wang et al., 2009). In solid tumor cells, the HDACI panobinostat was shown to downregulate EGFR/HER2 kinase and to interact synergistically with the EGFR/HER2 inhibitor lapatinib in human colon cancer cells (LaBonte et al., 2011). Of note, this interaction was associated with the marked downregulation of the MEK1/2/ERK1/2 and AKT signal transduction pathways. In lung cancer cells, HDACIs interacted synergistically with the tyrosine kinase inhibitor erlotinib in association with upregulation of E-cadherin expression (Witta et al., 2006).

5.2.2.4 Interactions with CDK inhibitors

CDK inhibitors were originally developed as inhibitors of transformed cell proliferation, but were subsequently shown to exert multiple other functions. These include inhibition of transcription as a consequence of inhibition of CDK9, leading to disruption of the pTEFb complex, and by extension, RNA PolII (Shapiro, 2006). Flavopiridol (alvocidib) was the first of the CDK inhibitors to enter the clinic (Senderowicz, 1999). In preclinical studies, exposure of human leukemia cells to flavopiridol blocked induction of p21CIP1 by HDACIs, culminating in a marked increase in apoptosis (Almenara et al., 2002). Similar interactions have been reported in human epithelial tumor cells exposed to flavopiridol in combination with the HDACI depsipeptide (Nguyen et al., 2004) as well as in neuroblastoma cells (Huang, Sheard, Ji, Sposto, & Keshelava, 2010). Trials combining HDACIs such as vorinostat with flavopiridol have been initiated in AML (Grant et al.,

2008), and the emergence of newer generation CDK inhibitors (e.g., SCH727965, CYC202 etc.) (MacCallum et al., 2005; Parry et al., 2010) offers the opportunity for further exploration of this strategy.

5.2.2.5 Aurora kinase inhibitors

Among their numerous actions, HDACIs can induce mitotic slippage and interfere with the mitotic spindle apparatus (Stevens et al., 2008). They also disrupt multiple proteins implicated in the chromosome passenger complex, including survivin and aurora A (Gabrielli, Chia, & Warrener, 2011). Furthermore, deacetylation of histone tails by HDAC3 is important for the proper regulation of mitosis by aurora B and interference with this process, for example, by HDAC3 inhibitors, disrupts progression of cells through mitosis (Li et al., 2006). In addition, HDACIs (e.g., LBH-589) have been shown to downregulate both aurora A and aurora B protein levels in transformed cells, for example, renal carcinoma cells (Cha et al., 2009). Finally, HDACIs interfere with kinetocore assembly by disrupting pericentromeric heterochromatin (Robbins et al., 2005). Collectively, these observations provide a theoretical basis for interactions between HDAC and aurora kinase inhibitors in transformed cells. In support of this notion, in human Bcr/abl^{+} leukemia cells, coadministration of HDACIs and aurora kinase inhibitors resulted in the synergistic induction of cell death (Dai, Chen, Venditti, et al., 2008; Fiskus et al., 2008). This combination regimen has also shown significant activity against Philadelphia-positive acute lymphoblastic leukemia cells (Okabe, Tauchi, & Ohyashiki, 2010). More recently, synergistic interactions between aurora kinase inhibitors have been described in human NHL cells in association with downregulation of c-Myc and c-TERT (Kretzner et al., 2011).

5.2.2.6 Hsp90 antagonists

Hsp90 antagonists act by interfering with Hsp90 chaperone function, leading to loss of mutant oncoprotein client proteins (e.g., Raf, AKT, Bcr/Abl etc.) upon which many transformed cells depend for their survival (Workman et al., 2007). Analogously, HDACIs that function as tubulin acetylases also interfere with Hsp90 function and may act in a similar function (Bali et al., 2005). Multiple studies have shown that coadministration of HDACIs and Hsp90 antagonists interact synergistically to trigger transformed cell death (George et al., 2005; Rahmani, Reese, Dai, Bauer, Kramer, et al., 2005; Rao et al., 2009; Yu et al., 2011). Whether such interactions reflect cooperativity at the level of Hsp90 function, or other actions related to one or both of these agents (e.g., induction of DNA damage) remain to be determined.

5.2.2.7 TRAIL

Multiple mechanisms have been invoked to account for synergistic interactions between HDACIs and TRAIL or other activators of the extrinsic pathway. For example, dual activation of both the intrinsic (i.e., by HDACIs) and the extrinsic (i.e., by TRAIL or death receptor agonistic antibodies) apoptotic pathways may lead to amplification of the death process by promoting Bid degradation and activation (Rosato, Almenara, Dai, & Grant, 2003). Alternatively, in some transformed cell types (e.g., leukemia), exposure to HDACIs can trigger death receptor upregulation, and in so doing, can sensitize cells to death receptor agonists (Insinga et al., 2005; Nebbioso et al., 2005). In glioblastoma cells, sensitization to the effects of concomitant HDACI/TRAIL exposure was related to the c-Myc-related downregulation of c-FLIP (Bangert et al., 2012). A similar mechanism was described in pancreatic cancer cells exposed to TRAIL in conjunction with HDACIs (Kauh et al., 2010). Synergistic induction of apoptosis by HDACIs and agonistic anti-DR4 antibodies has also been described in T-cell lymphoblastic leukemia cells (Sung et al., 2010).Finally, HDACIs have been shown to sensitize TRAIL-resistant breast cancer cells to TRAIL (Srivastava, Kurzrock, & Shankar, 2010).

5.2.2.8 NF-κB antagonists

In some transformed cells, HDACIs elicit a cytoprotective NF-κB response through induction of the inside-out ATM/NEMO/SUMOylation pathway, which diminishes HDACI lethality (Rosato et al., 2010). In addition, in Hodgkin's lymphoma cells, HDACI-mediated activation of NF-κB was related functionally to induction of TNF-α (Buglio et al., 2010). As noted previously, the NF-κB protective response can be attenuated by proteasome inhibitors which prevent RelA nuclear translocation by sparing IκBα from degradation, allowing it to trap RelA in the cytoplasm (Dai, Chen, Kramer, et al., 2008). Alternatively, the same effect can be achieved by inhibiting IKK, which is responsible for phosphorylating IκBα, a step necessary for its degradation (Dai et al., 2005). In this regard, IKK has attracted attention as a possible therapeutic target for malignancies dependent upon NF-κB for maintenance or survival (Tergaonkar, Bottero,

Ikawa, Li, & Verma, 2003). Of note, the parthenolide analog LC-1 is a potent IKK inhibitor now entering clinical trials (Hewamana et al., 2008). Interestingly, parthenolide has been shown to increase HDACI lethality in human leukemia cells in association with NF-κB inhibition and JNK activation (Dai et al., 2010b).

5.2.2.9 BH3 mimetics

Several BH3-mimetics, which recapitulate the actions of proapoptotic members of the Bcl-2 family, including ABT-263 (navitoclax), obatoclax, and AT-101, have entered the clinical arena (Zeitlin, Zeitlin, & Nor, 2008). Several preclinical studies have shown that HDACIs potentiate the activity of such agents in various transformed cell types, particularly hematopoietic malignancies. For example, coadministration of HDACIs has been shown to promote the lethality of ABT-737, and inhibitor of Bcl-2 and Bcl-xL, but not Mcl-1 (Oltersdorf et al., 2005) in human leukemia or lymphoma cells through a process involving upregulation of the proapoptotic proteins Bim or Bmf (Chen, Dai, et al., 2009; Wiegmans et al., 2011). In diverse Hodgkin's lymphoma cell types, synergistic interactions were observed between the Class I HDACI entinostat (MS-275) and several BH3-mimetics (e.g., obatoclax, ABT-737) (Jona et al., 2011). Synergism between ABT-737 and HDACIs in human myeloid leukemia cells has been shown to proceed through a Bim-dependent mechanism (Chen, Dai, et al., 2009).

5.2.2.10 Sorafenib

Like HDACIs, the multikinase inhibitor sorafenib has been shown to exert its effects through multiple mechanisms, including antiangiogenic actions, induction of ER stress, and inhibition of Mcl-1 translation (Rahmani et al., 2007). The capacity of sorafenib to trigger ER stress through phosphorylation of eIF2α, which effectively inhibits translation, may contribute to the latter phenomenon (Rahmani et al., 2007). In addition, in solid tumor systems, sorafenib is a potent inducer of autophagy (Park, Reinehr, et al., 2010). Consistent with these observations, synergistic interactions have been described between sorafenib and HDACIs in human leukemia cells in association with downregulation of Mcl-1 and $p21^{CIP1}$ (Dasmahapatra, Yerram, Dai, Dent, & Grant, 2007). Furthermore, in epithelial tumors, HDACI/sorafenib lethality has been attributed to induction ER stress and autophagy through a ceramide and CD95-dependent process (Park, Mitchell, et al., 2010; Park et al., 2008).

5.2.2.11 MEK1/2 inhibitors

In human leukemia cells, synergistic interactions have been described between HDACIs and inhibitors of the MAPK pathway, most notably MEK1/2 inhibitors (Nishioka, Ikezoe, Yang, Koeffler, & Yokoyama, 2008). Synergism between HDACI and MEK1/2 inhibitors was also observed in BCR/ABL^{+} human leukemia cells and was associated with potentiation of oxidative injury (Yu, et al., 2005). These results were confirmed in other tumor cell types, and it was observed that potentaition of HDACI-mediated lethality by MEK1/2/ERK1/2 inhibitors primarily occurred in cells in which the latter pathway was constitutively activated (Ozaki, Minoda, Kishikawa, & Kohno, 2006). In human osteosarcoma cells, activation of the MEK1/2/ERK1/2 pathway was implicated in resistance to HDACIs (Matsubara et al., 2009). In view of the ability of both MEK1/2 (Ewings et al., 2007) and HDAC (Tan et al., 2006) inhibitors to upregulate expression of the proapoptotic protein Bim, it is tempting to speculate that induction of this protein contributes to interactions between these agents. In addition, the observation that HDACIs activate the MEK1/2/ERK1/2 pathway through an SP1-dependent process (Yang, Kawai, Hanson, & Arinze, 2001) raises the possibility that ERK1/2 activation represents a cytoprotective response to HDAC inhibition, and that disabling this response contributes to potentiation of HDACI lethality. Finally, the Ras/Raf/MEK1/2/ERK1/2 pathway signals downstream to induce multiple DNA repair proteins, including XRCC1 and ERCC1, among others (Lammering et al., 2001), as well as DNA damage response proteins such as ATM (Golding et al., 2007). In view of extensive evidence implicating DNA damage and interference with DNA repair events in HDACI lethality (Gaymes et al., 2006; Petruccelli et al., 2011; Rosato et al., 2010, 2006), the possibility that MEK1/2 inhibitors promote HDACI-mediated genotoxic injury, and/or interfere with the response to such damage, appear plausible.

5.2.2.12 PI3K/AKT inhibitors

Synergistic interactions have been described between AKT inhibitors such as perifosine and HDACIs in human leukemia cells through a process involving both ERK1/2 and AKT1/2 inactivation as well as the generation of ceramide and ROS (Rahmani, Reese, Dai, Bauer, Payne, et al., 2005). Analogously, PI3K inhibitors have been shown to interact synergistically with HDACIs in both leukemic and epithelial tumor types (Denlinger, Rundall, & Jones, 2005; Rahmani et al., 2003). Interestingly, in the former case, lethality was more dependent upon inhibition of ERK1/2 than AKT (Rahmani et al., 2003). In contrast, synergistic interactions between HDACIs and dual PI3K/mTOR inhibitors in head and neck squamous carcinoma cells correlated strongly with AKT inactivation (Erlich et al., 2011). Similar interactions have recently been observed in cutaneous T-cell lymphoma cells (Wozniak et al., 2010). Notably, blockade of PI3K/AKT signaling substantially increased the lethality of HDACIs toward tumor cells resistant to tyrosine kinase inhibitors such as gefitinib or imatinib (Ozaki et al., 2010). In human DLBCL cells, resistance to the mTOR inhibitor rapamycin was associated with rapamycin-induced activation of AKT signaling. However, this phenomenon was largely prevented by coadministration of LBH-589, resulting in a sharp increase in apoptosis (Gupta et al., 2009). In lung carcinoma cells, this interaction was related to enhanced induction of ROS (Koshikawa, Hayashi, Nakagawara, & Takenaga, 2009). Finally, several studies have suggested that AKT activation is necessary for phosphorylation and activation of the p300 histone acetyltransferase, which promotes acetylation and transcriptional activation of RelA/p65, leading in turn to induction of various prosurvival NF-κB-dependent genes (Kim et al., 2006; Liu, Denlinger, Rundall, Smith, & Jones, 2006). Thus, interruption of the PI3K/AKT pathway may act, at least in part, by blocking NF-κB activation in HDACI-treated cells, an action known to lower the threshold for HDACI-mediated lethality (Dai et al., 2005).

5.2.2.13 JAK2 inhibitors

Mutation of JAK2 (e.g., JAK2V617F) has been implicated in the pathogenesis of myeloproliferative neoplasms (e.g., myelofibrosis/myeloid metaplasia) (Kilpivaara & Levine, 2008), prompting the development JAK2 inhibitors for such disorders. Such agents have shown significant activity in this group of disorders and have now been approved for this indication, although their primary benefit lies in improving symptomatology rather than in prolonging survival or eradicating disease (Harrison et al., 2012). In light of evidence that mutant oncoproteins are particularly dependent upon chaperone protein function for their survival, and that Class IIb HDACIs effectively downregulate the expression of such proteins (Bali et al., 2005), the possibility arose that pan-HDACIs might enhance the activity of JAK2 inhibitors. Indeed, studies in JAK2V6117F MPN cells demonstrated that coadministration of panobinostat with the JAK2 inhibitor TG101209 resulted in pronounced mutant protein downregulation and inhibition, leading to a dramatic increase in lethality for the combination compared to effects of the combination (Wang et al., 2009). Based upon these and related findings, clinical trials combining JAK2 with HDAC inhibitors are now being implemented in patients with myeloproliferative neoplasms.

6. SUMMARY AND FUTURE DIRECTIONS

Despite their relatively narrow spectrum of approved activities to date, interest in HDACIs as antineoplastic agents continues unabated. This stems from multiple factors, including their capacity to modify the expression of genes implicated in transformation, continuing evidence of their selective activity toward transformed cells and their capacity to interact synergistically with a large number of conventional as well as targeted agents. The latter ability is particularly noteworthy, and it is likely that the future role of HDACIs will be in rational combination regimens with other agents. However, to achieve their full potential, several important questions will have to be addressed. For example, does the activity of HDACIs reflect their ability to modify gene expression, exert direct cytotoxic effects, or a combination of the two? The latter possibility appears highly likely, in that HDACIs may trigger DNA damage through ROS production (Ruefli et al., 2001), while at the same time downregulating or disrupting the function of DNA repair proteins (Subramanian et al., 2005). In addition, given the expanding list of nonhistone proteins acetylated by HDACIs, and whose function may be disrupted by this class of agents, it is becoming increasingly less likely that HDACIs function solely as epigenetic agents, for example, as modulators of gene transcription (Shubassi et al., 2012). Conversely, it is hard to imagine that perturbations in gene function and protein expression triggered by HDACIs would not have a significant impact on cellular responses to the lethal effects of disruption of the function of diverse proteins resulting from acetylation. Another important question is whether the selectivity shown by HDACIs in preclinical studies can translate into a favorable therapeutic index, either when these agents are administered alone or in combination with other agents. Certainly, the therapeutic selectivity of HDACIs is not absolute, in that various toxicities, for example, gastrointestinal, systemic (e.g., fatigue) limits their use. A related question is whether HDAC specificity is desirable goal, that is, will HDACIs targeting specific HDACs (e.g., HDAC6) or classes of HDACs (e.g., Class I nuclear HDACs) will prove superior to, for example, pan-HDACIs. The answer to this question may ultimately depend upon whether the antitumor effects of HDACIs can ultimately be attributed to inhibition of specific HDACs or HDAC classes. Finally, given the enormous number of candidate targeted agents with which HDACIs can be combined to achieve synergistic antitumor effects, it will be important to determine which of the many possibilities offers the greatest opportunity for success. This may depend upon whether one or more of their preclinical actions can be recapitulated in humans. For example, if acetylation and disruption of the function of Hsp90 can be achieved *in vivo*, then combination with inhibitors of mutant oncogenic kinases or agents such as proteasome inhibitors which induce proteotoxic stress may prove to be particularly effective. Given the continuing interest in the clinical development of HDACIs, it is likely that answers to these questions will be forthcoming in the years to come.

ACKNOWLEDGMENTS

This work was supported by awards CA093738, CA100866, 1 P50 CA130805-01, and 1 P50CA142509-01 from the National Cancer Institute; award 6181-10 from the Leukemia and Lymphoma Society of America; and awards from the V Foundation and the Multiple Myeloma Research Foundation.

REFERENCES

Almenara, J., Rosato, R., & Grant, S. (2002). Synergistic induction of mitochondrial damage and apoptosis in human leukemia cells by flavopiridol and the histone deacetylase inhibitor suberoylanilide hydroxamic acid (SAHA). *Leukemia*, *16*, 1331–1343.

An, W. G., Schulte, T. W., & Neckers, L. M. (2000). The heat shock protein 90 antagonist geldanamycin alters chaperone association with p210bcr-abl and v-src proteins before their degradation by the proteasome. *Cell Growth & Differentiation*, *11*, 355–360.

Aron, J. L., Parthun, M. R., Marcucci, G., Kitada, S., Mone, A. P., Davis, M. E., et al. (2003). Depsipeptide (FR901228) induces histone acetylation and inhibition of histone deacetylase in chronic lymphocytic leukemia cells concurrent with activation of caspase 8-mediated apoptosis and down-regulation of c-FLIP protein. *Blood*, *102*, 652–658.

Ashkenazi, A. (2002). Targeting death and decoy receptors of the tumour-necrosis factor superfamily. *Nature Reviews. Cancer*, *2*, 420–430.

Bachmann, P. S., Piazza, R. G., Janes, M. E., Wong, N. C., Davies, C., Mogavero, A., et al. (2010). Epigenetic silencing of BIM in glucocorticoid poor-responsive pediatric acute lymphoblastic leukemia, and its reversal by histone deacetylase inhibition. *Blood*, *116*, 3013–3022.

Badros, A., Burger, A. M., Philip, S., Niesvizky, R., Kolla, S. S., Goloubeva, O., et al. (2009). Phase I study of vorinostat in combination with bortezomib for relapsed and refractory multiple myeloma. *Clinical Cancer Research*, *15*, 5250–5257.

Bali, P., George, P., Cohen, P., Tao, J., Guo, F., Sigua, C., et al. (2004). Superior activity of the combination of histone deacetylase inhibitor LAQ824 and the FLT-3 kinase inhibitor PKC412 against human acute myelogenous leukemia cells with mutant FLT-3. *Clinical Cancer Research*, *10*, 4991–4997.

Bali, P., Pranpat, M., Bradner, J., Balasis, M., Fiskus, W., Guo, F., et al. (2005). Inhibiition of histone deacetylase 6 acetylates and disrupts the chaperone function of heat shock protein

90: A novel basis of antileukemia activity of histone deacetylase inhibitors. *The Journal of Biological Chemistry, 280,* 26729–26734.

Bangert, A., Cristofanon, S., Eckhardt, I., Abhari, B. A., Kolodziej, S., Hacker, S., et al. (2012). Histone deacetylase inhibitors sensitize glioblastoma cells to TRAIL-induced apoptosis by c-myc-mediated downregulation of cFLIP. *Oncogene,* (in press).

Batty, N., Malouf, G. G., & Issa, J. P. (2009). Histone deacetylase inhibitors as anti-neoplastic agents. *Cancer Letters, 280,* 192–200.

Baumeister, P., Dong, D., Fu, Y., & Lee, A. S. (2009). Transcriptional induction of GRP78/BiP by histone deacetylase inhibitors and resistance to histone deacetylase inhibitor-induced apoptosis. *Molecular Cancer Therapeutics, 8,* 1086–1094, 2007.

Belinsky, S. A., Grimes, M. J., Picchi, M. A., Mitchell, H. D., Stidley, C. A., Tesfaigzi, Y., et al. (2011). Combination therapy with vidaza and entinostat suppresses tumor growth and reprograms the epigenome in an orthotopic lung cancer model. *Cancer Research, 71,* 454–462.

Belinsky, S. A., Klinge, D. M., Stidley, C. A., Issa, J. P., Herman, J. G., March, T. H., et al. (2003). Inhibition of DNA methylation and histone deacetylation prevents murine lung cancer. *Cancer Research, 63,* 7089–7093.

Besson, A., Dowdy, S. F., & Roberts, J. M. (2008). CDK inhibitors: Cell cycle regulators and beyond. *Developmental Cell, 14,* 159–169.

Bhalla, S., Balasubramanian, S., David, K., Sirisawad, M., Buggy, J., Mauro, L., et al. (2009). PCI-24781 induces caspase and reactive oxygen species-dependent apoptosis through NF-kappaB mechanisms and is synergistic with bortezomib in lymphoma cells. *Clinical Cancer Research, 15,* 3354–3365.

Bhaskara, S., Chyla, B. J., Amann, J. M., Knutson, S. K., Cortez, D., Sun, Z. W., et al. (2008). Deletion of histone deacetylase 3 reveals critical roles in S phase progression and DNA damage control. *Molecular Cell, 30,* 61–72.

Bhaskara, S., Knutson, S. K., Jiang, G., Chandrasekharan, M. B., Wilson, A. J., Zheng, S., et al. (2010). Hdac3 is essential for the maintenance of chromatin structure and genome stability. *Cancer Cell, 18,* 436–447.

Bubici, C., Papa, S., Pham, C. G., Zazzeroni, F., & Franzoso, G. (2004). NF-kappaB and JNK: An intricate affair. *Cell Cycle, 3,* 1524–1529.

Buglio, D., Mamidipudi, V., Khaskhely, N. M., Brady, H., Heise, C., Besterman, J., et al. (2010). The class-I HDAC inhibitor MGCD0103 induces apoptosis in Hodgkin lymphoma cell lines and synergizes with proteasome inhibitors by an HDAC6-independent mechanism. *British Journal of Haematology, 151,* 387–396.

Cameron, E. E., Bachman, K. E., Myohanen, S., Herman, J. G., & Baylin, S. B. (1999). Synergy of demethylation and histone deacetylase inhibition in the re-expression of genes silenced in cancer. *Nature Genetics, 21,* 103–107.

Carew, J. S., Nawrocki, S. T., Kahue, C. N., Zhang, H., Yang, C., Chung, L., et al. (2007). Targeting autophagy augments the anticancer activity of the histone deacetylase inhibitor SAHA to overcome Bcr-Abl-mediated drug resistance. *Blood, 110,* 313–322.

Cerchietti, L. C., Hatzi, K., Caldas-Lopes, E., Yang, S. N., Figueroa, M. E., Morin, R. D., et al. (2010). BCL6 repression of EP300 in human diffuse large B cell lymphoma cells provides a basis for rational combinatorial therapy. *The Journal of Clinical Investigation, 120*(12), 4569–4582.

Cha, T. L., Chuang, M. J., Wu, S. T., Sun, G. H., Chang, S. Y., Yu, D. S., et al. (2009). Dual degradation of aurora A and B kinases by the histone deacetylase inhibitor LBH589 induces G2-M arrest and apoptosis of renal cancer cells. *Clinical Cancer Research, 15,* 840–850.

Chen, S., Dai, Y., Pei, X. Y., & Grant, S. (2009). Bim up-regulation by histone deacetylase inhibitors mediates interactions with the Bcl-2 antagonist ABT-737: Evidence for distinct roles for Bcl-2, Bcl-xL and Mcl-1. *Molecular and Cellular Biology, 29,* 6149–6169.

Chen, L. F., & Greene, W. C. (2004). Shaping the nuclear action of NF-kappaB. *Nature Reviews Molecular Cell Biology, 5,* 392–401.

Chen, L. F., Mu, Y., & Greene, W. C. (2002). Acetylation of RelA at discrete sites regulates distinct nuclear functions of NF-kappaB. *The EMBO Journal, 21*, 6539–6548.

Chen, H. Z., Tsai, S. Y., & Leone, G. (2009). Emerging roles of E2Fs in cancer: An exit from cell cycle control. *Nature Reviews. Cancer, 9*, 785–797.

Cohen, H. Y., Lavu, S., Bitterman, K. J., Hekking, B., Imahiyerobo, T. A., Miller, C., et al. (2004). Acetylation of the C terminus of Ku70 by CBP and PCAF controls Bax-mediated apoptosis. *Molecular Cell, 13*, 627–638.

Dai, Y., Chen, S., Kramer, L. B., Funk, V. L., Dent, P., & Grant, S. (2008). Interactions between bortezomib and romidepsin and belinostat in chronic lymphocytic leukemia cells. *Clinical Cancer Research, 14*, 549–558.

Dai, Y., Chen, S., Venditti, C. A., Pei, X. Y., Nguyen, T. K., Dent, P., et al. (2008). Vorinostat synergistically potentiates MK-0457 lethality in chronic myelogenous leukemia cells sensitive and resistant to imatinib mesylate. *Blood, 112*(3), 793–804.

Dai, Y., Chen, S., Wang, L., Pei, X. Y., Kramer, L. B., Dent, P., et al. (2011). Bortezomib interacts synergistically with belinostat in human acute myeloid leukaemia and acute lymphoblastic leukaemia cells in association with perturbation in NF-kappaB and Bim. *British Journal of Haematology, 153*(2), 222–235.

Dai, Y., Guzman, M. L., Chen, S., Wang, L., Yeung, S. K., Pei, X. Y., et al. (2010a). The NF (Nuclear factor)-kappaB inhibitor parthenolide interacts with histone deacetylase inhibitors to induce MKK7/JNK1-dependent apoptosis in human acute myeloid leukaemia cells. *British Journal of Haematology, 151*, 70–83.

Dai, Y., Rahmani, M., Dent, P., & Grant, S. (2005). Blockade of histone deacetylase inhibitor-induced RelA/p65 acetylation and NF-{kappa}B activation potentiates apoptosis in leukemia cells through a process mediated by oxidative damage, XIAP downregulation, and c-Jun N-terminal kinase 1 activation. *Molecular and Cellular Biology, 25*, 5429–5444.

Daley, G. Q. (2003). Towards combination target-directed chemotherapy for chronic myeloid leukemia: Role of farnesyl transferase inhibitors. *Seminars in Hematology, 40*, 11–14.

Dasmahapatra, G., Lembersky, D., Kramer, L., Fisher, R. I., Friedberg, J., Dent, P., et al. (2010). The pan-HDAC inhibitor vorinostat potentiates the activity of the proteasome inhibitor carfilzomib in human DLBCL cells in vitro and in vivo. *Blood, 115*, 4478–4487.

Dasmahapatra, G., Lembersky, D., Son, M. P., Attkisson, E., Dent, P., Fisher, R. I., et al. (2011). Carfilzomib interacts synergistically with histone deacetylase inhibitors in mantle cell lymphoma cells in vitro and in vivo. *Molecular Cancer Therapeutics, 10*, 1686–1697.

Dasmahapatra, G., Yerram, N., Dai, Y., Dent, P., & Grant, S. (2007). Synergistic interactions between vorinostat and sorafenib in chronic myelogenous leukemia cells involve Mcl-1 and p21CIP1 down-regulation. *Clinical Cancer Research, 13*, 4280–4290.

Davies, C., Hogarth, L. A., Dietrich, P. A., Bachmann, P. S., MacKenzie, K. L., Hall, A. G., et al. (2011). p53-independent epigenetic repression of the p21(WAF1) gene in T-cell acute lymphoblastic leukemia. *The Journal of Biological Chemistry, 286*, 37639–37650.

Debatin, K. M., & Krammer, P. H. (2004). Death receptors in chemotherapy and cancer. *Oncogene, 23*, 2950–2966.

Denlinger, C. E., Rundall, B. K., & Jones, D. R. (2005). Inhibition of phosphatidylinositol 3-kinase/Akt and histone deacetylase activity induces apoptosis in non-small cell lung cancer in vitro and in vivo. *The Journal of Thoracic and Cardiovascular Surgery, 130*, 1422–1429.

Deroanne, C. F., Bonjean, K., Servotte, S., Devy, L., Colige, A., Clausse, N., et al. (2002). Histone deacetylases inhibitors as anti-angiogenic agents altering vascular endothelial growth factor signaling. *Oncogene, 21*, 427–436.

Druker, B. J. (2008). Translation of the Philadelphia chromosome into therapy for CML. *Blood, 112*, 4808–4817.

Drummond, D. C., Noble, C. O., Kirpotin, D. B., Guo, Z., Scott, G. K., & Benz, C. C. (2005). Clinical development of histone deacetylase inhibitors as anticancer agents. *Annual Review of Pharmacology and Toxicology, 45*, 495–528.

Erlich, R. B., Kherrouche, Z., Rickwood, D., Endo-Munoz, L., Cameron, S., Dahler, A., et al. (2011). Preclinical evaluation of dual PI3K-mTOR inhibitors and histone deacetylase inhibitors in head and neck squamous cell carcinoma. *British Journal of Cancer, 106*(1), 107–115.

Ewings, K. E., Hadfield-Moorhouse, K., Wiggins, C. M., Wickenden, J. A., Balmanno, K., Gilley, R., et al. (2007). ERK1/2-dependent phosphorylation of Bim(EL) promotes its rapid dissociation from Mcl-1 and Bcl-x(L). *The EMBO Journal, 26*, 2856–2867.

Fan, T., Jiang, S., Chung, N., Alikhan, A., Ni, C., Lee, C. C., et al. (2011). EZH2-dependent suppression of a cellular senescence phenotype in melanoma cells by inhibition of p21/CDKN1A expression. *Molecular Cancer Research, 9*, 418–429.

Fantin, V. R., Loboda, A., Paweletz, C. P., Hendrickson, R. C., Pierce, J. W., Roth, J. A., et al. (2008). Constitutive activation of signal transducers and activators of transcription predicts vorinostat resistance in cutaneous T-cell lymphoma. *Cancer Research, 68*, 3785–3794.

Fiskus, W., Pranpat, M., Balasis, M., Bali, P., Estrella, V., Kumaraswamy, S., et al. (2006). Cotreatment with vorinostat (suberoylanilide hydroxamic acid) enhances activity of dasatinib (BMS-354825) against imatinib mesylate-sensitive or imatinib mesylate-resistant chronic myelogenous leukemia cells. *Clinical Cancer Research, 12*, 5869–5878.

Fiskus, W., Pranpat, M., Bali, P., Balasis, M., Kumaraswamy, S., Boyapalle, S., et al. (2006). Combined effects of novel tyrosine kinase inhibitor AMN107 and histone deacetylase inhibitor LBH589 against Bcr-Abl-expressing human leukemia cells. *Blood, 108*, 645–652.

Fiskus, W., Wang, Y., Joshi, R., Rao, R., Yang, Y., Chen, J., et al. (2008). Cotreatment with vorinostat enhances activity of MK-0457 (VX-680) against acute and chronic myelogenous leukemia cells. *Clinical Cancer Research, 14*, 6106–6115.

Fotheringham, S., Epping, M. T., Stimson, L., Khan, O., Wood, V., Pezzella, F., et al. (2009). Genome-wide loss-of-function screen reveals an important role for the proteasome in HDAC inhibitor-induced apoptosis. *Cancer Cell, 15*, 57–66.

Fournel, M., Bonfils, C., Hou, Y., Yan, P. T., Trachy-Bourget, M. C., Kalita, A., et al. (2008). MGCD0103, a novel isotype-selective histone deacetylase inhibitor, has broad spectrum antitumor activity in vitro and in vivo. *Molecular Cancer Therapeutics, 7*, 759–768.

Gabrielli, B., Chia, K., & Warrener, R. (2011). Finally, how histone deacetylase inhibitors disrupt mitosis!. *Cell Cycle, 10*, 2658–2661.

Gammoh, N., Lam, D., Puente, C., Ganley, I., Marks, P. A., & Jiang, X. (2012). Role of autophagy in histone deacetylase inhibitor-induced apoptotic and nonapoptotic cell death. *Proceedings of the National Academy of Sciences of the United States of America, 109*, 6561–6565.

Gao, L., Cueto, M. A., Asselbergs, F., & Atadja, P. (2002). Cloning and functional characterization of HDAC11, a novel member of the human histone deacetylase family. *The Journal of Biological Chemistry, 277*, 25748–25755.

Garcia-Manero, G., Assouline, S., Cortes, J., Estrov, Z., Kantarjian, H., Yang, H., et al. (2008). Phase I study of the oral isotype specific histone deacetylase inhibitor MGCD0103 in leukemia. *Blood, 112*, 981–989.

Garcia-Manero, G., Yang, H., Bueso-Ramos, C., Ferrajoli, A., Cortes, J., Wierda, W. G., et al. (2008). Phase 1 study of the histone deacetylase inhibitor vorinostat (suberoylanilide hydroxamic acid [SAHA]) in patients with advanced leukemias and myelodysplastic syndromes. *Blood, 111*, 1060–1066.

Gaymes, T. J., Padua, R. A., Pla, M., Orr, S., Omidvar, N., Chomienne, C., et al. (2006). Histone deacetylase inhibitors (HDI) cause DNA damage in leukemia cells: A mechanism for leukemia-specific HDI-dependent apoptosis? *Molecular Cancer Research, 4*, 563–573.

George, P., Bali, P., Annavarapu, S., Scuto, A., Fiskus, W., Guo, F., et al. (2005). Combination of the histone deacetylase inhibitor LBH589 and the hsp90 inhibitor 17-AAG is highly active against human CML-BC cells and AML cells with activating mutation of FLT-3. *Blood, 105*, 1768–1776.

Gillenwater, A. M., Zhong, M., & Lotan, R. (2007). Histone deacetylase inhibitor suberoylanilide hydroxamic acid induces apoptosis through both mitochondrial and Fas (Cd95) signaling in head and neck squamous carcinoma cells. *Molecular Cancer Therapeutics, 6*, 2967–2975.

Gillespie, S., Borrow, J., Zhang, X. D., & Hersey, P. (2006). Bim plays a crucial role in synergistic induction of apoptosis by the histone deacetylase inhibitor SBHA and TRAIL in melanoma cells. *Apoptosis, 11*, 2251–2265.

Glozak, M. A., Sengupta, N., Zhang, X., & Seto, E. (2005). Acetylation and deacetylation of non-histone proteins. *Gene, 363*, 15–23.

Glozak, M. A., & Seto, E. (2007). Histone deacetylases and cancer. *Oncogene, 26*, 5420–5432.

Golding, S. E., Rosenberg, E., Neill, S., Dent, P., Povirk, L. F., & Valerie, K. (2007). Extracellular signal-related kinase positively regulates ataxia telangiectasia mutated, homologous recombination repair, and the DNA damage response. *Cancer Research, 67*, 1046–1053.

Gore, S. D., Baylin, S., Sugar, E., Carraway, H., Miller, C. B., Carducci, M., et al. (2006). Combined DNA methyltransferase and histone deacetylase inhibition in the treatment of myeloid neoplasms. *Cancer Research, 66*, 6361–6369.

Grant, S., Easley, C., & Kirkpatrick, P. (2007). Vorinostat. *Nature Reviews. Drug Discovery, 6*, 1–2.

Grant, S., Kolla, S., Sirulnik, A., Shapiro, G. I., Supko, J. G., Cooper, B., et al. (2008). Phase I trial of vorinostat (SAHA) in combination with alvocidib (flavopiridol) in patients with refractory, relapsed, or (selected) poor-prognosis AML or refractory anemia with excess blasts-2 (RAEB-2). *Blood, 112*, 2986 (Abstract).

Griffiths, E. A., & Gore, S. D. (2008). DNA methyltransferase and histone deacetylase inhibitors in the treatment of myelodysplastic syndromes. *Seminars in Hematology, 45*, 23–30.

Gupta, M., Ansell, S. M., Novak, A. J., Kumar, S., Kaufmann, S. H., & Witzig, T. E. (2009). Inhibition of histone deacetylase overcomes rapamycin-mediated resistance in diffuse large B-cell lymphoma by inhibiting Akt signaling through mTORC2. *Blood, 114*, 2926–2935.

Haggarty, S. J., Koeller, K. M., Wong, J. C., Grozinger, C. M., & Schreiber, S. L. (2003). Domain-selective small-molecule inhibitor of histone deacetylase 6 (HDAC6)-mediated tubulin deacetylation. *Proceedings of the National Academy of Sciences of the United States of America, 100*, 4389–4394.

Harrison, C., Kiladjian, J. J., Al-Ali, H. K., Gisslinger, H., Waltzman, R., Stalbovskaya, V., et al. (2012). JAK inhibition with ruxolitinib versus best available therapy for myelofibrosis. *The New England Journal of Medicine, 366*, 787–798.

Harrison, S. J., Quach, H., Link, E., Seymour, J. F., Ritchie, D. S., Ruell, S., et al. (2011). A high rate of durable responses with romidepsin, bortezomib, and dexamethasone in relapsed or refractory multiple myeloma. *Blood, 118*, 6274–6283.

Hewamana, S., Lin, T. T., Jenkins, C., Burnett, A. K., Jordan, C. T., Fegan, C., et al. (2008). The novel nuclear factor-kappaB inhibitor LC-1 is equipotent in poor prognostic subsets of chronic lymphocytic leukemia and shows strong synergy with fludarabine. *Clinical Cancer Research, 14*, 8102–8111.

Hideshima, T., Bradner, J. E., Wong, J., Chauhan, D., Richardson, P., Schreiber, S. L., et al. (2005). Small-molecule inhibition of proteasome and aggresome function induces synergistic antitumor activity in multiple myeloma. *Proceedings of the National Academy of Sciences of the United States of America*, *102*, 8567–8572.

Hideshima, T., Richardson, P. G., & Anderson, K. C. (2011). Mechanism of action of proteasome inhibitors and deacetylase inhibitors and the biological basis of synergy in multiple myeloma. *Molecular Cancer Therapeutics*, *10*, 2034–2042.

Hu, Y., Lu, W., Chen, G., Zhang, H., Jia, Y., Wei, Y., et al. (2010). Overcoming resistance to histone deacetylase inhibitors in human leukemia with the redox modulating compound beta-phenylethyl isothiocyanate. *Blood*, *116*, 2732–2741.

Huang, J. M., Sheard, M. A., Ji, L., Sposto, R., & Keshelava, N. (2010). Combination of vorinostat and flavopiridol is selectively cytotoxic to multidrug-resistant neuroblastoma cell lines with mutant TP53. *Molecular Cancer Therapeutics*, *9*, 3289–3301.

Hug, B. A., & Lazar, M. A. (2004). ETO interacting proteins. *Oncogene*, *23*, 4270–4274.

Inoue, S., MacFarlane, M., Harper, N., Wheat, L. M., Dyer, M. J., & Cohen, G. M. (2004). Histone deacetylase inhibitors potentiate TNF-related apoptosis-inducing ligand (TRAIL)-induced apoptosis in lymphoid malignancies. *Cell Death and Differentiation*, *11*(Suppl 2), S193–S206.

Inoue, S., Riley, J., Gant, T. W., Dyer, M. J., & Cohen, G. M. (2007). Apoptosis induced by histone deacetylase inhibitors in leukemic cells is mediated by Bim and Noxa. *Leukemia*, *21*, 1773–1782.

Insinga, A., Monestiroli, S., Ronzoni, S., Gelmetti, V., Marchesi, F., Viale, A., et al. (2005). Inhibitors of histone deacetylases induce tumor-selective apoptosis through activation of the death receptor pathway. *Nature Medicine*, *11*, 71–76.

Ishii, S., Kurasawa, Y., Wong, J., & Yu-Lee, L. Y. (2008). Histone deacetylase 3 localizes to the mitotic spindle and is required for kinetochore-microtubule attachment. *Proceedings of the National Academy of Sciences of the United States of America*, *105*, 4179–4184.

Jenuwein, T., & Allis, C. D. (2001). Translating the histone code. *Science*, *293*, 1074–1080.

Jona, A., Khaskhely, N., Buglio, D., Shafer, J. A., Derenzini, E., Bollard, C. M., et al. (2011). The histone deacetylase inhibitor entinostat (SNDX-275) induces apoptosis in Hodgkin lymphoma cells and synergizes with Bcl-2 family inhibitors. *Experimental Hematology*, *39*, 1007–1017.

Jones, P. A., & Baylin, S. B. (2007). The epigenomics of cancer. *Cell*, *128*, 683–692.

Kachhap, S. K., Rosmus, N., Collis, S. J., Kortenhorst, M. S., Wissing, M. D., Hedayati, M., et al. (2010). Downregulation of homologous recombination DNA repair genes by HDAC inhibition in prostate cancer is mediated through the E2F1 transcription factor. *PloS One*, *5*, e11208.

Kalac, M., Scotto, L., Marchi, E., Amengual, J., Seshan, V. E., Bhagat, G., et al. (2011). HDAC inhibitors and decitabine are highly synergistic and associated with unique gene-expression and epigenetic profiles in models of DLBCL. *Blood*, *118*, 5506–5516.

Kastan, M. B., & Bartek, J. (2004). Cell-cycle checkpoints and cancer. *Nature*, *432*, 316–323.

Kauh, J., Fan, S., Xia, M., Yue, P., Yang, L., Khuri, F. R., et al. (2010). c-FLIP degradation mediates sensitization of pancreatic cancer cells to TRAIL-induced apoptosis by the histone deacetylase inhibitor LBH589. *PloS One*, *5*, e10376.

Kawaguchi, Y., Kovacs, J. J., McLaurin, A., Vance, J. M., Ito, A., & Yao, T. P. (2003). The deacetylase HDAC6 regulates aggresome formation and cell viability in response to misfolded protein stress. *Cell*, *115*, 727–738.

Kilpivaara, O., & Levine, R. L. (2008). JAK2 and MPL mutations in myeloproliferative neoplasms: Discovery and science. *Leukemia*, *22*, 1813–1817.

Kim, E. H., Kim, H. S., Kim, S. U., Noh, E. J., Lee, J. S., & Choi, K. S. (2005). Sodium butyrate sensitizes human glioma cells to TRAIL-mediated apoptosis through inhibition of Cdc2 and the subsequent downregulation of survivin and XIAP. *Oncogene*, *24*, 6877–6889.

Kim, Y. K., Seo, D. W., Kang, D. W., Lee, H. Y., Han, J. W., & Kim, S. N. (2006). Involvement of HDAC1 and the PI3K/PKC signaling pathways in NF-kappaB activation by the HDAC inhibitor apicidin. *Biochemical and Biophysical Research Communications, 347*, 1088–1093.

Kirschbaum, M., Frankel, P., Popplewell, L., Zain, J., Delioukina, M., Pullarkat, V., et al. (2011). Phase II study of vorinostat for treatment of relapsed or refractory indolent non-Hodgkin's lymphoma and mantle cell lymphoma. *Journal of Clinical Oncology, 29*, 1198–1203.

Koshikawa, N., Hayashi, J., Nakagawara, A., & Takenaga, K. (2009). Reactive oxygen species-generating mitochondrial DNA mutation up-regulates hypoxia-inducible factor-1alpha gene transcription via phosphatidylinositol 3-kinase-Akt/protein kinase C/histone deacetylase pathway. *The Journal of Biological Chemistry, 284*, 33185–33194.

Kretzner, L., Scuto, A., Dino, P. M., Kowolik, C. M., Wu, J., Ventura, P., et al. (2011). Combining histone deacetylase inhibitor vorinostat with aurora kinase inhibitors enhances lymphoma cell killing with repression of c-Myc, hTERT, and microRNA levels. *Cancer Research, 71*, 3912–3920.

LaBonte, M. J., Wilson, P. M., Fazzone, W., Russell, J., Louie, S. G., El-Khoueiry, A., et al. (2011). The dual EGFR/HER2 inhibitor lapatinib synergistically enhances the antitumor activity of the histone deacetylase inhibitor panobinostat in colorectal cancer models. *Cancer Research, 71*, 3635–3648.

Lammering, G., Hewit, T. H., Hawkins, W. T., Contessa, J. N., Reardon, D. B., Lin, P. S., et al. (2001). Epidermal growth factor receptor as a genetic therapy target for carcinoma cell radiosensitization. *Journal of the National Cancer Institute, 93*, 921–929.

Lee, J. H., Choy, M. L., Ngo, L., Foster, S. S., & Marks, P. A. (2010). Histone deacetylase inhibitor induces DNA damage, which normal but not transformed cells can repair. *Proceedings of the National Academy of Sciences of the United States of America, 107*(33), 14639–14644.

Levine, B., & Klionsky, D. J. (2004). Development by self-digestion: Molecular mechanisms and biological functions of autophagy. *Developmental Cell, 6*, 463–477.

Li, Y., Kao, G. D., Garcia, B. A., Shabanowitz, J., Hunt, D. F., Qin, J., et al. (2006). A novel histone deacetylase pathway regulates mitosis by modulating Aurora B kinase activity. *Genes & Development, 20*, 2566–2579.

Lindemann, R. K., Newbold, A., Whitecross, K. F., Cluse, L. A., Frew, A. J., Ellis, L., et al. (2007). Analysis of the apoptotic and therapeutic activities of histone deacetylase inhibitors by using a mouse model of B cell lymphoma. *Proceedings of the National Academy of Sciences of the United States of America, 104*, 8071–8076.

Ling, Y. H., Liebes, L., Zou, Y., & Perez-Soler, R. (2003). Reactive oxygen species generation and mitochondrial dysfunction in the apoptotic response to Bortezomib, a novel proteasome inhibitor, in human H460 non-small cell lung cancer cells. *The Journal of Biological Chemistry, 278*, 33714–33723.

Liu, Y., Denlinger, C. E., Rundall, B. K., Smith, P. W., & Jones, D. R. (2006). Suberoylanilide hydroxamic acid induces Akt-mediated phosphorylation of p300, which promotes acetylation and transcriptional activation of RelA/p65. *The Journal of Biological Chemistry, 281*, 31359–31368.

Liu, S., Liu, Z., Xie, Z., Pang, J., Yu, J., Lehmann, E., et al. (2008). Bortezomib induces DNA hypomethylation and silenced gene transcription by interfering with Sp1/NF-kappaB-dependent DNA methyltransferase activity in acute myeloid leukemia. *Blood, 111*, 2364–2373.

MacCallum, D. E., Melville, J., Frame, S., Watt, K., Anderson, S., Gianella-Borradori, A., et al. (2005). Seliciclib (CYC202, R-Roscovitine) induces cell death in multiple myeloma cells by inhibition of RNA polymerase II-dependent transcription and down-regulation of Mcl-1. *Cancer Research, 65*, 5399–5407.

Magnaghi-Jaulin, L., Eot-Houllier, G., Fulcrand, G., & Jaulin, C. (2007). Histone deacetylase inhibitors induce premature sister chromatid separation and override the mitotic spindle assembly checkpoint. *Cancer Research*, *67*, 6360–6367.

Marchion, D. C., Bicaku, E., Daud, A. I., Richon, V., Sullivan, D. M., & Munster, P. N. (2004). Sequence-specific potentiation of topoisomerase II inhibitors by the histone deacetylase inhibitor suberoylanilide hydroxamic acid. *Journal of Cellular Biochemistry*, *92*, 223–237.

Marks, P. A., & Jiang, X. (2005). Histone deacetylase inhibitors in programmed cell death and cancer therapy. *Cell Cycle*, *4*, 549–551.

Marks, P. A., Richon, V. M., Kelly, W. K., Chiao, J. H., & Miller, T. (2004). Histone deacetylase inhibitors: Development as cancer therapy. *Novartis Foundation Symposium*, *259*, 269–281.

Martin, B. P., Frew, A. J., Bots, M., Fox, S., Long, F., Takeda, K., et al. (2011). Antitumor activities and on-target toxicities mediated by a TRAIL receptor agonist following cotreatment with panobinostat. *International Journal of Cancer*, *128*, 2735–2747.

Matsubara, H., Watanabe, M., Imai, T., Yui, Y., Mizushima, Y., Hiraumi, Y., et al. (2009). Involvement of extracellular signal-regulated kinase activation in human osteosarcoma cell resistance to the histone deacetylase inhibitor FK228 [(1S,4S,7Z,10S,16E,21R)-7-ethylidene-4,21-bis(propan-2-yl)-2-oxa-12,13-dithia-5,8,20,23-tetraazabicyclo[8.7.6] tricos-16-ene-3,6,9,19,22-pentone]. *The Journal of Pharmacology and Experimental Therapeutics*, *328*, 839–848.

Miller, C. P., Ban, K., Dujka, M. E., McConkey, D. J., Munsell, M., Palladino, M., et al. (2007). NPI-0052, a novel proteasome inhibitor, induces caspase-8 and ROS-dependent apoptosis alone and in combination with HDAC inhibitors in leukemia cells. *Blood*, *110*, 267–277.

Miller, C. P., Singh, M. M., Rivera-Del, V. N., Manton, C. A., & Chandra, J. (2011). Therapeutic strategies to enhance the anticancer efficacy of histone deacetylase inhibitors. *Journal of Biomedicine and Biotechnology*, *2011*, 514261.

Miller, K. M., Tjeertes, J. V., Coates, J., Legube, G., Polo, S. E., Britton, S., et al. (2010). Human HDAC1 and HDAC2 function in the DNA-damage response to promote DNA nonhomologous end-joining. *Nature Structural and Molecular Biology*, *17*, 1144–1151.

Mitsiades, C. S., Mitsiades, N. S., McMullan, C. J., Poulaki, V., Shringarpure, R., Hideshima, T., et al. (2004). Transcriptional signature of histone deacetylase inhibition in multiple myeloma: Biological and clinical implications. *Proceedings of the National Academy of Sciences of the United States of America*, *101*, 540–545.

Munster, P. N., Thurn, K. T., Thomas, S., Raha, P., Lacevic, M., Miller, A., et al. (2011). A phase II study of the histone deacetylase inhibitor vorinostat combined with tamoxifen for the treatment of patients with hormone therapy-resistant breast cancer. *British Journal of Cancer*, *104*, 1828–1835.

Nawrocki, S. T., Carew, J. S., Douglas, L., Cleveland, J. L., Humphreys, R., & Houghton, J. A. (2007). Histone deacetylase inhibitors enhance lexatumumab-induced apoptosis via a p21Cip1-dependent decrease in survivin levels. *Cancer Research*, *67*, 6987–6994.

Nawrocki, S. T., Carew, J. S., Pino, M. S., Highshaw, R. A., Andtbacka, R. H., Dunner, K., Jr., et al. (2006). Aggresome disruption: A novel strategy to enhance bortezomib-induced apoptosis in pancreatic cancer cells. *Cancer Research*, *66*, 3773–3781.

Nawrocki, S. T., Carew, J. S., Pino, M. S., Highshaw, R. A., Dunner, K., Jr., Huang, P., et al. (2005). Bortezomib sensitizes pancreatic cancer cells to endoplasmic reticulum stress-mediated apoptosis. *Cancer Research*, *65*, 11658–11666.

Nebbioso, A., Clarke, N., Voltz, E., Germain, E., Ambrosino, C., Bontempo, P., et al. (2005). Tumor-selective action of HDAC inhibitors involves TRAIL induction in acute myeloid leukemia cells. *Nature Medicine*, *11*, 77–84.

Nguyen, T., Dai, Y., Attkisson, E., Kramer, L., Jordan, N., Nguyen, N., et al. (2011). HDAC inhibitors potentiate the activity of the BCR/ABL kinase inhibitor KW-2449 in imatinib-sensitive or -resistant BCR/ABL+ leukemia cells in vitro and in vivo. *Clinical Cancer Research, 17*, 3219–3232.

Nguyen, D. M., Schrump, W. D., Chen, G. A., Tsai, W., Nguyen, P., Trepel, J. B., et al. (2004). Abrogation of p21 expression by flavopiridol enhances depsipeptide-mediated apoptosis in malignant pleural mesothelioma cells. *Clinical Cancer Research, 10*, 1813–1825.

Nishioka, C., Ikezoe, T., Yang, J., Koeffler, H. P., & Yokoyama, A. (2008). Inhibition of MEK/ERK signaling synergistically potentiates histone deacetylase inhibitor-induced growth arrest, apoptosis and acetylation of histone H3 on p21waf1 promoter in acute myelogenous leukemia cell. *Leukemia, 22*, 1449–1452.

Nishioka, C., Ikezoe, T., Yang, J., Takeuchi, S., Koeffler, H. P., & Yokoyama, A. (2008). MS-275, a novel histone deacetylase inhibitor with selectivity against HDAC1, induces degradation of FLT3 via inhibition of chaperone function of heat shock protein 90 in AML cells. *Leukemia Research, 32*, 1382–1392.

Ocker, M., & Schneider-Stock, R. (2007). Histone deacetylase inhibitors: Signalling towards p21cip1/waf1. *The International Journal of Biochemistry & Cell Biology, 39*, 1367–1374.

Okabe, S., Tauchi, T., & Ohyashiki, K. (2010). Efficacy of MK-0457 and in combination with vorinostat against Philadelphia chromosome positive acute lymphoblastic leukemia cells. *Annals of Hematology, 89*, 1081–1087.

Oltersdorf, T., Elmore, S. W., Shoemaker, A. R., Armstrong, R. C., Augeri, D. J., Belli, B. A., et al. (2005). An inhibitor of Bcl-2 family proteins induces regression of solid tumours. *Nature, 435*, 677–681.

O'reilly, M. A. (2005). Redox activation of p21Cip1/WAF1/Sdi1: A multifunctional regulator of cell survival and death. *Antioxidants & Redox Signaling, 7*, 108–118.

Ott, M., Norberg, E., Zhivotovsky, B., & Orrenius, S. (2009). Mitochondrial targeting of tBid/Bax: A role for the TOM complex? *Cell Death and Differentiation, 16*, 1075–1082.

Ozaki, K., Kosugi, M., Baba, N., Fujio, K., Sakamoto, T., Kimura, S., et al. (2010). Blockade of the ERK or PI3K-Akt signaling pathway enhances the cytotoxicity of histone deacetylase inhibitors in tumor cells resistant to gefitinib or imatinib. *Biochemical and Biophysical Research Communications, 391*, 1610–1615.

Ozaki, K., Minoda, A., Kishikawa, F., & Kohno, M. (2006). Blockade of the ERK pathway markedly sensitizes tumor cells to HDAC inhibitor-induced cell death. *Biochemical and Biophysical Research Communications, 339*, 1171–1177.

Park, M. A., Mitchell, C., Zhang, G., Yacoub, A., Allegood, J., Haussinger, D., et al. (2010). Vorinostat and sorafenib increase CD95 activation in gastrointestinal tumor cells through a Ca(2+)-de novo ceramide-PP2A-reactive oxygen species-dependent signaling pathway. *Cancer Research, 70*, 6313–6324.

Park, M. A., Reinehr, R., Haussinger, D., Voelkel-Johnson, C., Ogretmen, B., Yacoub, A., et al. (2010). Sorafenib activates CD95 and promotes autophagy and cell death via Src family kinases in gastrointestinal tumor cells. *Molecular Cancer Therapeutics, 9*, 2220–2231.

Park, M. A., Zhang, G., Martin, A. P., Hamed, H., Mitchell, C., Hylemon, P. B., et al. (2008). Vorinostat and sorafenib increase ER stress, autophagy and apoptosis via ceramide-dependent CD95 and PERK activation. *Cancer Biology & Therapy, 7*, 1648–1662.

Parry, D., Guzi, T., Shanahan, F., Davis, N., Prabhavalkar, D., Wiswell, D., et al. (2010). Dinaciclib (SCH 727965), a novel and potent cyclin-dependent kinase inhibitor. *Molecular Cancer Therapeutics, 9*, 2344–2353.

Pei, X. Y., Dai, Y., & Grant, S. (2004). Synergistic induction of oxidative injury and apoptosis in human multiple myeloma cells by the proteasome inhibitor bortezomib and histone deacetylase inhibitors. *Clinical Cancer Research, 10*, 3839–3852.

Petruccelli, L. A., Dupere-Richer, D., Pettersson, F., Retrouvey, H., Skoulikas, S., & Miller, W. H., Jr. (2011). Vorinostat induces reactive oxygen species and DNA damage in acute myeloid leukemia cells. *PloS One*, *6*, e20987.

Piekarz, R. L., Frye, R., Prince, H. M., Kirschbaum, M. H., Zain, J., Allen, S. L., et al. (2011). Phase 2 trial of romidepsin in patients with peripheral T-cell lymphoma. *Blood*, *117*, 5827–5834.

Polo, J. M., Dell'Oso, T., Ranuncolo, S. M., Cerchietti, L., Beck, D., Da Silva, G. F., et al. (2004). Specific peptide interference reveals BCL6 transcriptional and oncogenic mechanisms in B-cell lymphoma cells. *Nature Medicine*, *10*, 1329–1335.

Qian, D. Z., Wang, X., Kachhap, S. K., Kato, Y., Wei, Y., Zhang, L., et al. (2004). The histone deacetylase inhibitor NVP-LAQ824 inhibits angiogenesis and has a greater antitumor effect in combination with the vascular endothelial growth factor receptor tyrosine kinase inhibitor PTK787/ZK222584. *Cancer Research*, *64*, 6626–6634.

Rahmani, M., Davis, E. M., Crabtree, T. R., Habibi, J. R., Nguyen, T. K., Dent, P., et al. (2007). The kinase inhibitor sorafenib induces cell death through a process involving induction of endoplasmic reticulum stress. *Molecular and Cellular Biology*, *27*, 5499–5513.

Rahmani, M., Reese, E., Dai, Y., Bauer, C., Kramer, L. B., Huang, M., et al. (2005). Cotreatment with suberanoylanilide hydroxamic acid and 17-allylamino 17-demethoxygeldanamycin synergistically induces apoptosis in Bcr-Abl+ Cells sensitive and resistant to STI571 (imatinib mesylate) in association with down-regulation of Bcr-Abl, abrogation of signal transducer and activator of transcription 5 activity, and Bax conformational change. *Molecular Pharmacology*, *67*, 1166–1176.

Rahmani, M., Reese, E., Dai, Y., Bauer, C., Payne, S. G., Dent, P., et al. (2005). Coadministration of histone deacetylase inhibitors and perifosine synergistically induces apoptosis in human leukemia cells through Akt and ERK1/2 inactivation and the generation of ceramide and reactive oxygen species. *Cancer Research*, *65*, 2422–2432.

Rahmani, M., Yu, C., Reese, E., Ahmed, W., Hirsch, K., Dent, P., et al. (2003). Inhibition of PI-3 kinase sensitizes human leukemic cells to histone deacetylase inhibitor-mediated apoptosis through p44/42 MAP kinase inactivation and abrogation of p21(CIP1/WAF1) induction rather than AKT inhibition. *Oncogene*, *22*, 6231–6242.

Rao, R., Lee, P., Fiskus, W., Yang, Y., Joshi, R., Wang, Y., et al. (2009). Co-treatment with heat shock protein 90 inhibitor 17-dimethylaminoethylamino-17-demethoxygeldanamycin (DMAG) and vorinostat: A highly active combination against human mantle cell lymphoma (MCL) cells. *Cancer Biology & Therapy*, *8*, 1273–1280.

Rao, R., Nalluri, S., Fiskus, W., Savoie, A., Buckley, K. M., Ha, K., et al. (2010). Role of C/EBP homologous protein (CHOP) in Panobinostat-mediated potentiation of bortezomib-induced lethal ER stress in mantle cell lymphoma cells. *Clinical Cancer Research*, *16*(19), 4742–4754.

Rascle, A., Johnston, J. A., & Amati, B. (2003). Deacetylase activity is required for recruitment of the basal transcription machinery and transactivation by STAT5. *Molecular and Cellular Biology*, *23*, 4162–4173.

Richon, V. M., Sandhoff, T. W., Rifkind, R. A., & Marks, P. A. (2000). Histone deacetylase inhibitor selectively induces p21WAF1 expression and gene-associated histone acetylation. *Proceedings of the National Academy of Sciences of the United States of America*, *97*, 10014–10019.

Richter-Larrea, J. A., Robles, E. F., Fresquet, V., Beltran, E., Rullan, A. J., Agirre, X., et al. (2010). Reversion of epigenetically mediated BIM silencing overcomes chemoresistance in Burkitt lymphoma. *Blood*, *116*, 2531–2542.

Robbins, A. R., Jablonski, S. A., Yen, T. J., Yoda, K., Robey, R., Bates, S. E., et al. (2005). Inhibitors of histone deacetylases alter kinetochore assembly by disrupting pericentromeric heterochromatin. *Cell Cycle*, *4*, 717–726.

Robert, T., Vanoli, F., Chiolo, I., Shubassi, G., Bernstein, K. A., Rothstein, R., et al. (2011). HDACs link the DNA damage response, processing of double-strand breaks and autophagy. *Nature*, *471*, 74–79.

Rosato, R. R., Almenara, J. A., Dai, Y., & Grant, S. (2003). Simultaneous activation of the intrinsic and extrinsic pathways by histone deacetylase (HDAC) inhibitors and tumor necrosis factor-related apoptosis-inducing ligand (TRAIL) synergistically induces mitochondrial damage and apoptosis in human leukemia cells. *Molecular Cancer Therapeutics*, *2*, 1273–1284.

Rosato, R. R., Almenara, J. A., & Grant, S. (2003). The histone deacetylase inhibitor MS-275 promotes differentiation or apoptosis in human leukemia cells through a process regulated by generation of reactive oxygen species and induction of p21CIP1/WAF1 1. *Cancer Research*, *63*, 3637–3645.

Rosato, R. R., Almenara, J. A., Kolla, S. S., Maggio, S. C., Coe, S., Gimenez, M. S., et al. (2007). Mechanism and functional role of XIAP and Mcl-1 down-regulation in flavopiridol/vorinostat antileukemic interactions. *Molecular Cancer Therapeutics*, *6*, 692–702.

Rosato, R. R., Almenara, J. A., Maggio, S. C., Coe, S., Atadja, P., Dent, P., et al. (2008). Role of histone deacetylase inhibitor-induced reactive oxygen species and DNA damage in LAQ-824/fludarabine antileukemic interactions. *Molecular Cancer Therapeutics*, *7*, 3285–3297.

Rosato, R. R., Kolla, S. S., Hock, S. K., Almenara, J. A., Patel, A., Amin, S., et al. (2010). Histone deacetylase inhibitors activate NF-kappaB in human leukemia cells through an ATM/NEMO-related pathway. *The Journal of Biological Chemistry*, *285*, 10064–10077.

Rosato, R. R., Maggio, S. C., Almenara, J. A., Payne, S. G., Atadja, P., Spiegel, S., et al. (2006). The histone deacetylase inhibitor LAQ824 induces human leukemia cell death through a process involving XIAP down-regulation, oxidative injury, and the acid sphingomyelinase-dependent generation of ceramide. *Molecular Pharmacology*, *69*, 216–225.

Rottenberg, S., Jaspers, J. E., Kersbergen, A., van der, B. E., Nygren, A. O., Zander, S. A., et al. (2008). High sensitivity of BRCA1-deficient mammary tumors to the PARP inhibitor AZD2281 alone and in combination with platinum drugs. *Proceedings of the National Academy of Sciences of the United States of America*, *105*, 17079–17084.

Ruefli, A. A., Ausserlechner, M. J., Bernhard, D., Sutton, V. R., Tainton, K. M., Kofler, R., et al. (2001). The histone deacetylase inhibitor and chemotherapeutic agent suberoylanilide hydroxamic acid (SAHA) induces a cell-death pathway characterized by cleavage of Bid and production of reactive oxygen species. *Proceedings of the National Academy of Sciences of the United States of America*, *98*, 10833–10838.

Saito, Y., Suzuki, H., Tsugawa, H., Nakagawa, I., Matsuzaki, J., Kanai, Y., et al. (2009). Chromatin remodeling at Alu repeats by epigenetic treatment activates silenced microRNA-512-5p with downregulation of Mcl-1 in human gastric cancer cells. *Oncogene*, *28*, 2738–2744.

Sampath, D., Liu, C., Vasan, K., Sulda, M., Puduvalli, V. K., Wierda, W. G., et al. (2012). Histone deacetylases mediate the silencing of miR-15a, miR-16, and miR-29b in chronic lymphocytic leukemia. *Blood*, *119*, 1162–1172.

Santo, L., Hideshima, T., Kung, A. L., Tseng, J. C., Tamang, D., Yang, M., et al. (2012). Preclinical activity, pharmacodynamic, and pharmacokinetic properties of a selective HDAC6 inhibitor, ACY-1215, in combination with bortezomib in multiple myeloma. *Blood*, *119*, 2579–2589.

Scaffidi, C., Fulda, S., Srinivasan, A., Friesen, C., Li, F., Tomaselli, K. J., et al. (1998). Two CD95 (APO-1/Fas) signaling pathways. *The EMBO Journal*, *17*, 1675–1687.

Senderowicz, A. M. (1999). Flavopiridol: The first cyclin-dependent kinase inhibitor in human clinical trials. *Investigational New Drugs*, *17*, 313–320.

Shao, W., Growney, J. D., Feng, Y., O'Connor, G., Pu, M., Zhu, W., et al. (2010). Activity of deacetylase inhibitor panobinostat (LBH589) in cutaneous T-cell lymphoma models: Defining molecular mechanisms of resistance. *International Journal of Cancer, 127*, 2199–2208.

Shapiro, G. I. (2006). Cyclin-dependent kinase pathways as targets for cancer treatment. *Journal of Clinical Oncology, 24*, 1770–1783.

Shubassi, G., Robert, T., Vanoli, F., Minucci, S., & Foiani, M. (2012). Acetylation: A novel link between double-strand break repair and autophagy. *Cancer Research, 72*, 1332–1335.

Smith, K. T., Martin-Brown, S. A., Florens, L., Washburn, M. P., & Workman, J. L. (2010). Deacetylase inhibitors dissociate the histone-targeting ING2 subunit from the Sin3 complex. *Chemistry & Biology, 17*, 65–74.

Srivastava, R. K., Kurzrock, R., & Shankar, S. (2010). MS-275 sensitizes TRAIL-resistant breast cancer cells, inhibits angiogenesis and metastasis, and reverses epithelial-mesenchymal transition in vivo. *Molecular Cancer Therapeutics, 9*, 3254–3266.

Stathis, A., Hotte, S. J., Chen, E. X., Hirte, H. W., Oza, A. M., Moretto, P., et al. (2011). Phase I study of decitabine in combination with vorinostat in patients with advanced solid tumors and non-Hodgkin's lymphomas. *Clinical Cancer Research, 17*, 1582–1590.

Stevens, F. E., Beamish, H., Warrener, R., & Gabrielli, B. (2008). Histone deacetylase inhibitors induce mitotic slippage. *Oncogene, 27*, 1345–1354.

Subramanian, C., Opipari, A. W., Jr., Bian, X., Castle, V. P., & Kwok, R. P. (2005). Ku70 acetylation mediates neuroblastoma cell death induced by histone deacetylase inhibitors. *Proceedings of the National Academy of Sciences of the United States of America, 102*, 4842–4847.

Sung, E. S., Kim, A., Park, J. S., Chung, J., Kwon, M. H., & Kim, Y. S. (2010). Histone deacetylase inhibitors synergistically potentiate death receptor 4-mediated apoptotic cell death of human T-cell acute lymphoblastic leukemia cells. *Apoptosis, 15*, 1256–1269.

Tan, J., Zhuang, L., Jiang, X., Yang, K. K., Karuturi, K. M., & Yu, Q. (2006). Apoptosis signal-regulating kinase 1 is a direct target of E2F1 and contributes to histone deacetylase inhibitor-induced apoptosis through positive feedback regulation of E2F1 apoptotic activity. *The Journal of Biological Chemistry, 281*, 10508–10515.

Tergaonkar, V., Bottero, V., Ikawa, M., Li, Q., & Verma, I. M. (2003). IkappaB kinase-independent IkappaBalpha degradation pathway: Functional NF-kappaB activity and implications for cancer therapy. *Molecular and Cellular Biology, 23*, 8070–8083.

Terui, T., Murakami, K., Takimoto, R., Takahashi, M., Takada, K., Murakami, T., et al. (2003). Induction of PIG3 and NOXA through acetylation of p53 at 320 and 373 lysine residues as a mechanism for apoptotic cell death by histone deacetylase inhibitors. *Cancer Research, 63*, 8948–8954.

Thurn, K. T., Thomas, S., Moore, A., & Munster, P. N. (2011). Rational therapeutic combinations with histone deacetylase inhibitors for the treatment of cancer. *Future Oncology, 7*, 263–283.

Ungerstedt, J. S., Sowa, Y., Xu, W. S., Shao, Y., Dokmanovic, M., Perez, G., et al. (2005). Role of thioredoxin in the response of normal and transformed cells to histone deacetylase inhibitors. *Proceedings of the National Academy of Sciences of the United States of America, 102*, 673–678.

Vrana, J. A., Decker, R. H., Johnson, C. R., Wang, Z., Jarvis, W. D., Richon, V. M., et al. (1999). Induction of apoptosis in U937 human leukemia cells by suberoylanilide hydroxamic acid (SAHA) proceeds through pathways that are regulated by Bcl-2/Bcl-XL, c-Jun, and p21CIP1, but independent of p53. *Oncogene, 18*, 7016–7025.

Wang, Y., Fiskus, W., Chong, D. G., Buckley, K. M., Natarajan, K., Rao, R., et al. (2009). Cotreatment with panobinostat and JAK2 inhibitor TG101209 attenuates JAK2V617F levels and signaling and exerts synergistic cytotoxic effects against human myeloproliferative neoplastic cells. *Blood, 114*, 5024–5033.

Wang, H., Zhou, W., Zheng, Z., Zhang, P., Tu, B., He, Q., et al. (2012). The HDAC inhibitor depsipeptide transactivates the p53/p21 pathway by inducing DNA damage. *DNA Repair, 11*, 146–156.

Warrener, R., Beamish, H., Burgess, A., Waterhouse, N. J., Giles, N., Fairlie, D., et al. (2003). Tumor cell-selective cytotoxicity by targeting cell cycle checkpoints. *The FASEB Journal, 17*, 1550–1552.

White, E., & DiPaola, R. S. (2009). The double-edged sword of autophagy modulation in cancer. *Clinical Cancer Research, 15*, 5308–5316.

Whittaker, S. J., Demierre, M. F., Kim, E. J., Rook, A. H., Lerner, A., Duvic, M., et al. (2010). Final results from a multicenter, international, pivotal study of romidepsin in refractory cutaneous T-cell lymphoma. *Journal of Clinical Oncology, 28*, 4485–4491.

Wiegmans, A. P., Alsop, A. E., Bots, M., Cluse, L. A., Williams, S. P., Banks, K. M., et al. (2011). Deciphering the molecular events necessary for synergistic tumor cell apoptosis mediated by the histone deacetylase inhibitor vorinostat and the BH3 mimetic ABT-737. *Cancer Research, 71*, 3603–3615.

Witta, S. E., Gemmill, R. M., Hirsch, F. R., Coldren, C. D., Hedman, K., Ravdel, L., et al. (2006). Restoring E-cadherin expression increases sensitivity to epidermal growth factor receptor inhibitors in lung cancer cell lines. *Cancer Research, 66*, 944–950.

Workman, P., Burrows, F., Neckers, L., & Rosen, N. (2007). Drugging the cancer chaperone HSP90: Combinatorial therapeutic exploitation of oncogene addiction and tumor stress. *Annals of the New York Academy of Sciences, 1113*, 202–216.

Wozniak, M. B., Villuendas, R., Bischoff, J. R., Aparicio, C. B., Martinez Leal, J. F., de La, C. P., et al. (2010). Vorinostat interferes with the signaling transduction pathway of T-cell receptor and synergizes with phosphoinositide-3 kinase inhibitors in cutaneous T-cell lymphoma. *Haematologica, 95*, 613–621.

Xargay-Torrent, S., Lopez-Guerra, M., Saborit-Villarroya, I., Rosich, L., Campo, E., Roue, G., et al. (2011). Vorinostat-induced apoptosis in mantle cell lymphoma is mediated by acetylation of proapoptotic BH3-only gene promoters. *Clinical Cancer Research, 17*, 3956–3968.

Xia, Z., Dickens, M., Raingeaud, J., Davis, R. J., & Greenberg, M. E. (1995). Opposing effects of ERK and JNK-p38 MAP kinases on apoptosis. *Science, 270*, 1326–1331.

Xu, W., Ngo, L., Perez, G., Dokmanovic, M., & Marks, P. A. (2006). Intrinsic apoptotic and thioredoxin pathways in human prostate cancer cell response to histone deacetylase inhibitor. *Proceedings of the National Academy of Sciences of the United States of America, 103*, 15540–15545.

Xu, X., Xie, C., Edwards, H., Zhou, H., Buck, S. A., & Ge, Y. (2011). Inhibition of histone deacetylases 1 and 6 enhances cytarabine-induced apoptosis in pediatric acute myeloid leukemia cells. *PloS One, 6*, e17138.

Yang, J., Kawai, Y., Hanson, R. W., & Arinze, I. J. (2001). Sodium butyrate induces transcription from the G alpha(i2) gene promoter through multiple Sp1 sites in the promoter and by activating the MEK-ERK signal transduction pathway. *The Journal of Biological Chemistry, 276*, 25742–25752.

Yi, C., Ma, M., Ran, L., Zheng, J., Tong, J., Zhu, J., et al. (2012). Function and molecular mechanism of acetylation in autophagy regulation. *Science, 336*, 474–477.

Yu, C., Dasmahapatra, G., Dent, P., & Grant, S. (2005). Synergistic interactions between MEK1/2 and histone deacetylase inhibitors in Bcr/abl+ human leukemia cells. *Leukemia, 19*, 1579–1589.

Yu, X., Guo, Z. S., Marcu, M. G., Neckers, L., Nguyen, D. M., Chen, G. A., et al. (2002). Modulation of p53, ErbB1, ErbB2, and Raf-1 expression in lung cancer cells by depsipeptide FR901228. *Journal of the National Cancer Institute, 94*, 504–513.

Yu, C., Rahmani, M., Almenara, J., Subler, M., Krystal, G., Conrad, D., et al. (2003). Histone deacetylase inhibitors promote STI571-mediated apoptosis in STI571-sensitive and -resistant Bcr/Abl+ human myeloid leukemia cells. *Cancer Research, 63*, 2118–2126.

Yu, C., Rahmani, M., Conrad, D., Subler, M., Dent, P., & Grant, S. (2003). The proteasome inhibitor bortezomib interacts synergistically with histone deacetylase inhibitors to induce apoptosis in Bcr/Abl+ cells sensitive and resistant to STI571. *Blood, 102*, 3765–3774.

Yu, C., Rahmani, M., Dent, P., & Grant, S. (2004). The hierarchical relationship between MAPK signaling and ROS generation in human leukemia cells undergoing apoptosis in response to the proteasome inhibitor Bortezomib. *Experimental Cell Research, 295*, 555–566.

Yu, W., Wang, J., Jin, J., Qian, W., Qian, J., Cheng, Y., et al. (2011). Heat shock protein 90 inhibition results in altered downstream signaling of mutant KIT and exerts synergistic effects on Kasumi-1 cells when combining with histone deacetylase inhibitor. *Leukemia Research, 35*, 1212–1218.

Zeitlin, B. D., Zeitlin, I. J., & Nor, J. E. (2008). Expanding circle of inhibition: Small-molecule inhibitors of Bcl-2 as anticancer cell and antiangiogenic agents. *Journal of Clinical Oncology, 26*, 4180–4188.

Zhao, Y., Tan, J., Zhuang, L., Jiang, X., Liu, E. T., & Yu, Q. (2005). Inhibitors of histone deacetylases target the Rb-E2F1 pathway for apoptosis induction through activation of proapoptotic protein Bim. *Proceedings of the National Academy of Sciences of the United States of America, 102*, 16090–16095.

CHAPTER SEVEN

HDAC Inhibitors and Chaperone Function

Rekha Rao, Warren Fiskus, Siddhartha Ganguly, Suman Kambhampati, Kapil N. Bhalla[1]
The University of Kansas Cancer Center, Kansas City, Kansas, USA
[1]Corresponding author: e-mail address: kbhalla@kumc.edu

Contents

Abstract

Cellular chaperones promote the folding and maturation of newly synthesized proteins and partially folded proteins in the cytosol and endoplasmic reticulum (ER) as well as prevent the aggregation of misfolded proteins. Histone deacetylases (HDACs) and histone acetyl transferases catalyze the reversible acetylation of histones and nonhistone substrates to control the epigenetic and transcriptomic landscape of normal and tumor cells. Treatment with HDAC inhibitors results in the hyperacetylation of chaperones including heat shock protein (hsp)90, hsp70, hsp40, and the ER-resident hsp70 homolog, glucose-regulated protein 78 (GRP78), which affects their function. HDAC

Advances in Cancer Research, Volume 116
ISSN 0065-230X
http://dx.doi.org/10.1016/B978-0-12-394387-3.00007-0

inhibitor-mediated deregulation of chaperone function, in turn, deregulates protein homeostasis and induces protein misfolding and proteotoxic stress. In the context of tumors which are particularly dependent on functional chaperones for maintaining protein homeostasis, HDAC inhibitors tip the balance toward lethal proteotoxic and ER stress. In this chapter, we describe HDAC inhibitor-induced hyperacetylation of major chaperones and its implication for the use of HDAC inhibitors in the treatment of solid and hematologic tumors.

1. INTRODUCTION

Histone deacetylases (HDACs) belong to an evolutionarily conserved family of proteins that catalyze the deacetylation of histones and a growing list of "nonhistone" substrates (Glozak & Seto, 2007). To date, 18 HDACs have been described in humans. HDACs are classified into Class I (HDAC1, 2, 3, and 8), Class IIa (HDAC4, 5, 7, and 9), Class IIb (HDAC6 and 10), Class III (also called sirtuins or SIRT1–7), and Class IV (HDAC11) enzymes, based on their sequence and cofactor requirement for catalysis (Minucci & Pelicci, 2006). Class I, II, and IV HDACs are dependent on Zn^{2+} whereas Class III HDACs are dependent on NAD^+ for catalysis. In contrast to Class I HDACs, Class II HDACs are cytosolic with some members shuttling between the cytosol and the nucleus (Minucci & Pelicci, 2006). HDACs and histone acetyl transferases, along with other chromatin-modifying enzymes, are involved in the epigenetic regulation of gene expression. HDAC-dependent deacetylation of lysine tails in histones leads to an increased association of histones with DNA resulting in a transcriptionally silenced or closed chromatin configuration (Yoo & Jones, 2006; Spiegel, Milstien, & Grant, 2012). Additionally, HDAC1 and 2 exist in a complex with transcriptional repressors and participate in deacetylating histones thus partnering with repressors in transcriptional silencing of genes. Further, HDACs deacetylate transcription factors (e.g., YY-1, HIF1α, etc.) and regulate transcription independent of their chromatin-modifying activity (Geng et al., 2011; Yao, Yang, & Seto, 2001).

Phylogenetic analyses have revealed that members of the HDAC family are present in primitive bacteria that do not express histones. This suggests that evolutionarily HDACs were meant to deacetylate nonhistone proteins (Peng & Seto, 2011). Consequently, treatment with HDAC inhibitors results in the hyperacetylation of numerous nonhistone HDAC substrates.

A proteomic study to identify global changes in the acetylation of proteins revealed that the acetylation status of as many as 1750 proteins is altered in response to HDAC inhibitors, SAHA (suberoylanilide hydroxamic acid) and MS-275, treatment (Choudhary et al., 2009). The identified proteins included cytosolic and nuclear proteins, which could be functionally categorized into kinases, helicases, chromatin-remodeling proteins, DNA repair proteins, transcription factors, cytoskeletal proteins as well as a number of molecular chaperones, including heat shock proteins (hsps). This chapter will describe the regulation of major cellular chaperones by HDAC-mediated deacetylation. The implications of HDAC inhibitor-induced chaperone hyperacetylation on the tumor cell protein homeostasis will also be discussed.

2. HDAC INHIBITORS

HDAC inhibitors can be classified into structurally diverse, natural, and synthetic compounds. Most HDAC inhibitors that have been used in the clinic are pan-HDAC inhibitors that inhibit Class I, II, and IV isoforms. There are currently no clinically active Class III HDAC inhibitors. Based on their structure and chemical composition, HDAC inhibitors are classified into hydroxamic acid analogs, cyclic tetrapeptides, benzamides, and short-chain fatty acids (Minucci & Pelicci, 2006). The hydroxamic acid analogs act as Zn^{2+} chelators and inhibit HDAC activity at nanomolar to micromolar concentrations. The hydroxamic acid analogs, vorinostat, panobinostat (also called LBH589), and LAQ824, which also inhibit the cytosolic HDAC6 isoform, are highly active against solid and hematologic malignancies and exhibit *in vivo* potency at nM concentrations. Recently, several HDAC6-specific inhibitors have been developed owing to the discovery that HDAC6 regulates the acetylation and function of chaperones involved in tumor cell survival, in addition to promoting cell invasion and tumorigenesis in experimental models, as well as regulating cellular responses to proteotoxic stress, as discussed later (see below for more details). A Class I-specific inhibitor RGFP1316 (an *ortho*-aminoaniline) that shows selectivity for HDAC3, has been used as a negative regulator of long-term memory formation (McQuown et al., 2011). Vorinostat (or SAHA) and the cyclic tetrapeptide romidepsin have been approved by FDA for use in CTCL. MGCD0103 (a benzamide) also shows *in vivo* potency at nM concentration, but preferentially inhibits Class I HDACs. Among all HDAC isoforms, HDAC8 appears to stand out as it is not inhibited by commonly used HDAC

Table 7.1 *In vitro* inhibition of recombinant HDAC6 following treatment with various HDAC inhibitors

HDAC inhibitor	Specificity	EC_{50} for HDAC6 mean ± SEM (nM)
Vorinostat (SAHA)	Class I, II, and IV	90 ± 26
Panobinostat (LBH589)	Class I, II, and IV	61 ± 1
Belinostat (PDX101)	Class I, II, and IV	82 ± 19
MS-275	Class I	> 10,000
MGCD103	Class I	> 10,000
Valproic acid	Class I and IIa	> 10,000
ACY-1215	HDAC6	5[a]
RGFP136	HDAC3	400[b]

[a]Adapted from Khan et al. (2008) and Santo et al. (2012)
[b]Adapted from McQuown et al. (2011).

inhibitors, suggesting that the structural requirements for its inhibition are dramatically different from all other isoforms (Khan et al., 2008). A list of HDAC inhibitors with their class specificity is summarized in Table 7.1.

3. CHAPERONES AND PROTEIN HOMEOSTASIS

Although the primary structure of a protein contains all the necessary information that dictates its secondary and tertiary structure, nascent polypeptides attain their active conformation with the help of a complex machinery of molecular chaperones (Hartl, Bracher, & Hayer-Hartl, 2011). In the cellular milieu, with a high local concentration of newly synthesized proteins, chaperones prevent the aggregation of misfolded proteins (Hartl et al., 2011). Hsps are a family of conserved chaperones that are induced under conditions of stress and upon heat shock (Morimoto, 2012). Hsp70 and its cochaperone, hsp40 chaperone newly synthesized proteins or misfolded proteins induced during stress (Hartl et al., 2011; Morimoto, 2012). Unsuccessful refolding by hsp70 directs the abnormal proteins to polyubiquitylation and degradation by the proteasomes (Goldberg, 2003). Hsp90, on the other hand, promotes the folding of numerous metastable polypeptides in an ATP-dependent manner (Hartl & Hayer-Hartl, 2002; Taipale, Jarosz, & Lindquist, 2010). It accepts partially folded substrates from hsp70 and stabilizes its "client" proteins. Since hsp90 and hsp70 are not known to interact with each other directly, the hsp70/hsp90 organizing protein (HOP) mediates optimal interactions between hsp70 and hsp90 and promotes hsp90-mediated protein refolding (Johnson, Schumacher, Ross, & Toft, 1998; Walsh et al., 2011). There are two isoforms of hsp90, with hsp90β being the constitutively expressed isoform and hsp90α, the heat shock- and stress-inducible isoform. Hsp27 is a small hsp which exists in the cell, in an oligomerized form, and is dissociated into monomers following phosphorylation. Hsp27 regulates cytoskeletal stability, cell motility, protein folding, and promotes survival in response to heat shock and oxidative stress (Arrigo, 2007).

Proteasomes are integral components of cellular protein homeostasis; their main function being the degradation of misfolded, polyubiquitylated proteins, especially the stress-induced misfolded proteins. Inhibition of the chaperone machinery, or treatment with proteasome inhibitors (bortezomib, carfilzomib) results in the accumulation of misfolded proteins in the cell, leading to *proteotoxic stress*. In this chapter, we describe the regulation of the function of major cellular chaperones, including hs90, hsp70, hsp40, and glucose-regulated protein 78 (GRP 78), by reversible acetylation and describe specific cytosolic and/or nuclear HDACs that deacetylate the molecular chaperones. The implications of HDAC inhibitor-induced chaperone hyperacetylation on the tumor cell protein homeostasis will also be discussed.

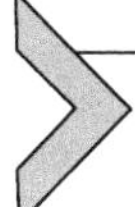

4. REGULATION OF HSP90 CHAPERONE FUNCTION BY REVERSIBLE ACETYLATION

4.1. The hsp90 chaperone complex

Hsp90 is an ATP-dependent molecular chaperone that promotes the stabilization of several partially folded client proteins that include steroid hormone receptors, kinases, signal transducers, DNA repair proteins, and structural proteins (Picard, 2002). The chaperoning of hsp90 clients is facilitated by cochaperones that associate with hsp90 in a cyclic manner to promote various stages of the client protein maturation. Hsp90-bound cochaperones include other hsps like hsp70, hsp40, as well as proteins that promote the chaperone activity of hsp90 such as p23, HOP, Cdc37, Aha1, and peptidyl prolyl isomerases, FKBP51, FKBP52, and CyP40 (Picard, 2002; Siligardi et al., 2004, 2002). The components of the chaperone complex formed depend on the kind of client protein that is being chaperoned, with Cdc37 being involved in the maturation of kinases and p23 in the activation of steroid hormone receptors. Hsp90 is organized into three functional domains—the N-terminus cochaperone, ATP- and drug-binding domain, the middle domain that binds cochaperones, client proteins and forms the active ATPase site, and the C-terminal dimerization and cochaperone-binding domain. Dimerization of hsp90 is required for its ATPase activity and chaperone function (Trepel, Mollapour, Giaccone, & Neckers, 2010; Whitesell & Lindquist, 2005). Geldanamycin and its analogs compete for the same ATP binding site on hsp90, thus inhibiting its chaperone function.

4.2. The hsp90 chaperone cycle

Hsp90-based chaperone complex interacts with its client proteins in an iterative manner, in which a homodimer of hsp90 cycles between ATP- or ADP-bound *closed* and *open* conformations respectively, mediated through multiple rounds of ATP binding and hydrolysis (Ali et al., 2006; Shiau, Harris, Southworth, & Agard, 2006). Upon binding of the client protein to the middle domain, the N-terminus and the middle domains of hsp90 undergo dramatic conformational changes that promote the ATP-bound *closed confirmation* of hsp90 (Street, Lavery, & Agard, 2011). Recent studies have demonstrated that substrate binding in the middle domain of hsp90 results in an intrinsically unfavorable twist in the N-terminal domain of hsp90 which then causes a secondary set of substrate binding sites in the middle domain to induce cross-monomer conformation changes that favor the ATP-bound closed conformation (Street et al., 2012). Replacement of ADP by ATP in hsp90 alters hsp90 conformation, thereby releasing cochaperones HOP and hsp70/hsp40 complex, while simultaneously recruiting another set of cochaperones including p23 and cyclophilin 40 (when the client protein is a steroid nuclear hormone receptor) or p50cdc37 (when the client protein is a signaling protein kinase; Trepel et al., 2010). The chaperone cycle is completed when ATP hydrolysis promotes the dissociation of the fully mature client protein from the complex. Conversely, ATP hydrolysis due to its intrinsic ATPase activity creates the ADP-bound conformation of hsp90, which directs the misfolded client protein to a covalent linkage with polyubiquitin and subsequent degradation by the 26S proteasome (Whitesell & Lindquist, 2005).

4.3. Hsp90 and HSF1 in a repressive complex

Hsp90 is also present in the cytosol as part of a repressive complex that contains hsp90–HDAC6–heat shock factor 1 (HSF1) and p97, an HDAC6-binding chaperone with segregase activity (Boyault et al., 2006). Accumulation of misfolded, polyubiquitylated proteins disrupts this complex, leading to the phosphorylation, trimerization, and nuclear localization of HSF1 where it promotes the transcription of genes containing a heat shock response element. HSF1-induced genes include hsp90, hsp70, hsp40, hsp27, and the hypoxia-inducible factor (HIF) 1α-inducible genes such as vascular endothelial growth factor (VEGF) and the HIF-1α target CAIX (carbonic anhydrase 9) (Gabai et al., 2012).

4.4. Hsp90 complex in cancer cells

Hsp90 client proteins often include mutated and overexpressed oncoproteins that promote the growth and survival of cancer cells (Cowen & Lindquist, 2005; Trepel et al., 2010; Whitesell & Lindquist, 2005). As a result, tumors may be addicted to oncogenes for their growth and survival. An updated list of hsp90 client proteins can be found at http://www.picard.ch/downloads/Hsp90interactors.pdf. Compared to normal cells, a major fraction of hsp90 in cancer cells is present in complex with cochaperones and exhibits higher ATPase activity. This suggests that it is actively involved in chaperoning client proteins in cancer cells (Kamal et al., 2003). These multiprotein complexes also bind with higher affinity to geldanamycin and its analog 17-*N*-allylamino-17-demethoxygeldanamycin (17-AAG), an observation that explains why tumor cells are more sensitive to hsp90 inhibition in comparison with normal cells (Kamal et al., 2003). PU-H71 is a purine scaffold hsp90 inhibitor with improved structural characteristics and better pharmacological profiles than the geldanamycin analogs (Chiosis, 2006). Further, a recent screen to identify PU-H71-bound protein complexes revealed that not only does hsp90 bind to its client oncoproteins in cancer cells but it also acts as a scaffolding protein to stabilize numerous signaling complexes, thus extending the multitude of functions that are influenced by hsp90 (Moulick et al., 2011). The heightened chaperone function in cancer cells forms the basis for the concept of *nononcogenic addiction* in tumors (Solimini, Luo, & Elledge, 2007). By virtue of their increased dependency on chaperones for stabilizing client oncoproteins and preventing proteotoxic stress-induced cell death, the tumor cells are addicted to the chaperone machinery for survival (Jego, Hazoumé, Seigneuric, & Garrido, 2010;

Luo, Solimini, & Elledge, 2009). Therefore, inhibition of the cellular chaperone machinery sensitizes tumor cells to agents that increase intracellular misfolded proteins and proteotoxic stress.

4.5. HDAC6 is the hsp90 deacetylase

The chaperone function of hsp90 is modulated by a variety of posttranslational modifications such as phosphorylation, acetylation, and S-nitrosylation (Mollapour & Neckers, 2012). Hyperacetylation of hsp90 has been demonstrated to inhibit its chaperone function. Treatment of cancer cells with pan-HDAC inhibitors such as LAQ824, LBH589 (or panobinostat), and vorinostat (which inhibit both Class I and II HDACs), but not sodium butyrate or trapoxin (which inhibits Class I HDACs) induces hyperacetylation of hsp90 and decreases its ATP and p23 (cochaperone) binding (Bali et al., 2005; Kovacs et al., 2005; Rao et al., 2008). This is associated with a decrease in the association of client proteins, for example, BCR-ABL, c-RAF, and AKT with hsp90 and a concomitant increase in the binding of client proteins to hsp70 (Bali et al., 2005; Rao et al., 2008). Shifting of client proteins to a complex containing hsp70 induces their polyubiquitylation by specific E3 ubiquitin ligases and their subsequent proteasomal degradation. Studies to identify the HDAC isoform that deacetylates hsp90 revealed that the predominantly cytosolic isoform HDAC6 was the major hsp90 deacetylase. Specifically, knocking down HDAC6 by siRNA resulted in hyperacetylation of hsp90 with a concomitant depletion of hsp90 client proteins including BCR-ABL, AKT, c-RAF, and glucocorticoid receptor (Bali et al., 2005; Kovacs et al., 2005; Fig. 7.1). Further, not only is HDAC6 the deacetylase for hsp90, but it is also an hsp90 client protein. Therefore, treatment with HDAC inhibitors and siRNA-mediated knockdown of HDAC6 enhances 17-AAG-mediated loss of hsp90 chaperone function as evidenced by greater decline in ATP and p23 binding to hsp90. This is associated with a greater depletion of hsp90 client proteins and loss of clonogenic survival in human leukemia (George et al., 2005; Rao et al., 2008).

To determine the acetylated lysine residues on hsp90 following panobinostat treatment, FLAG-tagged hsp90α protein was overexpressed in cells and treated with panobinostat. Hsp90α was affinity captured from the transfected cells utilizing anti-FLAG affinity beads. Following this, acetylated hsp90 was immunoprecipitated using acetyl lysine agarose beads. The

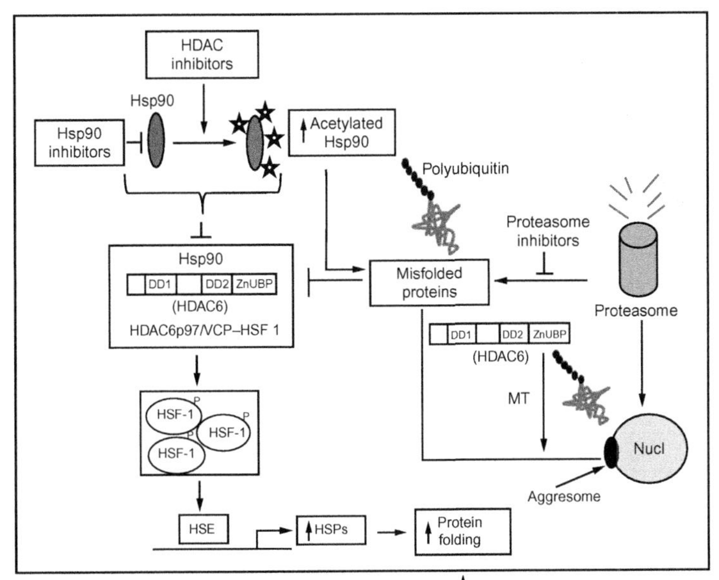

Figure 7.1 Cellular responses to misfolded proteins: HDAC and Hsp90 inhibitors induce acetylation of hsp90 which results in misfolding, polyubiquitylation, and proteasomal degradation of hsp90 client proteins. Treatment with proteasome inhibitors induce the accumulation of misfolded proteins which are bound to HDAC6 through the Zn-UBP domain located at the C-terminus of HDAC6 and shuttled into perinuclear aggresomes, in a dynein motor-dependent manner along MTs. Misfolded proteins also disrupt the hsp90-HDAC6–HSF1–p97 complex leading to the phosphorylation, trimerization, and nuclear translocation of HSF1 to promote the transcription of hsps, which ameliorate proteotoxic stress.

immunoprecipitated proteins were resolved by SDS-PAGE, hsp90α was excised from the gel, and an in-gel tryptic digestion to create peptides was carried out. The tryptic digests were further analyzed using nano high-performance liquid chromatography/MS/MS (tandem mass spectroscopy). Treatment with panobinostat resulted in the acetylation of hsp90 at lysines K69, K100, K292, K327, K478, K546, and K558 spread across the N-terminus and middle domain of the protein (Yang et al., 2008). A three-dimensional space-filling molecular structural model of hsp90α showed that all of the identified lysine residues that are acetylated reside on the surface and, thus, are accessible for modification. Increased acetylation of hsp90α at all of the above mentioned residues (except K292) decreased its ATP binding. Acetylated K292, K478, K546, and K558 were identified as residues that decrease binding of hsp90 to cochaperones p23, hsp40, hsp70, as well as with the client protein c-RAF. Binding to CHIP (COOH-terminal-interacting protein of hsp70) (Xu et al., 2002), which is an E3-ubiquitin ligase was also disrupted. An independent study reported the identification of acetylated lysine residues including K294 in cells treated with the HDAC inhibitor trichostatin (TSA; Scroggins et al., 2007). However, this was not confirmed in a separate study (Choudhary et al., 2009).

4.6. Acetylated hsp90α goes places

Interestingly, it has been demonstrated that HDAC-inhibitor-induced acetylation of hsp90α promotes its localization on the cell surface. Residues K69, K100, and K558 have been identified to be critical for the extracellular localization of hsp90. Extracellular hsp90 has been implicated in chaperoning client proteins such as MMP-2 and in promoting tumor invasiveness (Eustace et al., 2004; Lagarrigue et al., 2010; Sims, McCready, & Jay, 2011; Yang et al., 2008).

4.7. Deacetylase-independent functions of HDAC6

HDAC6 comprises two modular HDAC domains separated by a spacer domain and a unique C-terminal ubiquitin-binding domain, also known as the Zn-UBP (Ubiquitin C-terminus hydrolase-like zinc finger). HDAC6 binds to polyubiquitylated proteins through its Zn-UBP domain which accumulate for instance, following inhibition of proteasome. This has two consequences:

i. *Dissociation of HSF1 from the hsp90–HDAC6–HSF1–p97 (repressive) complex* and the concomitant activation of HSF1 resulting in the upregulation of hsps (Boyault, Sadoul, Pabion, & Khochbin, 2007; Boyault, Zhang, et al., 2007). Elevated levels of hsps, in turn, promote protein refolding and inhibit cell death due to proteotoxic stress (Fig. 7.1).

ii. *Association of HDAC6 with polyubiquitylated proteins in a p97-dependent manner* and the shuttling of misfolded proteins along microtubules in a dynein motor-dependent manner, into protective, perinuclear structures called *aggresomes* (Boyault et al., 2006; Kawaguchi et al., 2003; Fig. 7.1).

Therefore, HDAC6 acts as a master regulator of cellular responses to eliminate toxic protein aggregates (Boyault, Sadoul, et al., 2007; Boyault, Zhang, et al., 2007; Fig. 7.1). Inhibiting HDAC6 consequently enhances proteasome inhibitor-induced proteotoxic stress which forms the basis for the synergistic antitumor activity of HDAC inhibitors and proteasome inhibitors in multiple myeloma, mantle cell lymphoma, activated B-cell-like diffuse large B-cell lymphoma, and pancreatic ductal adenocarcinoma (Catley et al., 2006; Dasmahapatra et al., 2010; Hideshima et al., 2005; Hideshima, Richardson, & Anderson, 2011; Nawrocki et al., 2006; Rao, Nalluri, Fiskus, et al., 2010). While failure to resolve proteotoxic stress is one of the major mechanisms that can be ascribed to the synergistic induction of apoptosis in combination with HDAC and proteasome inhibitors, the involvement of other apoptotic mechanisms cannot be discounted (Hideshima et al., 2011).

4.8. Targeting HDAC6 in cancer

HDAC6 expression was found to be high in ovarian and breast carcinoma (Zhang et al., 2004; Saji et al., 2005; Zhang et al., 2007), as well as in a variety of other hematologic malignancies. Regardless of the level of their expression, increased dependency of tumor cells on hsp90 chaperone function makes them dependent on HDAC6 for maintaining protein homeostasis. Addiction of tumors to nononcogenic pathways (such as hsp90 chaperone function) therefore offers a unique mechanism to specifically target cancer cells without affecting normal cells. Further, HDAC6 promotes tumor cell migration and invasiveness (Rey, Irondelle, Waharte, Lizarraga, & Chavrier, 2011; Zhang et al., 2007). Studies using HDAC6 null mice revealed that HDAC6 promotes Ras- and carcinogen-induced oncogenic transformation (Lee et al., 2008). Collectively, these observations make HDAC6 an attractive target for developing antitumor agents. Consequently, several HDAC6-specific inhibitors have been developed and tested for their antitumor activity both in preclinical and early phase clinical trials. Tubacin, a relatively specific HDAC6 inhibitor was developed in a chemical screen to identify small molecules that induced α-tubulin (an HDAC6 substrate) acetylation but exhibited no effects on histone acetylation (Haggarty, Koeller, Wong, Grozinger, & Schreiber, 2003). More specific hydroxamate-based HDAC6 inhibitors have been developed that exhibit antitumor activity *in vitro* (Kozikowski, Tapadar, Luchini, Kim, & Billadeau, 2008). ACY-1215 (a hydroxamate with a low nanomolar IC50 for HDAC6) is being currently tested in the clinic for its antimultiple myeloma activity in combination with the proteasome inhibitor, bortezomib (Santo et al., 2012)

5. HDAC3 IS THE NUCLEAR HSP90 DEACETYLASE

While hsp90 is localized predominantly in the cytosol, about 3% of hsp90 has also been reported to reside in the nucleus (Csermely, Schnaider, Soti, Prohaszka, & Nardai, 1998). Several nuclear proteins including checkpoint kinase 1 (CHK1), ataxia-telangiectasia mutated and Rad3-related (ATR) are chaperoned by hsp90 (Arlander et al., 2006; Brazelle et al., 2010; Ha et al., 2011). Studies to determine the isoform of HDAC(s) responsible for nuclear hsp90 deacetylation demonstrated that genetic knockdown of HDAC3 but not of HDAC1 or 2, caused hyperacetylation of nuclear hsp90 and depletion of ATR and CHK1, while ectopic overexpression of HDAC3 inhibited nuclear hsp90 acetylation in transformed cells (Ha et al., 2012). These findings demonstrate that HDAC3 is the nuclear hsp90 lysine deacetylase. Genetic knockdown of HDAC3, or its inhibition by pan-HDAC inhibitors, induces hyperacetylation of hsp90 and mediates loss of hsp90 chaperone association and depletion of nuclear hsp90 client proteins, ATR and CHK1. Consequently, this results in the abrogation of γ-irradiation-induced double-stranded DNA repair, and results in sensitization of transformed cells to DNA damage due to γ-irradiation or replication stress (Ha et al., 2012).

6. HDAC INHIBITOR-MEDIATED REGULATION OF HSP70

pan-HDAC inhibitors induce hsp70 expression by disrupting the hsp90–HDAC6–HSF1–p97 complex and activating HSF1-dependent gene transcription. Elevated levels of hsp70 and hsp90 promote proper protein folding and inhibit both the intrinsic and extrinsic pathways of apoptosis. Additionally, treatment with panobinostat or siRNA-mediated knockdown of HDAC6 leads to the hyperacetylation of hsp70 (Rao et al., 2008). It remains unclear whether the chaperoning ability of hsp70 is affected by its acetylation status, although hyperacetylation of hsp70 following HDAC inhibitor treatment may favor the polyubiquitylation and proteasomal degradation of misfolded hsp90 client proteins such as BCR-ABL, c-RAF, and AKT (Rao et al., 2008; Wang et al., 2007). Further studies are required to assess the effect of HDAC6-mediated hyperacetylation of hsp70 on its chaperone function.

Recent studies show that pan-HADC inhibitor-induced hsp70 acetylation promotes autophagy, which is a conserved catabolic pathway in which cytoplasmic macromolecules and damaged organelles and nonfunctional long-lived proteins are sequestered in autophagosomes. These bags of organelles and proteins fuse with lysosomes for protein degradation and amino acid recycling. Initial steps of autophagy include vesicle nucleation and expansion of an isolation membrane, creating a phagophore (Levine & Kroemer, 2008). The edges of the phagophore fuse to form a double membrane structure called the autophagosome (Levine & Kroemer, 2008). The initiation of the phagophore formation is linked to the activation of the lipid kinase PI3KC3 complex, which consists of beclin 1 (Atg6), Atg14, Vps (vacuolar protein sorting) 34, and Vps15 (Funderburk, Wang, & Yue, 2010). HDAC inhibitor-induced hyperacetylated hsp70 has been demonstrated to significantly enhance the stability and activity of Vps34. This leads to binding of Vps34 to beclin1, thus promoting the initial steps of autophagosome formation (Rao et al., 2009). These findings suggest an hsp70-based mechanism by which treatment with HDAC inhibitors promotes autophagy.

The phagophore expansion and completion of autophagosome formation is mediated by the two ubiquitin-like Atg12–Atg5 and LC3-phosphatidylethanolamine (PE) conjugation pathways (Funderburk, et al., 2010). During autophagy, a cytosolic form of LC3B (LC3B-I) is conjugated to PE to form the LC3–PE conjugate (LC3B-II), which is recruited to autophagosomal membranes. LC3B-II is degraded by lysosomal hydrolases after the fusion of autophagosomes with lysosomes (Kroemer, Marino, & Levine, 2010). Previous studies demonstrated that hsp70 inhibition with phenylacetylenylsulfonamide (PES) inhibits association of hsp70 cochaperones including hsp40, CHIP, and BAG-1 with hsp70 as well as inhibits the hsp70-dependent chaperoning of hsp90 client proteins. Treatment with PES also inhibited the autophagy-lysosome pathways and the proteasomes thereby promoting the accumulation of misfolded proteins (Leu, Pimkina, Frank, Murphy, & George, 2009; Leu et al., 2011). Collectively, these findings suggest that inhibition of hsp70 function inhibits the induction of autophagy.

Autophagy has been generally regarded as a cytoprotective mechanism employed by tumor cells in response to chemotherapeutic agents. However, inhibition of autophagy with inhibitors such as chloroquine (which inhibits lysosomal function) leads to the accumulation of damaged organelles, ultimately leading to tumor cell death. This strategy of inducing tumor cell death by cotreatment with HDAC inhibitors and autophagy inhibitors has been tested against triple negative breast cancer cells, malignant nerve sheath tumors, and imatinib-resistant chronic myelogenous leukemia (Carew, Nawrocki, & Cleveland, 2007; Lopez, Torres, & Lev, 2011; Rao et al., 2012).

7. HDAC REGULATION OF HSP40

Hsp40 is an hsp70 cochaperone. It belongs to the DNAJ family of proteins, which possesses a J domain that promotes the ATPase activity of hsp70. Hsp40 stabilizes the interaction between hsp70 and its unfolded substrates. Failure of hsp70–hsp40-mediated protein refolding machinery is known to cause aggregation of poly-glutamine (poly-Q) protein commonly found in neurodegenerative disorders. A recent study has identified the hsp40 isoforms DNAJB6 and DNAJB8 as potent suppressors of misfolded poly-Q protein aggregation and cytotoxicity *in vitro* and *in vivo*. Further, DNAJB6/8 suppressed poly-Q protein aggregation independent of their ATPase activity and hsp70 binding. Inhibition of HDAC activity with TSA (a pan-HDAC inhibitor) or specifically the knockdown of HDAC4, a Class IIa HDAC abrogates the antiaggregation activity of DNAJ6/8. The study also identified lysine residues K16, K216, and K223 to be acetylated by TSA treatment. The acetylation of K216 was found to be critical for the suppression of protein aggregation (Hageman et al., 2010).

8. REGULATION OF GRP78 FUNCTION BY HDAC6

GRP78 or immunoglobulin heavy chain-binding protein is an endoplasmic reticulum (ER)-resident, ATP-dependent homologue of the molecular chaperone hsp70 (Dudek et al., 2009). Recent reports have demonstrated the presence of GRP78 on the surface of tumor cells and in the mitochondria while a splice variant is reported to be present in the cytoplasm (Ni, Zhang, & Lee, 2011). GRP78 is primarily involved in the folding and assembly of newly synthesized polypeptides in the ER and chaperoning of improperly folded proteins. In a process also referred to as ER-associated degradation, GRP78 recognizes hydrophobic regions of misfolded proteins and retains them in a soluble conformation, thus preventing their aggregation. The misfolded substrates are released following ATP hydrolysis, retrotranslocated into the cytosol, and degraded by the proteasomes (Smith, Ploegh, & Weissman, 2011; Vembar & Brodsky, 2008)

8.1. GRP78: A master regulator of the unfolded protein response

The unfolded protein response (UPR) ensues following accumulation of misfolded proteins in the ER after exposure to several forms of cellular stress, for example, alterations in Ca^{2+} homeostasis, inhibition of protein glycosylation, or alterations in the redox status which compromise ER protein-folding capacity (Jager, Bertrand, Gorman, Vandenabeele, & Samali, 2012). In unstressed cells, GRP78 binds to the luminal domain of the three documented ER transmembrane receptors (mediators of UPR), that is, inositol-requiring enzyme 1α (IRE1), protein kinase RNA-like ER kinase (PERK), and activating transcription factor-6α (ATF6α), and keeps them inactive. Accumulation of misfolded proteins in the ER leads to the dissociation of GRP78 from the three receptors, leading to their activation (Jager et al., 2012). Activated PERK phosphorylates eukaryotic initiation factor (eIF2α) which blocks "cap-dependent" protein synthesis but results in the preferential translation of ATF4, a transcription factor that upregulates the proapoptotic transcription factor CHOP (CAAT/enhancer-binding protein homologous protein) and genes required to restore normal ER function (Ron & Walter, 2007). Activated IRE1 catalyzes the splicing of XBP1 to generate a frameshift splice variant of XBP1, called XBP1s, that transactivates and upregulates GRP78, which participates in restoring normal ER function (Lin et al., 2007; Ron & Walter, 2007). Consequently, elevated levels of protective GRP78 have been shown to have an important role in the pathogenesis of several types of cancers; moreover, it confers poor prognosis (Ni et al., 2011; Sato, Yao, Arap, & Pasqualini, 2010). Unresolved UPR and protracted ER stress result in sustained activation of eIF2α phosphorylation and CHOP induction, resulting in ER stress-triggered apoptosis involving the induction of BH3 domain-only proteins BIK and BIM (Lin et al., 2007; Puthalakath et al., 2007) as well as the induction of the death receptor DR5 (Yamaguchi & Wang, 2004) and activation of the intrinsic pathway of apoptosis involving the activity of caspase-9, caspase-3, and caspase-7 (Masud et al., 2007). Further, prolonged IRE1 activation induces apoptosis by activating the proapoptotic Jun amino terminal kinase 1 (Urano et al., 2000). Alternatively, CHOP could induce apoptosis in ER-stressed cells by downregulating Bcl2 (McCullough, Martindale, Klotz, Aw & Holbrook, 2001). Prolonged ER stress also results in the induction of the CHOP-induced gene ER oxidase ERO1α, which promotes the hyperoxidation of the ER proteins, thus causing cell death (Li et al., 2009;

Tabas and Ron, 2011). These and other interlinked mechanisms convert the protective ER stress response into a lethal ER stress response, especially when the protective role of GRP78 is undermined by its acetylation (Rao, Nalluri, Fiskus, et al., 2010; Tabas and Ron, 2011). While the regulation of GRP78 function by its acetylation has been demonstrated, the role of HDAC inhibitor-induced acetylation of GRP94, the ER homolog of hsp90 has not been elucidated (Eletto, Dersh, & Argon, 2010; Marzec, Eletto, & Argon, 2012).

8.2. HDAC6 is the GRP78 deacetylase

Much like its cytosolic homolog hsp70, treatment with panobinostat resulted in the acetylation of GRP78 in human breast cancer cells. At least 11 lysine residues (K118, K122, K123, K125, K138, K152, K154, K352, K353, K376, and K633) were acetylated following treatment with panobinostat, although a detailed characterization of the importance of these sites was not determined (Rao, Nalluri, Kolhe, et al., 2010). The HDAC isoform involved in the deacetylation of GRP78 was identified to be HDAC6, as knockdown of HDAC6 resulted in the hyperacetylation of GRP78. While GRP78 is an ER-resident protein and HDAC6 is predominantly cytosolic, the question arises as to where in the cell might the two proteins actually interact. Immunofluorescence microscopy of cells stained with GRP78, HDAC6, and Calnexin (an ER transmembrane protein) revealed that HDAC6 interacts with GRP78 at the ER membrane. This finding is strengthened by the observation that GRP78 is capable of forming four transmembrane helices and a fraction of cellular GRP78 can exist as an ER transmembrane protein (Reddy et al., 2003). It is highly likely that the recently reported surface GRP78 that is present on the plasma membrane of tumor cells is derived from the ER membrane. Further studies have confirmed that the surface expression of GRP78 also adopts a similar conformation as the ER transmembrane GRP78 (Ni et al., 2011; Zhang, Liu, Ni, Gill, & Lee, 2010). While the significance of GRP78 acetylation in relation to its localization has not been determined, panobinostat-induced acetylation of GRP78 reduces binding of GRP78 to PERK, increased levels of p-eIF2α, ATF4, and CHOP in human breast cancer cells. Although panobinostat also induces GRP78 protein levels, it undermines the function of GRP78 by inducing its hyperacetylation, leading to the lethal consequences of ER stress. Additionally, knockdown of GRP78 with short hairpin RNA sensitizes breast cancer cells to panobinostat-induced as well as to the proteasome inhibitor bortezomib-induced UPR and cell death (Rao, Nalluri, Kolhe, et al., 2010). In summary, inhibition of GRP78 function by HDAC inhibitors results in the activation PERK–eIF2α–CHOP-mediated lethal ER stress. There is also accumulating evidence that ER stress can trigger autophagy in cells that do not experience lethal UPR and escape apoptosis (Hoyer-Hansen & Jaattela, 2007; True & Matthias, 2012; Yorimitsu, Nair, Yang, & Klionsky, 2006). Moreover, GRP78 has been shown to be necessary for ER stress-induced autophagy. Knockdown of GRP78, as expected, induces UPR, but results in the reduced formation of autophagosomes, in response to ER stress induction (Li et al., 2008; Fig. 7.2). Collectively, these observations suggest that depletion of GRP78 levels or inhibition of GRP78 function by hyperacetylation might also contribute to the

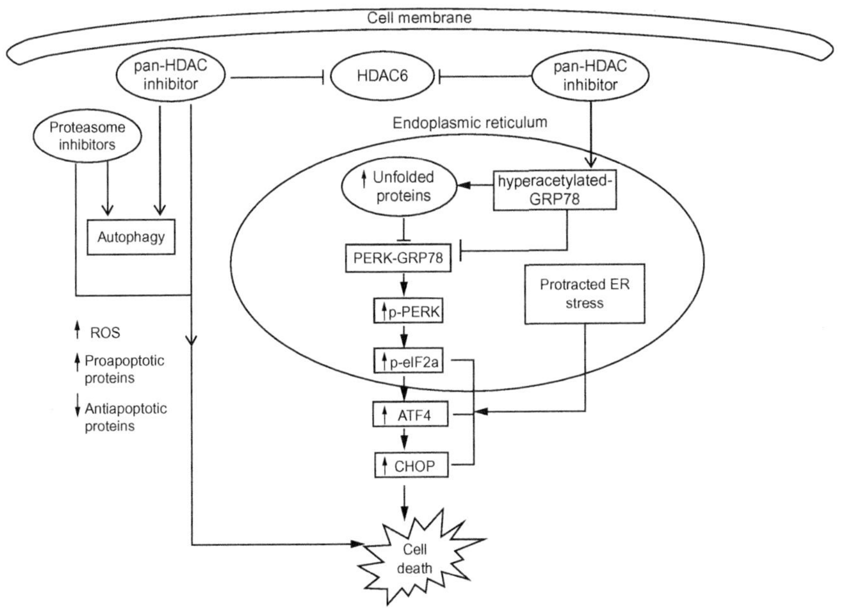

Figure 7.2 Mechanisms responsible for the induction of lethal UPR: PS-mediated inhibition of HDAC6 leads to the acetylation of GRP78, which dissociates GRP78 from the ER stress mediator, PERK, leading to the activation of p-eIf2α–ATF4–CHOP pathway. Increased accumulation of unfolded proteins and inhibition of GRP78 function by PS-induced acetylation lead to protracted ER stress, which results in the lethal outcome of UPR. PS and proteasome inhibitor-induced ROS, upregulation of proapoptotic proteins and downregulation of antiapoptotic proteins, as well as the induction of autophagy are additional mechanisms responsible for proteasome inhibitor/PS-induced cell death.

Table 7.2 Acetylated lysine residues in molecular chaperones and their biological functions

Protein	Acetylated lysine residues	Biological function
Hsp90	K292, K478, K546, and K558	Cochaperone, CHIP, and client protein binding
	K69, K100, K327, K478, K546, and K558	ATP binding
	K69, K100, and K558	Extracellular localization of hsp90, invasiveness
Hsp40	K216	Antiaggregation
Hsp70	K88,K126, K159, 523, K558, and K560	Not determined
GRP78	K118, K122, K123, K125, K138, K152, K154, K352, K353, K376, and K633	Not determined

induction of autophagy in response to HDAC inhibitor treatment. A comprehensive list of lysine residues identified in hsp90, hsp70, hsp40, and GRP78 is summarized in Table 7.2.

9. CONCLUDING REMARKS

In transformed cells, the presence of aneuploidy or gene copy number changes and the corresponding alterations in the protein stoichiometry in multiprotein complexes result in misfolding of proteins and the adaptive heat shock and ER stress response (Luo et al., 2009). Molecular chaperones perform a central role in mediating the heat shock and ER stress response and in sustaining the transformed phenotype. This supportive role may be especially heightened in those transformed cells that harbor activating mutations in hsp90 client oncoprotein kinases. The extended "interactome" of hsp90 overlaps with signaling pathways that are aberrantly activated by additional mechanisms in tumors (Moulick et al., 2011). HDAC inhibitor-induced hyperacetylation and inhibition of molecular chaperones and cochaperones abrogate the protective heat shock and ER stress responses, thereby undermining the transformed phenotype. HDAC inhibitors also attenuate the protein-folding ER quality control resulting from the inhibition of GRP78 function and make cancer cells more susceptible to proteasome inhibition. While the preclinical antitumor efficacy of HDAC inhibitor-mediated sensitization to proteasome inhibitor has been demonstrated, its clinical efficacy and the susceptible tumor types remain to be fully elucidated. Overall, in conclusion, the emerging understanding of the impact of lysine acetylation of molecular chaperones and cochaperones is defining novel strategies which exploit this mechanism and can be harnessed for cancer therapy.

REFERENCES

Ali, M. M., Roe, S. M., Vaughan, C. K., Meyer, P., Panaretou, B., Piper, P. W., et al. (2006). Crystal structure of an Hsp90-nucleotide-p23/Sba1 closed chaperone complex. *Nature, 440*, 1013–1017.

Arlander, S. J., Felts, S. J., Wagner, J. M., Stensgard, B., Toft, D. O., & Karnitz, L. M. (2006). Chaperoning checkpoint kinase 1 (Chk1), an Hsp90 client, with purified chaperones. *Journal of Biological Chemistry, 281*, 2989–2998.

Arrigo, A. P. (2007). The cellular "networking" of mammalian Hsp27 and its functions in the control of protein folding, redox state and apoptosis. *Advances in Experimental Medicine and Biology, 594*, 14–26.

Bali, P., Pranpat, M., Bradner, J., Balasis, M., Fiskus, W., Guo, F., et al. (2005). Inhibition of histone deacetylase 6 acetylates and disrupts the chaperone function of heat shock protein 90: A novel basis for antileukemia activity of histone deacetylase inhibitors. *Journal of Biological Chemistry, 280*, 26729–26734.

Boyault, C., Gilquin, B., Zhang, Y., Rybin, V., Garman, E., Meyer-Klaucke, W., et al. (2006). HDAC6-p97/VCP controlled polyubiquitin chain turnover. *EMBO Journal, 25*, 3357–3366.

Boyault, C., Sadoul, K., Pabion, M., & Khochbin, S. (2007). HDAC6, at the crossroads between cytoskeleton and cell signaling by acetylation and ubiquitination. *Oncogene, 26*, 5468–5476.

Boyault, C., Zhang, Y., Fritah, S., Caron, C., Gilquin, B., Kwon, S. H., et al. (2007). HDAC6 controls major cell response pathways to cytotoxic accumulation of protein aggregates. *Genes & Development, 21*, 2172–2181.

Brazelle, W., Kreahling, J. M., Gemmer, J., Ma, Y., Cress, W. D., Haura, E., et al. (2010). Histone deacetylase inhibitors downregulate checkpoint kinase 1 expression to induce cell death in non-small cell lung cancer cells. *PLoS One, 5*, e14335.

Carew, J. S., Nawrocki, S. T., & Cleveland, J. L. (2007). Modulating autophagy for therapeutic benefit. *Autophagy, 3*, 464–467.

Catley, L., Weisberg, E., Kiziltepe, T., Tai, Y. T., Hideshima, T., Neri, P., et al. (2006). Aggresome induction by proteasome inhibitor bortezomib and alpha-tubulin hyperacetylation by tubulin deacetylase (TDAC) inhibitor LBH589 are synergistic in myeloma cells. *Blood, 108*, 3441–3449.

Chiosis, G. (2006). Discovery and development of purine-scaffold Hsp90 inhibitors. *Current Topics in Medicinal Chemistry, 6*, 1183–1191.

Choudhary, C., Kumar, C., Gnad, F., Nielsen, M. L., Rehman, M., Walther, T. C., et al. (2009). Lysine acetylation targets protein complexes and co-regulates major cellular functions. *Science, 325*, 834–840.

Cowen, L. E., & Lindquist, S. (2005). Hsp90 potentiates the rapid evolution of new traits: Drug resistance in diverse fungi. *Science, 309*, 2185–2189.

Csermely, P., Schnaider, T., Soti, C., Prohaszka, Z., & Nardai, G. (1998). The 90-kDa molecular chaperone family: Structure, function, and clinical applications. A comprehensive review. *Pharmacology & Therapeutics, 79*, 129–168.

Dasmahapatra, G., Lembersky, D., Kramer, L., Fisher, R. I., Friedberg, J., Dent, P., et al. (2010). The pan-HDAC inhibitor vorinostat potentiates the activity of the proteasome inhibitor carfilzomib in human DLBCL cells in vitro and in vivo. *Blood, 115*, 4478–4487.

Dudek, J., Benedix, J., Cappel, S., Greiner, M., Jalal, C., Muller, L., et al. (2009). Functions and pathologies of BiP and its interaction partners. *Cellular and Molecular Life Sciences, 66,* 1556–1569.

Eletto, D., Dersh, D., & Argon, Y. (2010). GRP94 in ER quality control and stress responses. *Seminars in Cell & Developmental Biology, 21,* 479–485.

Eustace, B. K., Sakurai, T., Stewart, J. K., Yimlamai, D., Unger, C., Zehetmeier, C., et al. (2004). Functional proteomic screens reveal an essential extracellular role for hsp90 alpha in cancer cell invasiveness. *Nature Cell Biology, 6,* 507–514.

Funderburk, S. F., Wang, Q. J., & Yue, Z. (2010). The Beclin 1-VPS34 complex—At the crossroads of autophagy and beyond. *Trends in Cell Biology, 20,* 355–362.

Gabai, V. L., Meng, L., Kim, G., Mills, T. A., Benjamin, I. J., & Sherman, M. Y. (2012). Heat shock transcription factor Hsf1 is involved in tumor progression via regulation of hypoxia-inducible factor 1 and RNA-binding protein HuR. *Molecular and Cellular Biology, 32,* 929–940.

Geng, H., Harvey, C. T., Pittsenbarger, J., Liu, Q., Beer, T. M., Xue, C., et al. (2011). HDAC4 protein regulates HIF1alpha protein lysine acetylation and cancer cell response to hypoxia. *Journal of Biological Chemistry, 286,* 38095–38102.

George, P., Bali, P., Annavarapu, S., Scuto, A., Fiskus, W., Guo, F., et al. (2005). Combination of the histone deacetylase inhibitor LBH589 and the hsp90 inhibitor 17-AAG is highly active against human CML-BC cells and AML cells with activating mutation of FLT-3. *Blood, 105,* 1768–1776.

Glozak, M. A., & Seto, E. (2007). Histone deacetylases and cancer. *Oncogene, 26,* 5420–5432.

Goldberg, A. L. (2003). Protein degradation and protection against misfolded or damaged proteins. *Nature, 426,* 895–899.

Ha, K., Fiskus, W., Balusu, R., Rao, R., Venkannagari, S., & Bhalla, K. (2012). Inhibition of histone deacetylase (HDAC) 3 induces hyperacetylation and inhibition of nuclear heat shock protein (hsp) 90 leading to depletion of ATR and CHK1 with sensitization to DNA damage in breast and cervical cancer cells. In: *Proceedings of the 103rd annual meeting of the American association of cancer research* Abstract No: 4698.

Ha, K., Fiskus, W., Rao, R., Balusu, R., Venkannagari, S., Nalabothula, N. R., et al. (2011). Hsp90 inhibitor-mediated disruption of chaperone association of ATR with hsp90 sensitizes cancer cells to DNA damage. *Molecular Cancer Therapeutics, 10,* 1194–1206.

Hageman, J., Rujano, M. A., van Waarde, M. A., Kakkar, V., Dirks, R. P., Govorukhina, N., et al. (2010). A DNAJB chaperone subfamily with HDAC-dependent activities suppresses toxic protein aggregation. *Molecular Cell, 37,* 355–369.

Haggarty, S. J., Koeller, K. M., Wong, J. C., Grozinger, C. M., & Schreiber, S. L. (2003). Domain-selective small-molecule inhibitor of histone deacetylase 6 (HDAC6)-mediated tubulin deacetylation. *Proceedings of the National Academy of Sciences of the United States of America, 100,* 4389–4394.

Hartl, F. U., Bracher, A., & Hayer-Hartl, M. (2011). Molecular chaperones in protein folding and proteostasis. *Nature, 475,* 324–332.

Hartl, F. U., & Hayer-Hartl, M. (2002). Molecular chaperones in the cytosol: From nascent chain to folded protein. *Science, 295,* 1852–1858.

Hideshima, T., Bradner, J. E., Wong, J., Chauhan, D., Richardson, P., Schreiber, S. L., et al. (2005). Small-molecule inhibition of proteasome and aggresome function induces synergistic antitumor activity in multiple myeloma. *Proceedings of the National Academy of Sciences of the United States of America, 102,* 8567–8572.

Hideshima, T., Richardson, P. G., & Anderson, K. C. (2011). Mechanism of action of proteasome inhibitors and deacetylase inhibitors and the biological basis of synergy in multiple myeloma. *Molecular Cancer Therapeutics, 10,* 2034–2042.

Hoyer-Hansen, M., & Jaattela, M. (2007). Connecting endoplasmic reticulum stress to autophagy by unfolded protein response and calcium. *Cell Death and Differentiation, 14*, 1576–1582.

Jager, R., Bertrand, M. J., Gorman, A. M., Vandenabeele, P., & Samali, A. (2012). The unfolded protein response at the crossroads of cellular life and death during ER stress. *Biology of the Cell, 104*, 259–270.

Jego, G., Hazoumé, A., Seigneuric, R., & Garrido, C. (2010). Targeting heat shock proteins in cancer. *Cancer Letters*, http://dx.doi.org/10.1016/j.canlet.2010.10.014.

Johnson, B. D., Schumacher, R. J., Ross, E. D., & Toft, D. O. (1998). Hop modulates Hsp70/Hsp90 interactions in protein folding. *Journal of Biological Chemistry, 273*, 3679–3686.

Kamal, A., Thao, L., Sensintaffar, J., Zhang, L., Boehm, M. F., Fritz, L. C., et al. (2003). A high-affinity conformation of Hsp90 confers tumour selectivity on Hsp90 inhibitors. *Nature, 425*, 407–410.

Kawaguchi, Y., Kovacs, J. J., McLaurin, A., Vance, J. M., Ito, A., & Yao, T. P. (2003). The deacetylase HDAC6 regulates aggresome formation and cell viability in response to misfolded protein stress. *Cell, 115*, 727–738.

Khan, N., Jeffers, M., Kumar, S., Hackett, C., Boldog, F., Khramtsov, N., et al. (2008). Determination of the class and isoform selectivity of small-molecule histone deacetylase inhibitors. *Biochemical Journal, 409*, 581–589.

Kovacs, J. J., Murphy, P. J., Gaillard, S., Zhao, X., Wu, J. T., Nicchitta, C. V., et al. (2005). HDAC6 regulates Hsp90 acetylation and chaperone-dependent activation of glucocorticoid receptor. *Molecular Cell, 18*, 601–607.

Kozikowski, A. P., Tapadar, S., Luchini, D. N., Kim, K. H., & Billadeau, D. D. (2008). Use of the nitrile oxide cycloaddition (NOC) reaction for molecular probe generation: A new class of enzyme selective histone deacetylase inhibitors (HDACIs) showing picomolar activity at HDAC6. *Journal of Medicinal Chemistry, 51*, 4370–4373.

Kroemer, G., Marino, G., & Levine, B. (2010). Autophagy and the integrated stress response. *Molecular Cell, 40*, 280–293.

Lagarrigue, F., Dupuis-Coronas, S., Ramel, D., Delsol, G., Tronchere, H., Payrastre, B., et al. (2010). Matrix metalloproteinase-9 is upregulated in nucleophosmin-anaplastic lymphoma kinase-positive anaplastic lymphomas and activated at the cell surface by the chaperone heat shock protein 90 to promote cell invasion. *Cancer Research, 70*, 6978–6987.

Lee, Y. S., Lim, K. H., Guo, X., Kawaguchi, Y., Gao, Y., Barrientos, T., et al. (2008). The cytoplasmic deacetylase HDAC6 is required for efficient oncogenic tumorigenesis. *Cancer Research, 68*, 7561–7569.

Leu, J. I., Pimkina, J., Frank, A., Murphy, M. E., & George, D. L. (2009). A small molecule inhibitor of inducible heat shock protein 70. *Molecular Cell, 36*, 15–27.

Leu, J. I., Pimkina, J., Pandey, P., Murphy, M. E., & George, D. L. (2011). HSP70 inhibition by the small-molecule 2-phenylethynesulfonamide impairs protein clearance pathways in tumor cells. *Molecular Cancer Research, 9*, 936–947.

Levine, B., & Kroemer, G. (2008). Autophagy in the pathogenesis of disease. *Cell, 132*, 27–42.

Li, G., Mongillo, M., Chin, K. T., Harding, H., Ron, D., Marks, A. R., et al. (2009). Role of ERO1-alpha-mediated stimulation of inositol 1,4,5-triphosphate receptor activity in endoplasmic reticulum stress-induced apoptosis. *The Journal of Cell Biology, 186*, 783–792.

Li, J., Ni, M., Lee, B., Barron, E., Hinton, D. R., & Lee, A. S. (2008). The unfolded protein response regulator GRP78/BiP is required for endoplasmic reticulum integrity and stress-induced autophagy in mammalian cells. *Cell Death and Differentiation, 15*, 1460–1471.

Lin, J. H., Li, H., Yasumura, D., Cohen, H. R., Zhang, C., Panning, B., et al. (2007). IRE1 signaling affects cell fate during the unfolded protein response. *Science*, *318*, 944–949.

Lopez, G., Torres, K., & Lev, D. (2011). Autophagy blockade enhances HDAC inhibitors' pro-apoptotic effects: Potential implications for the treatment of a therapeutic-resistant malignancy. *Autophagy*, 7, 440–441.

Luo, J., Solimini, N. L., & Elledge, S. J. (2009). Principles of cancer therapy: Oncogene and non-oncogene addiction. *Cell*, *136*, 823–837.

Marzec, M., Eletto, D., & Argon, Y. (2012). GRP94: An HSP90-like protein specialized for protein folding and quality control in the endoplasmic reticulum. *Biochimica et Biophysica Acta*, *1823*, 774–787.

Masud, A., Mohapatra, A., Lakhani, S. A., Ferrandino, A., Hakem, R., & Flavell, R. A. (2007). Endoplasmic reticulum stress-induced death of mouse embryonic fibroblasts requires the intrinsic pathway of apoptosis. *The Journal of Biological Chemistry*, *282*, 14132–14139.

McCullough, K. D., Martindale, J. L., Klotz, L. O., Aw, T. Y., & Holbrook, N. J. (2001). Gadd153 sensitizes cells to endoplasmic reticulum stress by downregulating Bcl2 and perturbing the cellular redox state. *Molecular Cell Biology*, *21*, 1249–1259.

McQuown, S. C., Barrett, R. M., Matheos, D. P., Post, R. J., Rogge, G. A., Alenghat, T., et al. (2011). HDAC3 is a critical negative regulator of long-term memory formation. *Journal of Neuroscience*, *31*, 764–774.

Minucci, S., & Pelicci, P. G. (2006). Histone deacetylase inhibitors and the promise of epigenetic (and more) treatments for cancer. *Nature Reviews. Cancer*, *1*, 38–51.

Mollapour, M., & Neckers, L. (2012). Post-translational modifications of Hsp90 and their contributions to chaperone regulation. *Biochimica et Biophysica Acta*, *1823*, 648–655.

Morimoto, R. I. (2012). The Heat Shock Response: Systems Biology of Proteotoxic Stress in Aging and Disease. *Cold Spring Harbor Symposia on Quantitative Biology*, http://dx.doi.org/10.1101/sqb.2012.76.010637.

Moulick, K., Ahn, J. H., Zong, H., Rodina, A., Cerchietti, L., Gomes DaGama, E. M., et al. (2011). Affinity-based proteomics reveal cancer-specific networks coordinated by Hsp90. *Nature Chemical Biology*, 7, 818–826.

Nawrocki, S. T., Carew, J. S., Pino, M. S., Highshaw, R. A., Andtbacka, R. H., Dunner, K., Jr., et al. (2006). Aggresome disruption: A novel strategy to enhance bortezomib-induced apoptosis in pancreatic cancer cells. *Cancer Research*, *66*, 3773–3781.

Ni, M., Zhang, Y., & Lee, A. S. (2011). Beyond the endoplasmic reticulum: Atypical GRP78 in cell viability, signalling and therapeutic targeting. *The Biochemical Journal*, *434*, 181–188.

Peng, L., & Seto, E. (2011). Deacetylation of nonhistone proteins by HDACs and the implications in cancer. *Handbook of Experimental Pharmacology*, *206*, 39–56.

Picard, D. (2002). Heat-shock protein 90, a chaperone for folding and regulation. *Cellular and Molecular Life Sciences*, *59*, 1640–1648.

Puthalakath, H., O'Reilly, L. A., Gunn, P., Lee, L., Kelly, P. N., Huntington, N. D., et al. (2007). ER stress triggers apoptosis by activating BH3-only protein Bim. *Cell*, *129*, 1337–1349.

Rao, R., Balusu, R., Fiskus, W., Mudunuru, U., Venkannagari, S., Chauhan, L., et al. (2012). Combination of Pan-Histone Deacetylase Inhibitor and Autophagy Inhibitor Exerts Superior Efficacy against Triple-Negative Human Breast Cancer Cells. *Molecular Cancer Therapeutics*, *11*, 973–983.

Rao, R., Fiskus, W., Yang, Y., Lee, P., Joshi, R., Fernandez, P., et al. (2008). HDAC6 inhibition enhances 17-AAG-mediated abrogation of hsp90 chaperone function in human leukemia cells. *Blood*, *112*, 1886–1893.

Rao, R., Fiskus, W., Yang, Y., Mudunuru, U., Wang, Y., Atadja, P., et al. (2009). Targeting autophagy induced by pan-HDAC inhibitor panobinostat and promoted by acetylated hsp70: A novel therapy for breast cancer. *Molecular Cancer Therapeutics*, *8*(Issue 12) Supplement 1.

Rao, R., Nalluri, S., Fiskus, W., Savoie, A., Buckley, K. M., Ha, K., et al. (2010). Role of CAAT/enhancer binding protein homologous protein in panobinostat-mediated potentiation of bortezomib-induced lethal endoplasmic reticulum stress in mantle cell lymphoma cells. *Clinical Cancer Research*, *16*, 4742–4754.

Rao, R., Nalluri, S., Kolhe, R., Yang, Y., Fiskus, W., Chen, J., et al. (2010). Treatment with panobinostat induces glucose-regulated protein 78 acetylation and endoplasmic reticulum stress in breast cancer cells. *Molecular Cancer Therapeutics*, *9*, 942–952.

Reddy, R. K., Mao, C., Baumeister, P., Austin, R. C., Kaufman, R. J., & Lee, A. S. (2003). Endoplasmic reticulum chaperone protein GRP78 protects cells from apoptosis induced by topoisomerase inhibitors: Role of ATP binding site in suppression of caspase-7 activation. *The Journal of Biological Chemistry*, *278*, 20915–20924.

Rey, M., Irondelle, M., Waharte, F., Lizarraga, F., & Chavrier, P. (2011). HDAC6 is required for invadopodia activity and invasion by breast tumor cells. *European Journal of Cell Biology*, *90*, 128–135.

Ron, D., & Walter, P. (2007). Signal integration in the endoplasmic reticulum unfolded protein response. *Nature Reviews. Molecular Cell Biology*, *8*, 519–529.

Saji, S., Kawakami, M., Hayashi, S., Yoshida, N., Hirose, M., Horiguchi, S., et al. (2005). Significance of HDAC6 regulation via estrogen signaling for cell motility and prognosis in estrogen receptor-positive breast cancer. *Oncogene*, *24*, 4531–4539.

Santo, L., Hideshima, T., Kung, A. L., Tseng, J. C., Tamang, D., Yang, M., et al. (2012). Preclinical activity, pharmacodynamic and pharmacokinetic properties of a selective HDAC6 inhibitor, ACY-1215, in combination with bortezomib in multiple myeloma. *Blood*, *119*, 2579–2589.

Sato, M., Yao, V. J., Arap, W., & Pasqualini, R. (2010). GRP78 signaling hub a receptor for targeted tumor therapy. *Advances in Genetics*, *69*, 97–114.

Scroggins, B. T., Robzyk, K., Wang, D., Marcu, M. G., Tsutsumi, S., Beebe, K., et al. (2007). An acetylation site in the middle domain of Hsp90 regulates chaperone function. *Molecular Cell*, *25*, 151–159.

Shiau, A. K., Harris, S. F., Southworth, D. R., & Agard, D. A. (2006). Structural Analysis of *E. coli* hsp90 reveals dramatic nucleotide-dependent conformational rearrangements. *Cell*, *127*, 329–340.

Siligardi, G., Hu, B., Panaretou, B., Piper, P. W., Pearl, L. H., & Prodromou, C. (2004). Co-chaperone regulation of conformational switching in the Hsp90 ATPase cycle. *The Journal of Biological Chemistry*, *279*, 51989–51998.

Siligardi, G., Panaretou, B., Meyer, P., Singh, S., Woolfson, D. N., Piper, P. W., et al. (2002). Regulation of Hsp90 ATPase activity by the co-chaperone Cdc37p/p50cdc37. *The Journal of Biological Chemistry*, *277*, 20151–20159.

Sims, J. D., McCready, J., & Jay, D. G. (2011). Extracellular heat shock protein (Hsp)70 and Hsp90alpha assist in matrix metalloproteinase-2 activation and breast cancer cell migration and invasion. *PLoS One*, *6*, e18848.

Smith, M. H., Ploegh, H. L., & Weissman, J. S. (2011). Road to ruin: Targeting proteins for degradation in the endoplasmic reticulum. *Science*, *334*, 1086–1090.

Solimini, N. L., Luo, J., & Elledge, S. J. (2007). Non-oncogene addiction and the stress phenotype of cancer cells. *Cell*, *130*, 986–988.

Spiegel, S., Milstien, S., & Grant, S. (2012). Endogenous modulators and pharmacological inhibitors of histone deacetylases in cancer therapy. *Oncogene*, *3*, 537–551.

Street, T. O., Lavery, L. A., & Agard, D. A. (2011). Substrate binding drives large-scale conformational changes in the Hsp90 molecular chaperone. *Molecular Cell*, *42*, 96–105.

Street, T. O., Lavery, L. A., Verba, K. A., Lee, C. T., Mayer, M. P., & Agard, D. A. (2012). Cross-monomer substrate contacts reposition the Hsp90 N-terminal domain and prime the chaperone activity. *Journal of Molecular Biology, 415*, 3–15.

Tabas, I., & Ron, D. (2011). Integrating the mechanisms of apoptosis induced by endoplasmic reticulum stress. *Nature Cell Biology, 13*, 184–190.

Taipale, M., Jarosz, D. F., & Lindquist, S. (2010). HSP90 at the hub of protein homeostasis: Emerging mechanistic insights. *Nature Reviews. Molecular Cell Biology, 11*, 515–528.

Trepel, J., Mollapour, M., Giaccone, G., & Neckers, L. (2010). Targeting the dynamic HSP90 complex in cancer. *Nature Reviews. Cancer, 10*, 537–549.

True, O., & Matthias, P. (2012). Interplay between histone deacetylases and autophagy—From cancer therapy to neurodegeneration. *Immunology & Cell Biology, 90*, 78–84.

Urano, F., Wang, X., Bertolotti, A., Zhang, Y., Chung, P., Harding, H. P., et al. (2000). Coupling of stress in the ER to activation of JNK protein kinases by transmembrane protein kinase IRE1. *Science, 287*, 664–666.

Vembar, S. S., & Brodsky, J. L. (2008). One step at a time: Endoplasmic reticulum-associated degradation. *Nature Reviews. Molecular Cell Biology, 9*, 944–957.

Walsh, N., Larkin, A., Swan, N., Conlon, K., Dowling, P., McDermott, R., et al. (2011). RNAi knockdown of Hop (Hsp70/Hsp90 organising protein) decreases invasion via MMP-2 down regulation. *Cancer Letters, 306*, 180–189.

Wang, Y., Wang, S. Y., Zhang, X. H., Zhao, M., Hou, C. M., Xu, Y. J., et al. (2007). FK228 inhibits Hsp90 chaperone function in K562 cells via hyperacetylation of Hsp70. *Biochemical and Biophysical Research Communications, 356*, 998–1003.

Whitesell, L., & Lindquist, S. L. (2005). HSP90 and the chaperoning of cancer. *Nature Reviews. Cancer, 5*, 761–772.

Xu, W., Marcu, M., Yuan, X., Mimnaugh, E., Patterson, C., & Neckers, L. (2002). Chaperone-dependent E3 ubiquitin ligase CHIP mediates a degradative pathway for c-ErbB2/Neu. *Proceedings of the National Academy of Sciences of the United States of America, 99*, 12847–12852.

Yamaguchi, H., & Wang, H. G. (2004). CHOP is involved in endoplasmic reticulum stress-induced apoptosis by enhancing DR5 expression in human carcinoma cells. *Journal of Biological Chemistry, 279*, 45495–45502.

Yang, Y., Rao, R., Shen, J., Tang, Y., Fiskus, W., Nechtman, J., et al. (2008). Role of acetylation and extracellular location of heat shock protein 90alpha in tumor cell invasion. *Cancer Research, 68*, 4833–4842.

Yao, Y. L., Yang, W. M., & Seto, E. (2001). Regulation of transcription factor YY1 by acetylation and deacetylation. *Molecular and Cellular Biology, 21*, 5979–5991.

Yoo, C. B., & Jones, P. A. (2006). Epigenetic therapy of cancer: Past, present and future. *Nature Reviews. Drug Discovery, 5*, 37–50.

Yorimitsu, T., Nair, U., Yang, Z., & Klionsky, D. J. (2006). Endoplasmic reticulum stress triggers autophagy. *The Journal of Biological Chemistry, 281*, 30299–30304.

Zhang, Y., Liu, R., Ni, M., Gill, P., & Lee, A. S. (2010). Cell surface relocalization of the endoplasmic reticulum chaperone and unfolded protein response regulator GRP78/BiP. *The Journal of Biological Chemistry, 285*, 15065–15075.

Zhang, Z., Yamashita, H., Toyama, T., Sugiura, H., Omoto, Y., Ando, Y., et al. (2004). HDAC6 expression is correlated with better survival in breast cancer. *Clinical Cancer Research, 10*, 6962–6968.

Zhang, X., Yuan, Z., Zhang, Y., Yong, S., Salas-Burgos, A., Koomen, J., et al. (2007). HDAC6 modulates cell motility by altering the acetylation level of cortactin. *Molecular Cell, 27*(2), 197–213.

INDEX

Note: Page numbers followed by "*f*" indicate figures, and "*t*" indicate tables.

F

G

H

I

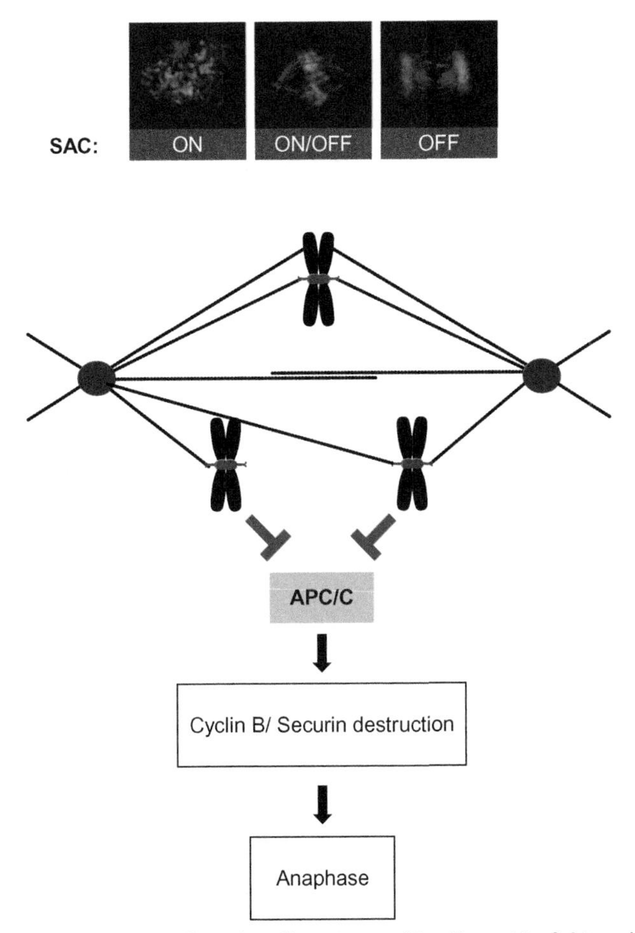

Figure 1.4, Brian Gabrielli and Mellissa Brown (See Page 12 of this volume.)

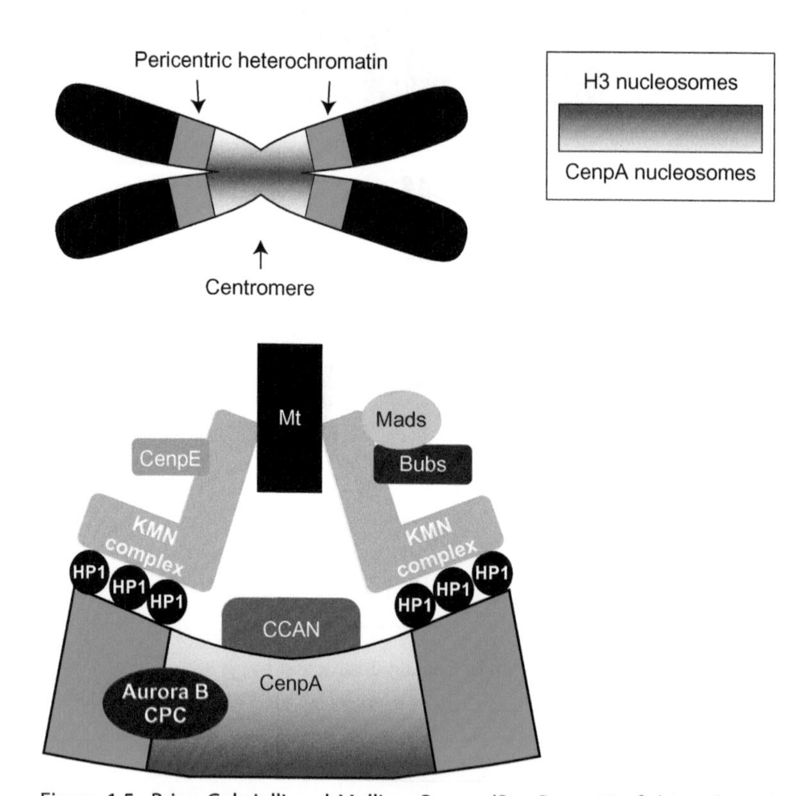

Figure 1.5, Brian Gabrielli and Mellissa Brown (See Page 15 of this volume.)

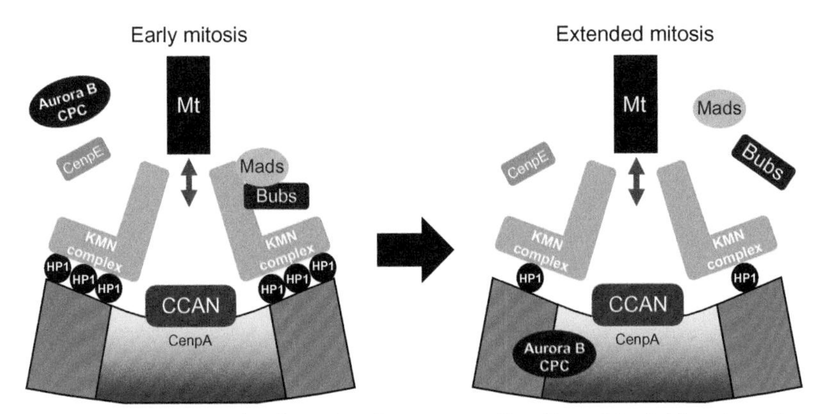

Figure 1.6, Brian Gabrielli and Mellissa Brown (See Page 20 of this volume.)

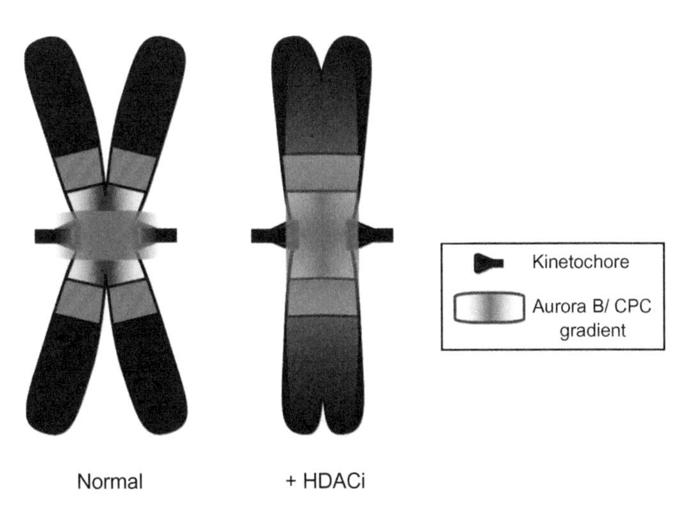

Figure 1.7, Brian Gabrielli and Mellissa Brown (See Page 22 of this volume.)

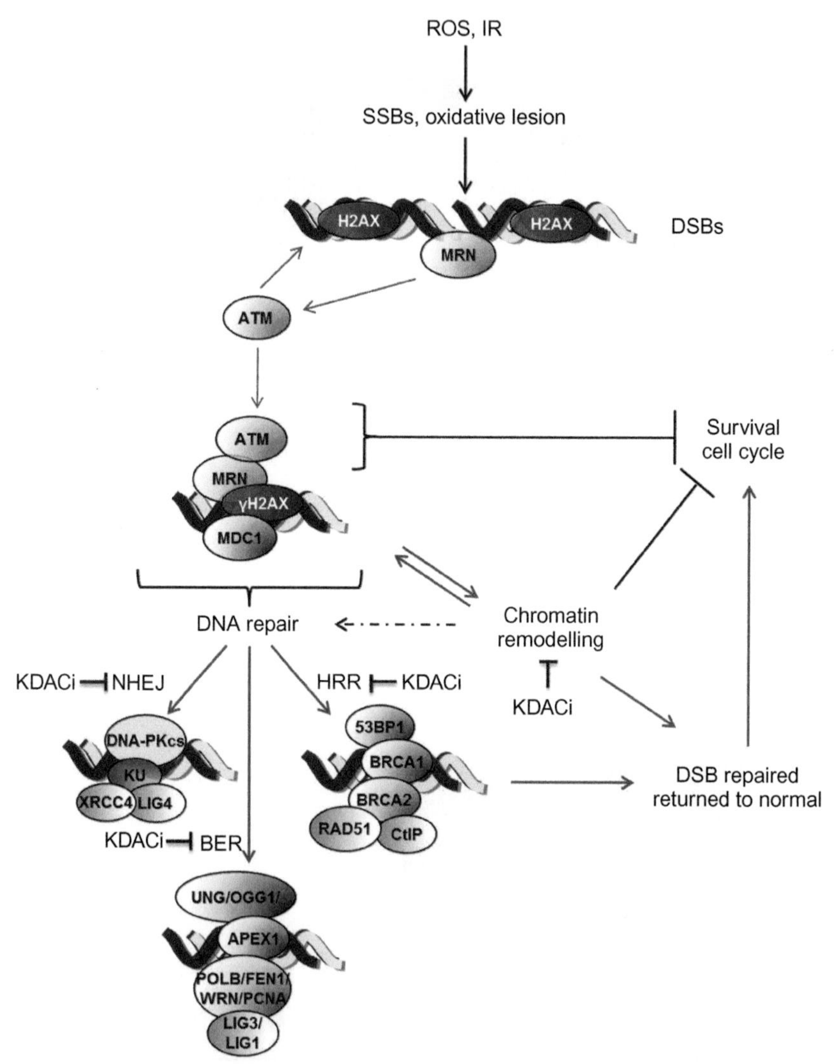

Figure 3.1, Carine Robert and Feyruz V. Rassool (See Page 94 of this volume.)

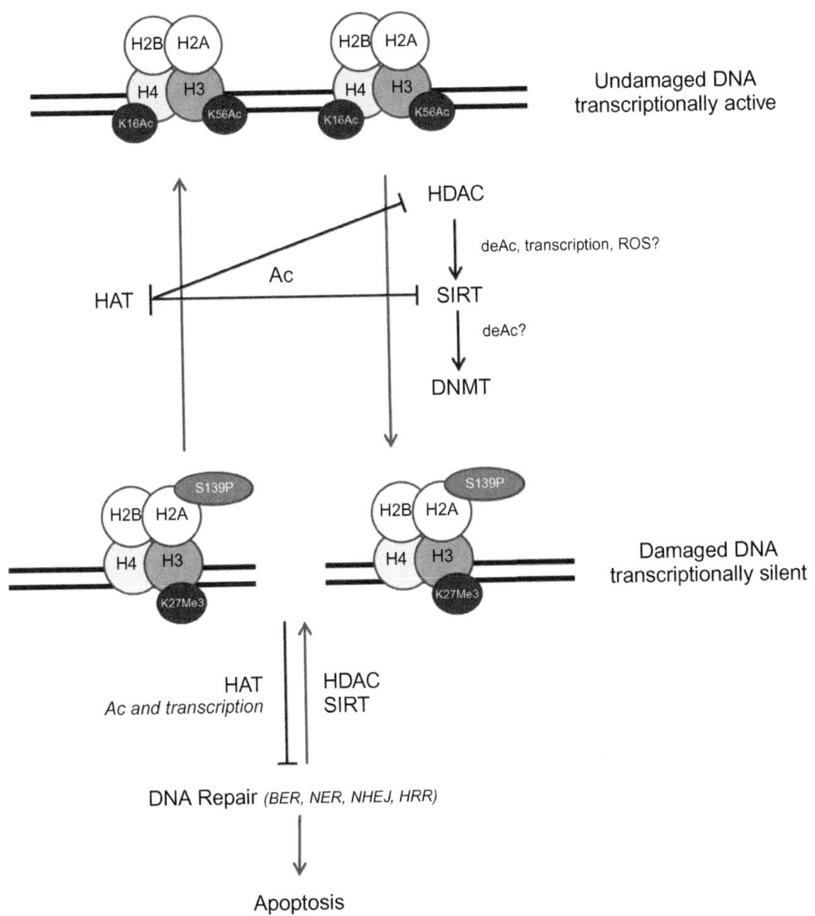

Figure 3.2, Carine Robert and Feyruz V. Rassool (See Page 104 of this volume.)

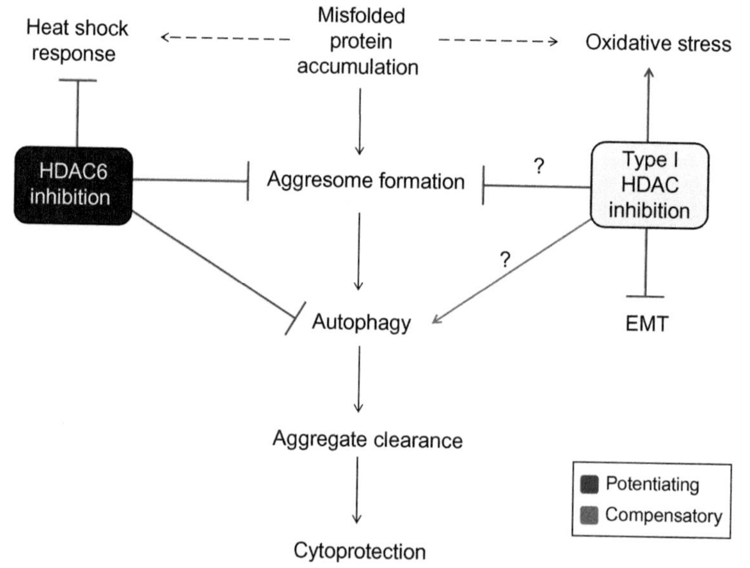

Figure 4.4, David J. McConkey *et al.* (See Page 145 of this volume.)

Printed and bound by CPI Group (UK) Ltd, Croydon, CR0 4YY
08/05/2025
01864955-0001